Ichthyology

Illustrated by William L. Brudon

John Wiley & Sons, Inc., New York · London · Sydney

Ichthyology

Second Edition

KARL F. LAGLER
School of Natural Resources
University of Michigan

JOHN E. BARDACH
Hawaii Institute of Marine Biology
University of Hawaii

ROBERT R. MILLER
Museum of Zoology
University of Michigan

DORA R. MAY PASSINO
Great Lakes Fishery Laboratory
U.S. Fish and Wildlife Service

JOHN WILEY & SONS
New York • Santa Barbara • London • Sydney • Toronto

Library of Congress Cataloging in Publication Data
Main entry under title:

Ichthyology.

   First ed. by K. F. Lagler, J. E. Bardach, and R. R. Miller.
   Includes bibliographies and indexes.
   1. Fishes. I. Lagler, Karl Frank, 1912-   Ich-
thyology.
QL615.L3   1977   597   •   76-50114
ISBN 0-471-51166-8

Printed in the United States of America

10 9 8 7 6 5 4 3 2

Dedicated to
Carl L. Hubbs
Scholar, Teacher,
and Friend

Preface

In the fourteen some years since the first edition was published, many studies have improved knowledge in ichthyology. A good proportion of these studies have been done elsewhere than in North America and we have drawn on them in this work; but again we have continued to emphasize the use of American examples.

Dr. Dora R. May Passino, fish biochemist, joined us as an author in the current effort, with the kind permission of her employer, the U.S. Fish and Wildlife Service.

We acknowledge with gratitude our indebtedness to our students, colleagues, and other users of the book who have helped us with corrections, criticisms, and ideas. We are especially indebted to Dr. Gerald P. Smith for his help in the revision of Chapter 2 on major fish groups and to Dr. Paul R. Webb for his suggestions for Chapter 6 on locomotion. Most competent technical assistance in the preparation of the revision was given by Elsie Goode, Frances Hubbs Miller, Joanna Schmidt, Mary Lou Lagler, and Donna Brown, and by our Wiley editor, Robert L. Rogers, and production supervisor, Christine Pines.

Karl F. Lagler
John E. Bardach
Robert R. Miller
Dora R. May Passino

Preface to First Edition

Not since David Starr Jordan's venerable *Guide to the Study of Fishes* appeared in 1905 has there been a text reference in ichthyology that drew its examples primarily from the American fauna. We have tried to fill this void with the hope that additional vigorous American students will be stimulated to work in the science of ichthyology and that informed fishermen may further their knowledge of fishes.

Needless to say, there are too many fishes too widely distributed in the vast water areas of the earth for this book to be comprehensive. Although it deals with the broad principles of ichthyology, the treatment of many interesting and important details is very brief or omitted entirely. It is hoped that the progressive student will understand the need for much additional reading. To help in accomplishing this, we have included selected references at the end of each chapter.

The table of contents discloses our efforts to make this book widely adaptable in collegiate courses. We planned it for core reading in both introductory and advanced presentations of ichthyology. However, care may be taken through choice of chapters and sections assigned to make it valuable also for use in courses of comparative anatomy of the vertebrates, vertebrate natural history, comparative physiology, and evolution. Throughout, we have emphasized comparisons among the major groups of fishes both in structure and function.

Early chapters (1 through 3) introduce the diversity of fishes and show the position and content of the major groups, their classification, relationships, and basic structure, with emphasis on living fishes. Chapters 4 through 11 describe and discuss the comparative anatomy and physiology of the classical ten body systems and their integration into the whole fish. Principles of genetics, evolution, systematics, ecology, and ichthyogeography comprise Chapters

12 through 14. Thus a college course in systematic and ecological ichthyology might use primarily Chapters 1 through 3 and 12 through 14, whereas one with emphasis on fish behavior and physiology might use primarily Chapters 4 through 11.

In the preparation of this book we have become greatly indebted to many of our students, colleagues, the publisher, and others. To them we offer our sincere thanks and the wish that they continue to give us the benefit of their indispensable help and criticism. Acknowledgement of illustrations not our own is made beneath each such illustration.

KARL F. LAGLER
JOHN E. BARDACH
ROBERT R. MILLER

Contents

Ichthyology

1

Fish, Animals, and Man

WHAT FISH ARE

Fishes are cold-blooded animals, typically with backbones, gills, and fins (rather than pentadactyl limbs), and are primarily dependent on water as a medium in which to live. Their study composes the pure and applied aspects of the science of ichthyology. Obviously not included in this field of learning are mammals, such as whales, seals, and porpoises; reptiles, such as aquatic turtles; and invertebrates, such as clams, shrimp, and lobsters ("shellfish").

Fishes are the most numerous of the vertebrates (Fig. 1.1), with estimates of around 20,000 recent species, although guesses range as high as 40,000. In contrast, it is commonly assumed that bird species number about 8600, mammals, 4500 (of which living man is only one), reptiles, 6000, and amphibians, 2500. Thus, not only are there many different fishes but they come in many different shapes and sizes. Included are pygmies such as the American percid least darter (*Etheostoma microperca*) which matures sexually at a length of 27 mm. and a dwarf pygmy goby (*Eviota*) of the Philippines which breeds at sizes less than 15 mm. There are giants, too, such as the whale shark (*Rhincodon*), which has been judged to attain lengths near 21 m. and weights of 25 tons or more. Most fishes are torpedo-shaped, but some are round, others are flat, and still others are angular.

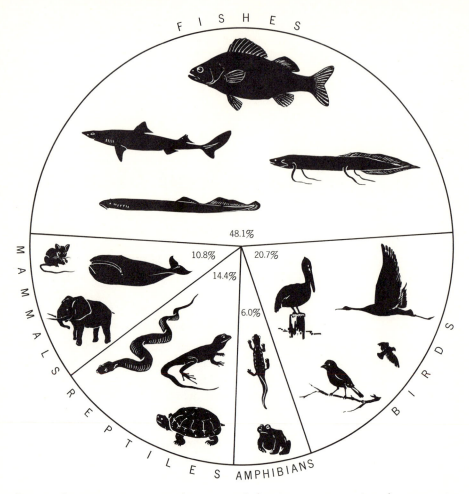

Fig. 1.1   Percentage composition by groups of the some 41,600 species of recent vertebrates.

FISH AS ANCESTORS TO MAN

According to evolutionary theory, which is based on evidence including fossils, comparative anatomy, embryology, and genetics, fishes have a distant place in the ancestry of man. Their presence on earth antedates man's ape-like ancestors by some 500 million years, and all other vertebrates by more than 100 million years. Without piscine ancestry, man might never have evolved. Many features of life ways and structure of man were originated or were already present aeons of time ago in fishy ancestors! These features include the ground plans and basic functions of the ten organ systems, including such striking features as sight, internal fertilization, intrauterine nourishment (including placenta), live birth, and, presumably, learning and memory.

WHERE FISH LIVE

It is really no wonder that there are so many different kinds of fishes when their antiquity and the extent and variety of their habitat are considered. At present, more than 70 percent of the earth's surface is covered with water (Fig. 1.2). When the fish group was in its evolutionary youth, there was even more because much that is now land was ocean bottom then. Development of diversity of living conditions in water might be expected to accelerate the rate of speciation.

Fishes seem to have been able to keep pace with the development of variety in places of abode and now live almost wherever there is water, both on the surface and in the surface-connected subterranean waters. They occupy everything from Antarctic waters below freezing to hot springs of more than 40°C, and from soft, fresh water to water saltier than the seas. They are present in sunlit mountain streams so torrential that neither man nor dog can wade or swim them, and in waters so quiet, deep, and dark that they have never been inhabited by other vertebrates or thoroughly explored by man. Their vertical range of distribution exceeds that of any of the other vertebrates. Fishes range from approximately 5 km above sea level to some 11 km beneath it (Fig. 1.2).

HOW FISH LIVE

Water is highway, byway, communications medium, nursery, playground, school, room, bed, board, drink, toilet, and grave for a fish. All of the fishes' vital functions of feeding, digestion, assimilation, growth, responses to stimuli, and reproduction are dependent on water. For the fish, the most important aspects of water are dissolved oxygen, dissolved salts, light penetration, temperature, toxic substances, concentrations of disease organisms, and opportunity to escape enemies.

Although humans are able to absorb oxygen directly from air through the vascularized walls of the lungs, few fish have lungs or other devices for utilizing oxygen from air. Most fish, including those with lungs, depend mainly upon gills to extract oxygen dissolved in water. Fish cannot live long in a habitat rare or deficient in dissolved oxygen any more than humans can survive in the upper atmosphere or the space beyond unless they carry an oxygen supply with them.

The pasturage that the sea, lakes, and streams afford to fish depends

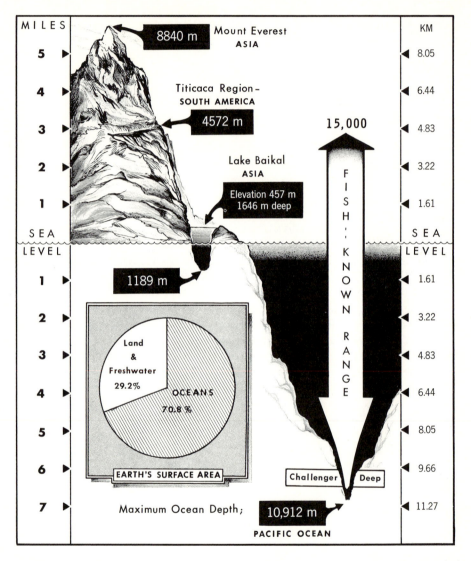

Fig. 1.2   Water area and vertical extent of fish distribution in relation to maximum relief on surface of the earth.

initially on the penetration of light into water, even as growth of grass on the open range relies upon the sun. The "grass" of the waters is microscopic plant life—diatoms and algae, collectively termed phytoplankton.

The beginning of the chain of life leading to fish production is generally in the bodies of the phytoplankters. They utilize light energy and dissolved carbon dioxide to manufacture organic matter that eventually becomes food for fish. Besides providing energy for food production for all fishes, light is also known to trigger mechanisms of reproduction, growth, and many kinds of behavior, including that of feeding.

Unwanted materials such as toxins produced in nature and pollution from human activities are serious menaces to fish life. The aquatic habitat provides no places of escape from damaging substances in solution. The threat to fish of water-borne toxic materials is comparable to that of air-borne pollutants to human beings. Although fish are able to detect many such chemical contaminants, they are often unable to avoid them.

Like all animals, fishes have a very full complement of diseases with which to contend. Many of these are due to external agencies; others arise internally. From outside come viruses, fungi, bacteria, parasitic protozoans, worms, crustaceans, and lampreys. From within arise almost all the common organic and degenerative disorders that plague man himself. Included are cancer, rickets, degeneration of the liver, blindness, and a host of developmental anomalies such as Siamese twinning and spinal flexure. And even if not killed by a disease or disorder such as the foregoing, the fish must still survive periodic adverse chemical conditions in water, predators, and the capturing devices of fishermen.

HOW AND WHY FISH ARE STUDIED

Growth in knowledge of fishes has resulted from our lasting curiosity about nature and from our need for information concerning species used for commerce or recreation. At least ten centuries before Christ, the Chinese were trying to find out enough about fishes successfully to propagate them. Ancient Egyptians, Greeks, and Romans recorded observations on the varieties, habits, and qualities of various fishes. The symbol of the early Christian underground movement in the Catacombs of Rome was the fish.

The study of fishes (ichthyology) was hardly scientific until the eighteenth century in Europe. Since that time it has grown the world over along the following major lines:

Classification—the continuing, long-term effort to arrange all kinds of living and fossil fishes into various groups or taxa and to determine their natural relationships.

Anatomy—the structure of fishes from microscopic, embryological, and comparative points of view, including fossils where available and pertinent.

Evolution and genetics—the origins of fishes and the sequence and manner in which modern fishes evolved from previous ones and the mechanisms by which changes have come about; speciation; origin, transmission, and changes of the characters by which kinds of fishes are recognized.

Natural history and ecology—life ways and habitats and the interactions of fish with each other and their environment.

Physiology and biochemistry—functions of organs and systems, the metabolism and integration of systems at the molecular level, and tolerances to changes in physical conditions of surroundings.

Conservation—wise use and management of fish resources for the benefit of man by means of fishery statistics, fishery technology and marketing, laws, population manipulation, fish culture and stocking, and environmental improvement.

The foregoing areas of work are carried on by international organizations, governmental agencies, museums, universities, and industries. The Food and Agriculture Organization (FAO) of the United Nations has an active Fish Division. Most countries have federal fishery units comparable in function to the Fish and Wildlife Service and National Marine Fisheries Service in the United States, and individual states or provinces often have specialized fish sections in their natural resource organizations. Museums and academies including science in their programs have fish divisions. Exemplary among these are the British Museum (Natural History), the U.S. National Museum, the California Academy of Sciences, and the Museum of Zoology at The University of Michigan. A number of institutions of higher learning offer instruction in ichthyology or fisheries, or both curriculums.

OPPORTUNITIES IN ICHTHYOLOGY

Opportunities to study fish are limitless; they exist for everyone whether or not they are professional ichthyologists. Many contributions to the knowledge of fishes have come from philosophers, clergymen, doctors of medicine, fishermen, and hobby aquarists. Research opportunities are boundless. In all of the aspects of ichthyology, much more is unknown than is known.

Teaching positions in ichthyology are not numerous; fewest are those in which the primary responsibility is to study and to teach ichthyology. Curatorial opportunities, which involve the development, care, and study of collections in museums are fewer still. There are only some 20 museums in North America, for example, that have staff positions for ichthyologists, amounting to between 35 and 50 curatorial jobs. Duties involve the development, care, and study of old and new collections, supervision of public exhibits, answering of questions and correspondence from the public, and the preparation for publication of scientific research. Many curators (all those at museums associated with educational institutions) are involved in graduate training programs and other forms of teaching. Openings for curators are infrequent, perhaps averaging only one per year, and there is keen competition for these positions, which typically require a PhD degree. In recent years, however, a technical position titled Collection Manager requires far less formal training.

The great commercial fisheries, professional fishery management, and the propagation and sale of food, game, ornamental, and bait fishes afford vast opportunities for those with training or experience in ichthyology. Such jobs may require no more than a master's degree. Trained personnel are needed in managing the fisheries of marine and inland waters and also those of reservoirs and hatcheries. The expanding nature of the fishery field affords many opportunities for employment in areas not previously investigated. In the 1970s there has been a great upsurge in positions with private and governmental agencies involved in environmental impact assessment. And most popular of all opportunities for students of fishes are the hobbies of recreational fishing and aquarium care. Here the science of ichthyology has much to offer in the enrichment of living.

GENERAL REFERENCES

American Fisheries Society. 1970. Fisheries as a profession. A career guide for the field of fisheries science. Distributed by the Society.

American Society of Ichthyologists and Herpetologists. 1973. Career opportunities for the ichthyologists. Prepared by Victor G. Springer and distributed by the Society.

Boulenger, G. A. 1904. Fishes (systematic account of Teleostei). In The Cambridge natural history. Vol. VII. The Macmillan Company, New York. Reprint edition, 1958, Hafner Publishing Company, New York.

Bridge, T. W. 1904. Fishes (exclusive of the systematic account of Teleostei). In The Cambridge natural history. Vol. VII. The Macmillan Company, New York. Reprint edition, 1958, Hafner Publishing Company, New York.

Brown, M. E. (Ed.). 1957. The physiology of fishes. Vol. 1 Metabolism. Vol. 2 Behavior. Academic Press, New York.

Curtis, B. 1949. The life story of the fish his morals and manners. Harcourt, Brace and Company, New York.

Dean, B. 1895. Fishes, living and fossil. The Macmillan Company, New York.

Goodrich, E. S. 1909. Cyclostomes and fishes. In A treatise on zoology. Pt. IX, Fas. 1. A. and C. Black, London.

Grassé, P.-P. (Ed.). 1958. Agnathes et poissons anatomie, éthologie, systematique. Traité de zoologie. Tome XIII, 3 Vols. Masson et Cie., Paris.

Günther, A. C. L. G. 1880. An introduction to the study of fishes. A. and C. Black, Edinburgh.

Herald, E. S. 1961. Living fishes of the world. Doubleday & Company Inc., Garden City, New York.

Hoar, W. S., and D. J. Randall. 1969–1976. Fish physiology. Academic Press, New York and London. 8 volumes.

Jordan, D. S. 1905. A guide to the study of fishes. 2 Vols. Henry Holt and Company, New York.

Kyle, H. M. 1926. The biology of fishes. Sidgwick and Jackson, London; the Macmillan Company, New York.

Lanham, U. 1962. The fishes. Columbia University Press, New York.

Marshall, N. B. 1965. The life of fishes. Weidenfeld and Nicolson, London.

Matsubara, K. 1955. Fish morphology and hierarchy. Ishizaki Shoten, Tokyo.

Myers, G. W. 1970. How to become an ichthyologist. T. F. H. Publications, Jersey City, N.J.

Nikolski, G. W. 1957. Spezielle Fischkunde. VEB Deutscher Verlag der Wissenschaften, Berlin. (Also: 1961, Special ichthyology. Israel Program for Scientific Translations, Jerusalem)

Norman, J. R. 1975. A history of fishes. Third edition by P. H. Greenwood. John Wiley and Sons, New York.

Pincher, C. 1948. A study of fish. Duell, Sloan and Pearce, New York.

Roule, L. 1935. Fishes and their ways of life. W. W. Norton and Company, New York.

Schultz, L. P., and E. M. Stern. 1948. The ways of fishes. D. Van Nostrand Company, Princeton, N.J.

Sterba, G. 1959. Süsswasserfische aus aller Welt. Verlag Zimmer & Herzog, Berchtesgaden.

Young, J. Z. 1962. The life of vertebrates. (2nd ed.) Oxford University Press, London.

2

The Major Groups
of Fishes

The subject of ichthyology has been defined as the study of fishes, and fishes
have been defined primarily as aquatic and gill-breathing vertebrates equipped
with fins. Beyond this generality, the diversity of shapes, sizes, and varieties
of fishes, and the different distributions of the many kinds in time and space
are so great that many difficulties have been encountered in trying to classify
them. In addition, relatively few contemporary scientists have attained the
breadth and depth of knowledge needed to make major efforts in classification. The lack of agreement among the schemes that have been proposed in
the twentieth century points up the many imperfections in existing knowledge, as well as the challenge of this group of animals.

SCOPE OF CLASSIFICATION

The materials to be classified are the specimens of all of the thousands of
living and fossil kinds of fishes that have been discovered and all of those
yet to become known to science. Almost any area of the earth may be expected to yield new examples to be described and categorized. As commonly
conceived these fishes include some without jaws, of which a few are living
and unarmored (the lampreys and hagfishes), along with others that are
archaic and armored, and the many fishes with jaws, some extinct and some
living (sharks and bony fishes and their relatives).

9

AIMS OF CLASSIFICATION

In order to obtain an overview of the many, many kinds of fishes that have appeared on earth, a student must immediately contend with the problem of dividing them into groups for study. Solution of this problem requires the establishment of concepts that will enable grouping. The formulation of these concepts is aided by a definition of possible aims of classification. At least two aims exist: (1) practical; and (2) ideal.

The practical aim is simply to arrange the kinds of fishes into some system for the pure convenience of finding again something that has been handled before, such as in the alphabetical arrangement by name of specimens in jars on a shelf, or of information on species on sheets of paper in a drawer. Such a functional arrangement might also be adopted to make it as easy as possible to distinguish one kind from another. Thus, the basic theme of a practical classification would be for mechanical convenience—it might be done alphabetically, by size, or by any other useful means.

If we add to the utilitarian aspects of the practical aim, the thought that the arrangement should also depict the evolutionary history of fishes, we shall have expressed the ideal aim. Here the system, while affording convenience since it is systematic, is also conceptual and imparts a condensed expression of the phylogeny of the entire major kind of organism. The ideal aim, then, is for classification to be evolutionary, genetic, or natural in character rather than merely artificial or mechanical. Gaps in knowledge, however, make it difficult to achieve a purely ideal aim. The result is that modern classifications of fishes, although based on ideal aims, are always partly artificial.

METHOD OF CLASSIFICATION

Simultaneous attainment of practical and ideal objectives in the classification of plants and animals, including fishes, early led to the acceptance of the following seven standard categories: kingdom, phylum, class, order, family, genus, and species. These are listed from largest (most inclusive) to smallest, hence also from the oldest to youngest in time. Each group encompasses one or more of the succeeding ones in it. As in a filing system, divisions progress from major to minor, with the smallest entry always being made under the species (or subspecies) name, which name is found under its appropriate genus, genus under appropriate family, and so on up to kingdom.

Standardization exists not only for the major categories of classification, but also for how family, generic, and specific names are to be made (Chapter 13). There are, however, no rules regarding the formation of phylum, class, or ordinal names. Some of the most confusing name situations exist at the class and order level (Table 2.1); lack of an established practice here offers an impediment to students who wish quickly to grasp a feeling for the natural groups of fishes.

VERTEBRATE ANCESTRY

The following animals are often treated with fishes in the Phylum Chordata on the basis of the common characteristic of a notochord (hence termed chordates):

Lacking both a cranium (Acrania) and vertebrae (Protochordata)
    Acorn worms (such as *Balanoglossus*)
    Sea squirts or tunicates (such as *Molgula*)
    Lancelets (such as amphioxus, *Branchiostoma*)
Possessing both a cranium (Craniata) and vertebrae (Vertebrata)
    Fishes (Pisces and relatives)
    Amphibians (Amphibia)—frogs, toads, salamanders, caecilians
    Reptiles (Reptilia)—lizards, snakes, turtles, crocodiles
    Birds (Aves)
    Mammals (Mammalia)

The living chordate groups leading up to and including the fishes have often been named as follows:

Phylum Chordata
    Subphylum Hemichordata (acorn worms)
    Subphylum Urochordata (sea squirts)   collectively,
    Subphylum Cephalochordata (lancelets)   protochordates
    Subphylum Vertebrata (fishes through mammals)

It is possible that the ancestors of the vertebrates were among the protochordates and generally were without hard skeletal structures. The remote metazoan ancestor of the protochordates and vertebrates may have been "a simple sedentary creature composed of little but a digestive tract and an apparatus for gathering food particles from the surrounding water . . ." (Romer, 1968). The lowest chordate types, which no doubt had bottom-dwelling adults and a free-swimming ciliate larva such as the tornaria of certain hemichordates, possess this type of structure. The next stage in development toward a higher chordate would be the appearance of a system of gill-slit, filter-feeding, which, indeed, is found among true living vertebrates in the lamprey larva, as well as others. From the ammocoete-like larval stage, the next step may have been to a fish-like form with direct development.

Table 2.1. Comparative Terminology for Major Groups that Include Living Fishes—Classes (and Subclasses)

Jordan (1923)	Regan (1929)	Berg (1940)	Traité de Zoologie (1958)[a]	Present Work
MARSIPOBRANCHII Lampreys and hagfishes	MARSIPOBRANCHII	PETROMYZONTES Lampreys MYXINI Hagfishes	CYCLOSTOMI (Cephalaspidomorphi) Lampreys (Pteraspidomorphi) Hagfishes	CEPHALASPIDOMORPHI Lampreys and hagfishes
ELASMOBRANCHII (Selachii) Sharks, rays and skates (Holocephali) Chimaeras	SELACHII (Euselachii) (Holocephali) Chimaeras	ELASMOBRANCHII Sharks, rays and skates HOLOCEPHALI Chimaeras	CHONDRICHTHYES (Euselachii) Sharks, rays and skates (Bradyodonti) Chimaeras	CHONDRICHTHYES (Elasmobranchii) Sharks, rays and skates (Holocephali) Chimaeras
PISCES Bony fishes (Dipneusti) Lungfishes (Crossopterygii) Lobefins (Actinopteri) Rayfins	PISCES (Dipneusti) Lungfishes (Crossopterygii) Lobefins (Paleopterygii) Primitive bony fishes (Neopterygii) Modern bony fishes (Actinopteri) Rayfins	DIPNOI TELEOSTOMI (Crossopterygii)[b] (Actinopterygii) Rayfins	OSTEICHTHYES (Dipneusti) Lungfishes (Crossopterygii) Lobefins (Brachiopterygii) Reedfishes (Actinopterygii) Rayfins	OSTEICHTHYES (Dipnoi) Lungfishes (Crossopterygii) Lobefins (Actinopterygii) Rayfins

[a] In "Traité de Zoologie, . . ." Tome XIII, Stensiö classified the agnathous fishes and Bertin and Arambourg, the rest.

[b] The classification of the lobefins as Teleostomi by Berg appears to have been an error; the positioning by the other authors cited appears to be more realistic.

The question of whether the earliest vertebrates were freshwater inhabitants or marine has been repeatedly debated. The problem is where did the Agnatha arise, since it is generally accepted that the first craniates belonged to this group. Recent support for the theory of a marine origin of agnathans is given by Lehtola (1973), who described remains of the jawless heterostracan, *Astraspis desiderata*, from shallow marine limestones of Middle Ordovician age in Ontario, Canada. This is the first record of an Ordovician vertebrate in limestone, a rock deposited in the open sea. Her paper reviews the literature on this controversial subject.

Many theories of chordate evolution have been proposed and most have been discarded. The remarkable structural similarity between certain Ordovician echinoderms (mitrates) and heterostracans has led to a new, controversial theory of vertebrate origins. The mitrates are nearly symmetrical and are believed to have had a notochord, a dorsal nerve chord, and a complex central nervous system like ostracoderms, in addition to gill slits in the pharynx like those of hemichordates. Early heterostracans (Fig. 2.2b) could have evolved directly from mitrates.

MAJOR GROUPS OF LIVING FISHES

Living fishes probably exceed 20,000 species, an abundance and diversity unequalled among all other vertebrates combined (Fig. 1.1). Their ancient ancestry, extending into the past for some 500 million years, has allowed a vast span of time for evolutionary divergence and for the origin and extinction of major phyletic lines. Even the modern bony fishes, or teleosts, had their origin as long ago as 200 million years, in the Triassic period, when the major groups (classes and subclasses) of fishes were firmly established. The evolution and relationships of fishes are still highly debatable, despite repeated and continuing investigation; partly because of the variety of philosophical approaches used by systematists, details of the higher classification of these animals remain evanescent.

Fully cognizant of these limitations, we have nevertheless decided to adopt a single classification of fishes in order to provide an operational base for student use and, we hope, a more holistic view of this difficult subject. The classification we present is a compromise chosen from various recent sources, all of which are cited either directly or indirectly at the end of this chapter. Pertinent literature references given there also provide a basis for historical review of the numerous fish classifications that have been widely used during the 1900s; four of these are compared with ours in Table 2.1. The primary division in vertebrate classification is between the jawless vertebrates (earli-

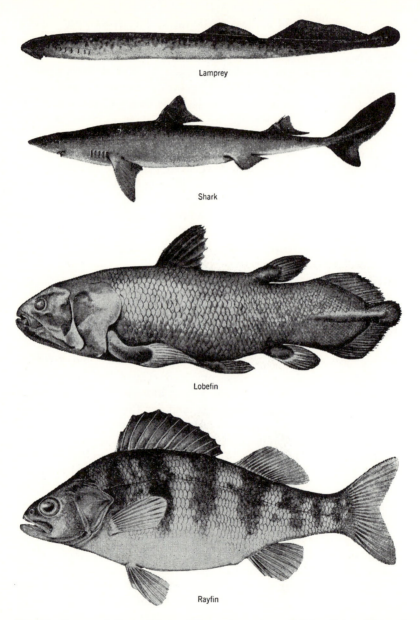

Fig. 2.1   Representatives of four of the major groups of living fishes. Shown from top to bottom are: Class Cephalaspidomorphi, lampreys and hagfishes (sea lamprey, *Petromyzon marinus*); Class Chondrichthyes, sharks, rays, skates, etc. (spiny dogfish, *Squalus acanthias*); Class Osteichthyes, the bony fishes—Subclass Crossopterygii, the lobefins (coelacanth lobefin, *Latimeria chalumnae*), and Subclass Actinopterygii, the rayfins (yellow perch, *Perca flavescens*). The actinopterygian Subclass Dipnoi, the lungfishes (Fig. 4.2), is not shown here.

est fishes), the Agnatha, and those vertebrates that possess jaws, the Gnathostomata.

We here adopt Pisces as a convenient name to include all fish and fish-like vertebrates, and use the terms given below for the major groups of living fishes. These constitute three important natural divisions (Fig. 2.1): (1) CEPHALASPIDOMORPHI—the hagfishes and lampreys, which are jawless (agnathous) and have pouched gills; (2) CHONDRICHTHYES—the chimaeras, sharks, rays, and their allies, which possess true jaws (gnathostomous), have gills borne on wall-like partitions between gill chambers (a single chamber in chimaeras), and have a cartilaginous skeleton (that may be calcified but not bony); and (3) OSTEICHTHYES—all higher fishes, which are gnathostomous, have a branchial chamber protected by a hyoidean opercular series, including the gill cover or operculum, and have a bony skeleton.

Pisces (Groups with living representatives)

I.   Agnatha—the jawless cyclostomes, a relict group derived from extinct ostracoderms.
1.   Class Cephalaspidomorphi—comprising the Subclass Cyclostomata—Myxiniformes, hagfishes and slime eels, and the Petromyzoniformes, lampreys.

II. Gnathostomata—jawed fishes*.
(Superclass Elasmobranchiomorphi—the extinct placoderms and the sharks, holocephalans, and their relatives)
2.   Class Chondrichthyes—the cartilaginous chimaeras, sharks, rays, and their allies.
(Superclass Teleostomi—the fossil acanthodians and others and the bony fishes)
3.   Class Osteichthyes—coelacanths, lungfishes, and all other bony fishes, which include more than 90 percent of the species of living fishes.

CHARACTERIZATION OF LIVING FISH GROUPS

In characterizing the living representatives of the major groups of fishes, it is necessary to generalize. It must be understood, therefore, that in the following lists of features the most frequent condition is stated and exceptions, even where they are known, are usually omitted. In addition to gross characterization, useful terms that have been applied to subgroups or to anatomical features are included.

---

* As here applied, the term includes all jawed vertebrates but only the jawed fishes are Pisces.

Class Cephalaspidomorphi (Hagfishes and Lampreys)

a.   Notochord unconstricted.
b.   Jawless (agnathous).
c.   Main axial skeleton (vertebrae) cartilaginous or fibrous.
d.   Two semicircular canals in ear on each side of head in lampreys but only one on each side in hagfishes.
e.   No true gill arches for the support and protection of gills; instead a branchial basket situated external to the gills, gill arteries (branchial arteries and truncus arteriosus), and branchial nerves; gills in pouches.
f.   Branchial basket firmly united to brain case (neurocranium).
g.   Paired fins (pectorals and pelvics) absent.
h.   Single median nostril (monorhinous).

Class Chondrichthyes (Chimaeras, Sharks, Rays, and Skates)

a.   Notochord constricted by vertebrae.
b.   Jawed (gnathostomous).
c.   Vertebrae cartilaginous (with some calcification but no ossification in living forms).
d.   Three semicircular canals in ear on each side of head.
e.   Gill arches cartilaginous and internal to gills and their supplying arteries and nerves.
f.   Gill arches not firmly united to brain case, but joined to it by connective tissue.
g.   Paired fins present.
h.   Paired nostrils (dirhinous).

Subclass Holocephali (Chimaeras)
a.   Gills in four pairs; gill clefts a single pair.
b.   Spiracle absent.
c.   Skin naked in adults.
d.   Cloaca absent.
e.   Males with pelvic intromittent organs (claspers) that may have prepelvic tenacula, and some (*Chimaera*) have a clasping organ (tenaculum) on forehead.

Subclass Elasmobranchii (Sharks, Rays, and Skates)
a.   Gills and gill clefts in five to seven pairs.
b.   Spiracle present.
c.   Placoid scales (dermal denticles) present or absent.
d.   Cloaca present.
e.   Males usually with pelvic intromittent organs (myxopterygia).

Class Osteichthyes (Bony Fishes)

a. Notochord unconstricted or constricted.
b. Jawed (gnathostomous).
c. Skeleton bony.
d. Three semicircular canals in ear on each side of head.
e. Gill arches bony and internal to gills and their supplying arteries and nerves.
f. Gill arches not firmly united to brain case.
g. Paired fins present.
h. Paired nostrils (dirhinous).

Subclass Dipnoi (Lungfishes)
a. Maxilla and premaxilla absent, three pairs of tooth plates.
b. Internal nares (in recent forms).
c. No movable joint between anterior and posterior part of skull.
d. Palatoquadrate fused to cranium.
e. Extension of radials and muscles into fin base (in recent forms); a single dorsal fin.
f. Cloaca present.

Subclass Crossopterygii (Lobefins)
a. Maxilla absent (except in some fossil forms); premaxilla present; teeth normal.
b. No internal nares (in recent forms).
c. Movable joint between anterior and posterior part of skull.
d. Palatoquadrate not fused to cranium.
e. Extension of radials and muscles into fin base; two separate dorsal fins.
f. Cloaca absent.

Subclass Actinopterygii (Rayfins, Higher Bony Fishes)
a. Maxilla and premaxilla present.
b. No internal nares.
c. No movable joint between anterior and posterior part of skull.
d. Palatoquadrate not fused to cranium.
e. No extension of radials and muscles into fin base (except in Polypteriformes); dorsal fin single or divided.
f. Cloaca absent.

MAJOR GROUPS OF EXTINCT FISHES

To a paleontologist living animals constitute but a brief (and incomplete) cross section of vertebrate history. This is readily understandable if we consider that although there may be as many as 5 or 10 million recent animal species; those that have become extinct may have numbered as high as a half a billion (Mayr, 1969; see Ch. 13). Except for mammals, the fossil record of living vertebrates is fraught with wide gaps and inadequate remains. Much fossil material is represented only by fragments that are often impossible to identify to species, mainly because the osteology of extant relatives is so

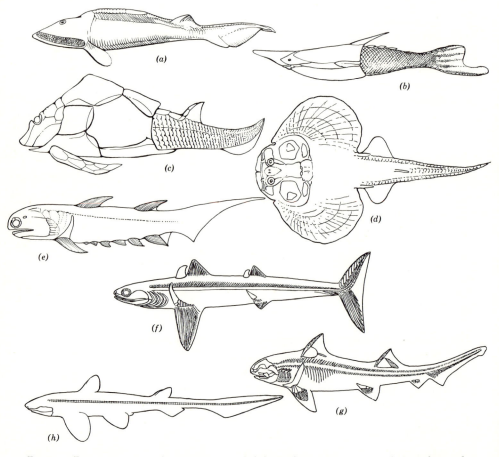

Fig. 2.2   Representatives of groups of fossil fishes. Shown are (a) a cephalaspidomorph (*Hemicyclaspis*); (b) a pteraspidomorph (*Pteraspis*); (c) an arthrodiran placoderm (*Remigolepis*), restored; (d) an antiarchan placoderm (*Gemuendina*), dorsal view; (e) an acanthodian (*Climatius*); (f) a cladodont-level shark, *Cladoselache*; (g) a hybodont-level shark (*Hybodus*); and (h) oldest known modern shark (*Paleospinax*). (Figs 2.2 f through h redrawn from Schaeffer, 1967).

poorly known. Thus the precise delimitation and positioning of extinct groups has proven to be particularly difficult, not only because of the imperfect record, but also because many stores of fossil materials have not as yet been diagnosed and much of the remains are scattered in study collections over the world.

Interpretation of fossils requires an imaginative mind that is carefully disciplined by thorough training in comparative morphology and systematics coupled with intuitive judgment. Although inferences regarding fossil relationships are being greatly enhanced by the incorporation of new principles and methods of study, there is warm debate among specialists concerning the classification of fossil groups.

All fossil and living fishes are included in either the agnathans or gnathostomes but, whereas the independent evolution of each of these fundamentally different groups in reasonably well known, the phylogeny of fishes as a whole can only be reconstructed in its broadest outlines. There are many groups of jawless and jawed fishes that have failed to survive geological time and are now extinct. The most ancient of the jawless forms are the Paleozoic heterostracans and thelodonts, which are the first vertebrates to appear in the fossil record. These and other Paleozoic agnathans are frequently combined as the ostracoderms, as Romer does (1966), although Miles (in Moy-Thomas and Miles, 1971) avoided this term because he believes it represents an artificial assemblage, and Romer did not use it in his formal classification.

The classification for Agnatha favored by Miles is adopted here, except that ostracoderms are retained. Two classes, Cephalaspidomorphi and Pteraspidomorphi, are used (corresponding to the Monorhina and Diplorhina of Romer). For Gnathostomata, the extinct groups are the Placodermi and the Acanthodii. Although there is no complete agreement on the interrelationships of Paleozoic fishes, the raw materials constitute a valuable record of the early evolution of fishes that involves the following major groups.

Class Cephalaspidomorphi (extinct osteostracans and anaspids, and living hagfishes and lampreys), Fig. 2.2a

In extinct forms,* head and anterior trunk covered with a continuous bony shield. Neurocranium of either mixed cartilage and bone, or (rarely) ossified throughout. A single median nasal opening not communicating with mouth cavity. Two semicircular canals. Gills in pouches as in living lampreys and hagfishes. Branchial skeleton external to the associated musculature. Ten pairs of external gill openings on lower surface of head. Body behind the cephalic

* Living forms characterized on pages 15–17.

shield covered with imbricating scales. Caudal fin heterocercal. Pectoral fins usually present, covered with scales and lacking dermal rays. No pelvic fins (but a paired ventral finfold is present). One or two dorsal fins. Nine orders are recognized by Miles.

Class Pteraspidomorphi (heterostracans, thelodonts), Fig. 2.2b

Head and anterior part of trunk covered with deck of solid bone (carapace) without true bone cells. External nasal apertures apparently absent, but an internal one opens into the mouth cavity. Two semicircular canals. A single external branchial opening on each side. Body behind carapace covered with scales. Caudal fin hypocercal. No other fins. Nine orders are recognized by Miles.

Class Placodermi (placoderms), Fig. 2.2c and d

A very diversified group of extinct, jawed bony fishes that are dorsoventrally compressed, with a head shield movably articulated with a trunk shield that covers the anterior part of the body, the body tapering posterior to the trunk shield, and with the tail heterocercal in most species. The arthrodires, which show the greatest adaptive radiation of all placoderm groups, have been intensively studied for some forty years by Stensiö (1969). The class may be polyphyletic in origin; furthermore it gave rise to both the Chondrichthyes and Osteichthyes. Miles recognizes six orders.

Class Acanthodii (acanthodians), Fig. 2.2e

Acanthodians are the earliest true jawed fishes with true bone in the endoskeleton, small, square-crowned scales, stout spines before the dorsal, anal, and paired fins, and a heterocercal caudal fin. They are the only Paleozoic fishes with paired-fin spines. The position of this group is debatable although recent opinion suggests that acanthodians are more closely related to osteichthyans than they are to chondrichthyans (Miles, 1974).

Two important Mesozoic fossil groups, the pholidophorid and leptolepid fishes, have been discussed at length by Patterson (1975), who concluded that they "stand at the base of the Teleostei" (p. 278). He further stated that "on the evidence of the braincase the living teleosts form a monophyletic group whose origins lie within the leptolepid grade, not the pholidophorid" (p. 562). In the present classification, we have emphasized this point by placing the leptolepid fishes within the Teleostei, although it is conceded that the pholidophoroids are forerunners of the Teleostei (Gosline, 1971, p. 96).

RELATIONSHIPS OF THE MAJOR GROUPS OF FISHES

The relationships of animals in time support the theory of evolution. This support is strong enough in groups where the fossil record is good to carry over into groups where the fossil record leaves much to be desired. Even for groups with scanty evidence, no thoughtful person could conclude that apparently isolated or bizarre forms arose *de novo*. For example, the fact that the earliest known fishes are heavily armored does not preclude the possibility that their ancestors were not so. Rather it seems that the armored forms persisted as fossils whereas their unarmored relatives did not. In general, the sequence of appearance of fish groups is suggestive of an orderly evolution, even though fossil links between major groups remain unknown or not understood.

Progress in understanding fish interrelationships has been hampered by the incompleteness of the fossil record as well as by a scarcity of primitive forms and an excess of highly evolved forms in both fossil and living faunas. It has also suffered in the past from the treatment of fossil and living fishes as two independent fields of investigation. That this dichotomy is being bridged, however, is well shown in a recent important volume on this subject (Greenwood, Miles, and Patterson, 1974).

Knowledge of the earliest known fishes is based on bony fragments of heterostracans, dating from deposits of middle Ordovician age, about 500 million years ago. By Silurian time the two major groups of vertebrates—the jawless fishes, or Agnatha, and the earliest jawed vertebrates, or Gnathostomata—were already well established.

The morphological and evolutionary gap that separates the Agnatha from the other classes of fishes is actually greater than that between the gnathostome fishes and the four classes of the tetrapods. Thus fishes, or Pisces, used in the broad sense, embrace a diverse assemblage of ancient, unrelated lineages that have been arranged by different authorities in from four to more than a dozen classes.

Primitive Fishes

Most primitive of living vertebrates are the cephalaspidomorph cyclostomes, comprising the lampreys and the hagfishes, which constitute greatly modified, degenerate descendants of armored agnathans. In addition to the absence of biting jaws, these slimy and superficially eel-like creatures, some of which have become adapted primarily for life as parasites or scavengers, also lack the paired appendages—fins or limbs—that are present (except when secondarily lost) in all of the more advanced fishes. They further possess a purely cartilaginous skeleton, once thought also to constitute a primitive characteristic. It has been well demonstrated, however, that bone, or bonelike tissue, and cartilage are both exceedingly ancient (Moss, 1964).

The ancestral line from which the cyclostomes sprang comprises the jawless, heavily armored, bottom-living agnathous ostracoderms, with their dorso-ventrally compressed and finless bodies encased in a bony shell. Perhaps in part because of their well ossified skeletons, these were the most conspicuous types of vertebrates during the oldest stages of known fish history—the late Silurian and the Devonian. They varied in length from only a few to thirty centimeters or so. Until the advent of their revolutionary study by the great Swedish paleontologist Erik Stensiö in the late 1920s, the ostracoderms had been superficially described and largely ignored. By painstaking and elaborate techniques, Stensiö demonstrated that there was fundamental anatomical agreement between the cephalaspids (Fig. 2.2a; one of the major groups of ostracoderms) and the lampreys (Petromyzonidae), especially with their ammocete larvae. However, the hagfishes (Myxinidae) differ from the lampreys in having the nasal opening at the extremity of the head, the pituitary canal opening into the pharynx, 5 to 15 (rather than 7) gill pouches on each side, a rudimentary branchial skeleton, one (rather than two distinct) semicircular canal, and the eggs larger with direct development (no larval stage). Although hagfishes are classified with lampreys and cephalaspidomorphs here, they were possibly derived from a different lineage (the pteraspids; Fig. 2.2b). A third evolutionary line (the thelodonts) comprises only extinct ostracoderms.

The striking external changes from the cephalaspid ancestral line to the living lampreys are partly the result of regressive development of the skeleton, which first brought about a disintegration and finally a total loss of the bony skeleton. Ostracoderms still flourished in the early Devonian but they had vanished by the end of that period, during which other, higher vertebrates had already appeared.

The presence of bone, covering early agnathans from snout to tail, has puzzled investigators. It has been suggested that this armor was evolved as a defense against the scorpion-like eurypterids, which were contemporaneous potential predators. There is an alternate idea that the armor "waterproofed" these early vertebrates (which were of salt-water ancestry) against osmotic flooding of their body tissues in fresh water.

Early Jawed Fishes

The Gnathostomata, which include the jaw-bearing fishes, comprise the basal stocks from which sprang the tetrapods (Fig. 2.3). Somewhat off the main line of bony fish evolution, near the base of the Elasmobranchiomorphi, lie the placoderms (Fig. 2.3). The remains of these fishes first appear during late

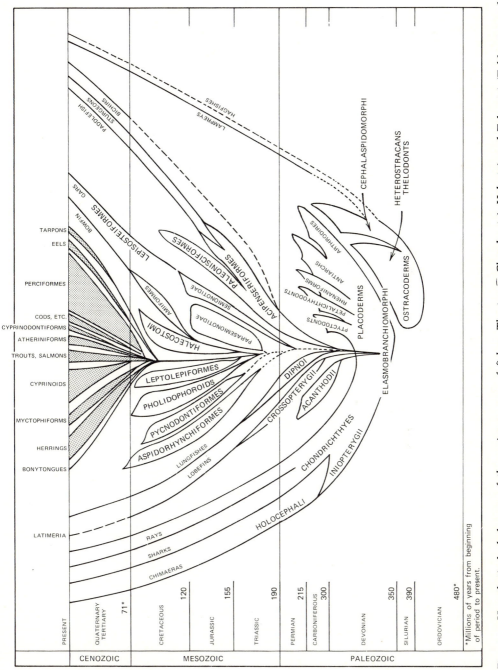

Fig. 2.3 Hypothetical phylogeny of the major groups of fishes. The terms Chondrostei, Holostei, and Teleostei (Tables 2.2 and 2.3) represent grades that cut across phylogenetic lines.

23

Silurian time, and they flourished in the Devonian period, the "Age of Fishes," and then became extinct in the Carboniferous. Their history is thus confined to the Paleozoic Era (Moy-Thomas and Miles, 1971). All have bony skeletal tissue and paired fins, although some of these are built on unusual plans.

The "spiny sharks" or acanthodians (Fig. 2.2f) were a fairly large and very successful group characterized by the possession of stout spines strengthening all fins except the caudal, which was strongly tilted upward in the heterocercal fashion of the sharks. Their body covering consisted of a series of small, flat bony scales, quite unlike shark denticles. The large spines, probably not movable, evidently served as stabilizers. They had exceptionally short snouts, as a result of reduction in the ethmoid region of the head and the large orbits. The gill region was apparently covered by opercular structures. Their bodies were encased in a complete armor of true scales, the microscopic structure of which is almost identical with that of the ganoid scales of certain actinopterygians. Acanthodians typically grew no larger than a few centimeters long, although some aberrant marine genera reached considerable size. In the Lower Devonian they had already attained their peak in numbers and variety and comprise one of the most abundant vertebrate groups in deposits of that age. These fishes show an interesting combination of primitive and advanced features, and although they do not appear to be along the direct line from which higher fishes were evolved, they are clearly close to the ancestral stock that gave rise to bony fishes and are placed by some recent workers at or near the base of the Osteichthyes (see discussion by Romer, 1966). *Acanthodes bronni* (the Permian species that reveals details of internal structure) and primitive osteichthyans show fundamental mutual resemblances in the skull that are not shared by chondrichthyans (Miles, 1974).

The placoderms appeared at about the same time as the acanthodians. However, except for a single genus that survived into the Carboniferous, they died out at the close of Devonian time, over 300 million years ago. Two major subgroups are recognized, the Arthrodira (Fig. 2.2e), and the Antiarchi (Fig. 2.2d).

The so-called joint-necked armored fishes, the arthrodires, were the largest of the placoderm assemblages and form the commonest remains of Devonian vertebrates. The vernacular refers to the arrangement of the bony armor in two rigid parts, one covering the head and gill region (the head shields) and the second enclosing much of the trunk (the thoracic shields); the latter segment articulated with the anterior shield by ball-and-socket joints. Thus the head was for the first time freely movable up and down on the trunk. Their rather depressed body form suggests a bottom-dwelling habit. These peculiar fishes, ranging from 0.5 to perhaps 10 meters in length, had odd jaw plates quite unlike the jaw structures of higher vertebrates. The general absence of armor on the posterior part of the body is indicative of improvements in locomotion that enabled these fishes to begin the abandonment of heavy armor.

Another, still more aberrant assemblage assigned to the Placodermi is the antiarchs, grotesque little fishes with jointed bony appendages that constituted neither spines nor fins. They were widely distributed in Devonian time and were mostly rather small, not more than 30 cm. Their basic features resembled those of the arthrodires, from which they differed chiefly in the very feeble, nibbling jaws, and in the extraordinary, jointed, and freely movable pectoral "fins"—perhaps derived from the fixed spines found in the previous two groups, and used to hold the animal to the substrate in rivers. Detailed anatomical studies have suggested the possibility that these animals had lungs.

Sharks and Their Relatives (Chondrichthyes)

The next major group of fishes, the class Chondrichthyes, comprises the sharks, rays, chimaeras, and their fossil relatives. Sharks and rays are the last major fish group to appear in the fossil record. Internal fertilization is characteristic of all living forms and is known in all Jurassic sharks; it may be inferred as far back as late Devonian times. The skeleton lacks bone; however, as we have noted above, this is believed to be a derived rather than an ancestral condition, and the sharks' ancestry is to be sought among the early bony fishes along the evolutionary path that gave rise to the placoderms and their relatives (Fig. 2.3). The main line of shark evolution took place in the sea.

Shark remains first appear in the Lower Devonian as teeth and spines of a type that are characteristic of a number of early elasmobranchs. By late Devonian time, two primitive shark types are known to have borne such teeth. One is the famous genus *Cladoselache* (Fig. 2.2f) in which the paired fins are broad based and no trace of claspers has been found. By Carboniferous time there were genera seemingly transitional between these Devonian types and the relatively primitive forms known from later Paleozoic to mid-Mesozoic strata. Sharks and rays of modern type are traceable back to the Jurassic period when their extensive adaptive radiation began. Thus the fossil history of elasmobranchs, from late Devonian to the present, is comparatively well known (Zangerl, 1974). After diversification during the Carboniferous period these fishes became greatly reduced in number and variety by the close of the Permian—a time that marked the extinction of many ancient forms. After expansion of groups again in the Triassic and Jurassic, nearly all the modern lineages of the sharks, rays, and skates were represented in early Cretaceous seas.

In their evolutionary history the elasmobranchs passed through two successive levels of organization or radiations, the "cladodont," and the "hybodont," in leading to the modern forms, which attained a new (hyostylic) jaw suspension and an axial skeleton with calcified vertebral centra (the galeoid, squaloid, and batoid lines). *Cladoselache* (Fig. 2.2f) is a representative of the cladodont level, *Hybodus* (Fig. 2.2g) of the hybodont level, and *Paleospinax* (Fig. 2.2h) is the oldest known modern shark.

Present-day chimaeras (Holocephali) are grotesque marine fishes that feed principally on shellfish. They possess cartilaginous skeletons and other shark-like attributes but it has been thought that the two groups are only distantly related. Recent fossil discoveries (Zangerl, 1974), however, indicate that the Carboniferous forerunners of this group, the Iniopterygii (Fig. 2.3), which have been interpreted as holocephalans, have dentitional features that link them to early elasmobranchs; thus the view (source cited in Zangerl) that holocephalans may have arisen from some branch of the placoderms is equivocal.

Bony Fishes (Osteichthyes)

The major groups of fishes considered thus far, although extremely important in Paleozoic strata, largely died out (except for sharks) by the close of the Permian. The major class of living forms, the Osteichthyes or bony fishes, has, in contrast, continually increased in importance since its inception in early Paleozoic times. Although the fossil record indicates that bony fishes appeared no earlier than the late Silurian, the major groups were well established by middle Devonian time and thus it is reasonable to infer that the beginnings of the group extend back into Ordovician times, over 500 million years ago.

Although the Osteichthyes possess a bony skeleton, we have already seen that this feature is not exclusive to them. The fossil evidence clearly points to the conclusion that their ancestors not only had bone but many of them were better ossified than their living descendants. Three major groups of bony fishes may be recognized (the Acanthodii, treated above, are by some regarded as a subclass of the Osteichthyes): (1) the lobefins or crossopterygians (Crossopterygii), important as the ancestral line leading to the tetrapods, (2) the lungfishes or dipnoans (Dipnoi), and (3) the rayfin fishes (Actinopterygii). Groups (1) and (2) equal the subclass Sarcopterygii of Romer (1966) but not of Nelson (1969) who believes this group should include also the Polypteriformes (bichirs and reedfishes), here placed within the grade Chondrostei under group (3).

Dipnoans range from Lower Devonian to recent with six species in three genera and two families surviving today; there are about 35 extinct genera in nine families.

Lobefins also range from Lower Devonian to recent with about 55 extinct genera and only one living species (*Latimeria chalumnae*, known since 1938).

Rayfins, ranging from Lower Devonian to recent, comprise about 20,000 living species, with many extinct groups.

The earliest rayfinned fishes were the mainly Paleozoic palaeoniscoids—a vast group tremendously rich in diversified forms. The majority of them were of modest size and bore a close superficial resemblance to their actinopterygian successors—for example, the gars (Lepisosteidae), herrings (Clupeidae), and minnows (Cyprinidae). There were, however, fundamental structural differences—the thick ganoid scales, heterocercal caudal fin, primitive cranial pattern, and a shoulder girdle that possessed a clavicle as well as a cleithrum. In *Cheirolepis*, one of the most primitive genera, the tiny, squarish scales are remarkably like those of the shark-like acanthodians (see above); other primitive traits included a long gape and maxilla and the broad-based, many-rayed, pelvic fins.

The paleoniscoids, ancestral to modern teleosts, were the dominant freshwater fishes from Carboniferous to early Triassic times. The adaptive radiation of this extinct group may have rivalled that of present-day bony fishes. Modern representatives of the paleoniscoids are a few specialized, degenerate forms such as the sturgeons (Acipenseridae) and the paddlefishes (*Polyodon, Psephurus*), with their almost wholly cartilaginous internal skeletons, archaic and shark-like fins, reduced scales (forming bony plates in sturgeons), etc. Actinopterygians of Mesozoic ancestry are the gars and bowfins, which persist today as only two genera, *Lepisosteus* and *Amia*, the gars and bowfin respectively. Ancestral gars were most abundant in the early Mesozoic, whereas the bowfins were especially well represented in middle Mesozoic strata.

Paleoniscoid ancestry seems, by passage through two intermediate stages, to have arrived at the great flowering point that is manifested by the true bony fishes of today. Descriptive terms, of historical significance in ichthyology have been applied to these stages, although it is recognized that these are artificial and that there is not complete agreement on the fish families to be included in each group. Nevertheless, the terms are descriptive of living representatives of the groups. Progressively from primitive through intermediate to advanced they are (*a*) Chondrostei, including the sturgeons (Acipenseridae) and the paddlefishes (Polyodontidae); (*b*) Holostei (gars, Lepisosteidae; bowfin, Amiidae); and (*c*) Teleostei, the vast array of commonly encountered, ordinary fishes. Recognized trends in evolution through these stages are summarized in Table 2.2. There is growing indication that the evolution through these groups was devious (not orthogenetic), with the advanced features independently derived along different lines, and with both the Holostei and Teleostei being of polyphyletic (multiple) origin. In general, the evolutionary history of the Actinopterygii has been one in which skeletal structures have undergone reduction; this has been true of the endocranium, the dermal bones, and the scales.

In the most numerous higher bony fishes, the Teleostei (Table 2.3), great

Table 2.2. Evolutionary Changes in Rayfin Fishes (Actinopterygii)—Modified from Colbert (1955) and Bailey and Cavender (1971)

Structure	Chondrostei	Holostei	Teleostei
Caudal fin	Typically heterocercal, with notochord extending into upper lobe	Abbreviate heterocercal, with notochord ending at fin base	Variously homocercal, rays attached to enlarged hypural bones (6 primitively) at end of vertebral column
Spiracle	Present	Absent	Absent
Upper jaw	Attached to cheek bones	Attached only at snout	Attached only at snout; upper jaw protractile
Notochord	Unreduced	Present; surrounded or replaced by bony vertebrae in some	Replaced by bony vertebrae
Skull roof	Nearly smooth	Nearly smooth	With crests and depressions
Fulcra[a]	Well developed (reduced in living forms)	Well developed (gars) or absent (bowfin)[b]	Absent
Cheek	Covered with bones	Covered with bones[c]	Open
Paired fins	Broad-based	Narrow-based	Narrow-based; modified in position from primitive condition to pelvic forward and pectoral elevated

[a] Spinelike scales along fronts of fins.
[b] Present in fossil amiids.
[c] Except in advanced pycnodonts.

Table 2.3. **Summary of Some Evolutionary Changes in the Modern Bony Fishes (Teleostei)**

Character	Soft-Rayed Species[a] (Malacopterygii)	Spiny-Rayed Species (Acanthopterygii)
Fin rays	Soft, composed of bilateral, segmented elements, and often branched (rarely secondarily hardened and pungent)	Comprising (1) medial, non-segmented, usually hard and pungent, but sometimes feeble spines, and (2) soft-rays
Scales	Cycloid (occasionally ctenoid)	Ctenoid (ctenii often lost)
Maxillary	Enters gape	Excluded from gape
Pelvic fins	Typically abdominal	Typically thoracic or jugular
Pelvic soft-rays	Usually more than 5, without spine	Typically 5, preceded by a spine
Pectoral-fin base	Chiefly more or less ventral and transverse	More or less lateral and vertical
Gas bladder	Physostomous	Physoclistous
Principal caudal rays	19 (not often reduced)	17 (often reduced)
Pancreas	Usually well developed as a separate organ	Usually incorporated in liver, to form hepato-pancreas
Branchiostegal rays	Variable, typically not folded like a fan beneath opercle	4 attached on outside or upper part of hyoid arch, 1 to 3 (rarely 0 to 7 or more) attached to inside of lower part of arch; folded like a fan under opercles
Orbitosphenoid	Usually present	Absent (except in Beryciformes)
Mesocoracoid	Usually present	Absent

[a] The soft-rayed fishes in general are orders Elopiformes through Notacanthiformes; the orders beyond the latter are essentially spiny-rayed.

divergence occurred among the generalized types of herring-like and salmon-like fishes (variously termed the Isospondyli or, approximately, Clupeiformes). They compose a large share of the soft-rayed fishes, the "Malacopterygii" or "Malacopteri" of early authors. Many families both of marine and freshwater habits have flowered in this group. Typically they are bony but soft-bodied, have fins supported only by soft-rays (not by spines), possess cycloid scales, have the maxillary bone forming a part of the gape (upper border of the mouth opening), and have the gas bladder (air bladder, swim bladder) opening into the esophagus (physostomous condition). In Cretaceous time the berycoids or Beryciformes provided a transition to spiny-rayed fishes (more or less the "Acanthopterygii" or "Acanthopteri" of various authors), and another great flowering of families occurred.

The largest group resulting from divergence of the mainstem of evolution is the percoid or percomorph fishes, the Perciformes. To this group belong thousands of marine and freshwater fishes, typified by the world-wide sea basses (Serranidae) and the freshwater perches (Percidae) of the Northern Hemisphere. In the living members, fins are supported both by spines and by soft-rays, the scales are most often ctenoid, the maxillae are excluded from the gape of the mouth by elongation of the premaxillae, and the pelvic fins, when present (as they usually are) are thoracic or jugular in position (with a reduced formula, typically of one spine and five soft-rays) and the gas bladder lacks a duct to the esophagus (physoclistous condition).

LIST OF COMMON AND REPRESENTATIVE FAMILIES OF LIVING FISHES

Because there has been a great flurry of recent activity on the classification of fishes (see references), the system of Berg (1940), heretofore used by us, is no longer tenable. "Fishes of the World" (Lindberg, 1971) contains helpful keys and characterizes all families of fishes (550 in 62 orders, as recognized by him), but is a compilation rather than a synthesis. The system used below follows no single, current authority, but does represent a concerted effort to select what seemed to us to be the current most reasonable choices.

Agnatha
    Class Cephalaspidomorphi
        Subclass Cyclostomata
            Order Myxiniformes
                Family Myxinidae—hagfishes (Fig. 2.4)
            Order Petromyzoniformes
                Family Petromyzonidae—lampreys (Figs. 2.1 and 2.5)
Gnathostomata

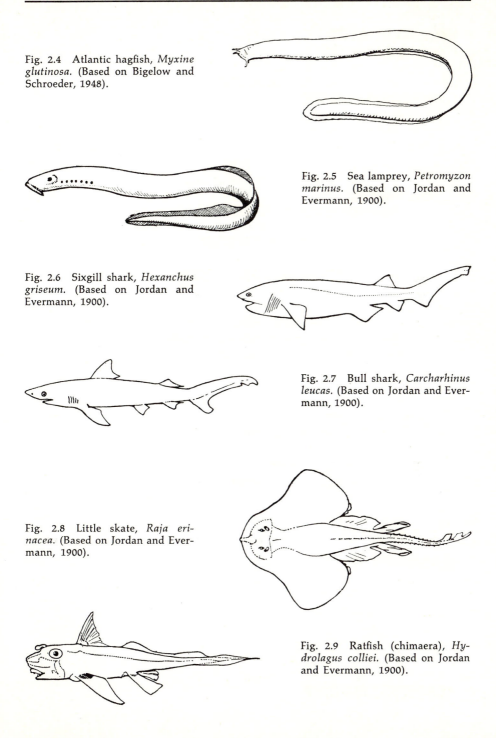

Fig. 2.4 Atlantic hagfish, *Myxine glutinosa*. (Based on Bigelow and Schroeder, 1948).

Fig. 2.5 Sea lamprey, *Petromyzon marinus*. (Based on Jordan and Evermann, 1900).

Fig. 2.6 Sixgill shark, *Hexanchus griseum*. (Based on Jordan and Evermann, 1900).

Fig. 2.7 Bull shark, *Carcharhinus leucas*. (Based on Jordan and Evermann, 1900).

Fig. 2.8 Little skate, *Raja erinacea*. (Based on Jordan and Evermann, 1900).

Fig. 2.9 Ratfish (chimaera), *Hydrolagus colliei*. (Based on Jordan and Evermann, 1900).

Class Chondrichthyes
  Subclass Holocephali
    Order Chimaeriformes
      Family Chimaeridae—chimaeras (Fig. 2.9)
  Subclass Elasmobranchii (Selachii)
    Order Heterodontiformes
      Family Heterodontidae—bullhead sharks
    Order Hexanchiformes
      Family Hexanchidae—cow sharks (Fig. 2.6)
      Family Chlamydoselachidae—frill sharks
    Order Squaliformes
      Family Odontaspidae—sand sharks
      Family Alopiidae—thresher sharks
      Family Lamnidae—mackerel sharks
      Family Orectolobidae—carpet sharks
      Family Rhincodontidae—whale sharks
      Family Cetorhinidae—basking sharks
      Family Scyliorhinidae—cat sharks
      Family Carcharhinidae—requiem sharks (Fig. 2.7)
      Family Sphyrnidae—hammerhead sharks
      Family Squalidae—dogfish sharks (Figs. 2.1 and 3.15a)
      Family Squatinidae—angel sharks
    Order Pristiophoriformes
      Family Pristiophoridae—saw sharks
    Order Rajiformes (Batoidei)
      Family Pristidae—sawfishes
      Family Rhinobatidae—guitarfishes
      Family Rajidae—skates (Figs. 2.8 and 6.2d)
      Family Dasyatidae—stingrays
      Family Myliobatidae—eagle rays
      Family Mobulidae—mantas
    Order Torpediniformes
      Family Torpedinidae—electric rays
Class Osteichthyes
  Subclass Crossopterygii
    Order Coelacanthiformes
      Family Coelacanthidae—coelacanths (Figs. 2.1 and 2.10)
  Subclass Dipnoi
    Order Dipteriformes
      Family Ceratodontidae—Australian lungfish (Fig. 2.11)
      Family Lepidosirenidae—South American and African lungfishes
  Subclass Actinopterygii

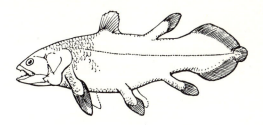

Fig. 2.10 Coelacanth, *Latimeria chalumnae*. (Based on E. London Mus., S. Africa).

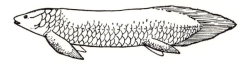

Fig. 2.11 Australian lungfish, *Neoceratodus forsteri*. (Based on Norman, 1931).

Fig. 2.12 Bichir, *Polypterus senegalus*. (Based on Bridge, 1904).

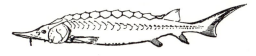

Fig. 2.13 Atlantic sturgeon, *Acipenser oxyrhynchus*. (Based on Jordan and Evermann, 1900).

Fig. 2.14 Bowfin, *Amia calva*. (Based on Jordan and Evermann, 1900).

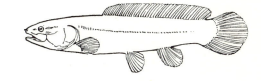

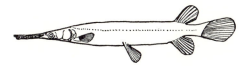

Fig. 2.15 Longnose gar, *Lepisosteus osseus*. (Based on Jordan and Evermann, 1900).

Fig. 2.16 Herring, *Clupea harengus*. (Based on Jordan and Evermann, 1900).

Order Polypteriformes
    Family Polypteridae—bichirs and reedfishes (Fig. 2.12)
Order Acipenseriformes
    Family Acipenseridae—sturgeons (Fig. 2.13)
    Family Polyodontidae—paddlefishes (Fig. 5.5g)
Order Amiiformes
    Family Amiidae—bowfin (Figs. 2.14 and 6.2a)
Order Lepisosteiformes   Semiodontiformes
    Family Lepisosteidae—gars (Fig. 2.15)
Order Elopiformes
  Suborder Elopoidei
    Family Elopidae—tenpounders
    Family Megalopidae—tarpons
  Suborder Albuloidei
    Family Albulidae—bonefishes
Order Anguilliformes
  Suborder Anguilloidei
    Family Anguillidae—freshwater eels (Fig. 2.25)
    Family Muraenidae—morays
    Family Dysommidae—arrowtooth eels
    Family Muraenesocidae—pike congers
    Family Nettastomidae—duckbill eels
    Family Congridae—conger eels
    Family Ophichthidae—snake eels
    Family Simenchelyidae—snubnose eels
    Family Derichthyidae—longneck eels
    Family Nemichthyidae—snipe eels (Fig. 3.1)
  Suborder Saccopharyngoidei
    Family Saccopharyngidae—swallowers (Fig. 2.22)
    Family Eurypharyngidae—gulpers
Order Notacanthiformes
    Family Halosauridae—halosaurid eels (Fig. 2.26)
    Family Notacanthidae—spiny eels (Fig. 2.27)
Order Clupeiformes
  Suborder Denticipitoidei
    Family Denticipitidae—African herring
  Suborder Clupeoidei
    Family Dussumieriidae—round herrings
    Family Clupeidae—herrings (Figs. 2.16 and 8.12)
    Family Engraulidae—anchovies
Order Osteoglossiformes
  Suborder Osteoglossoidei

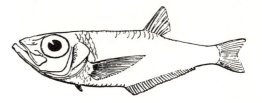

Fig. 2.17   Deepsea herring, *Bathyclupea argentea*. (Based on Jordan and Evermann, 1900).

Fig. 2.18   New Zealand trout, *Galaxias brevipinnis*. (Based on Boulenger, 1904).

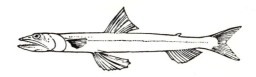

Fig. 2.19   Inshore lizardfish, *Synodus foetens*. (Based on Jordan and Evermann, 1900).

Fig. 2.20   Ateleopid (deepsea), *Ateleopus japonicus*. (Based on Goode and Bean, 1895).

Fig. 2.21   Giganturid (deepsea), *Gigantura vorax*. (Based on Regan, 1925).

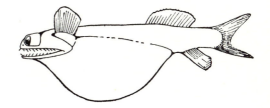

Fig. 2.22   Swallower, *Saccopharynx ampullaceus*. (Based on Jordan and Evermann, 1900).

Family Osteoglossidae—bonytongues
Family Pantodontidae—African mudskipper
Suborder Notopteroidei
Family Hiodontidae—mooneyes
Family Notopteridae—featherbacks
Suborder Mormyroidei
Family Mormyridae—elephantfishes (Fig. 2.23 and 5.5a)
Family Gymnarchidae—gymnarchids
Order Salmoniformes
Suborder Salmonoidei
Family Salmonidae—trouts, salmons, whitefishes, and graylings
(Figs. 3.15b and 10.4)
Family Plecoglossidae—ayu
Family Osmeridae—smelts
Suborder Argentinoidei
Family Argentinidae—argentines
Family Opisthoproctidae—spookfishes
Family Bathylagidae—deepsea smelts
Family Alepocephalidae—deepsea slickheads
Family Searsidae—searsids
Family Bathylaconidae—bathylaconids
Suborder Galaxioidei
Family Salangidae—salangids
Family Retropinnidae—southern smelts
Family Galaxiidae—galaxiids (New Zealand trout, Fig. 2.18)
Suborder Esocoidei
Family Umbridae—mudminnows
Family Esocidae—pikes
Suborder Stomiatoidei
Family Gonostomatidae—lightfishes
Family Sternoptychidae—deepsea hatchetfishes
Family Photichthyidae—photichthyids
Family Astronesthidae—deepsea snaggletooths (Fig. 8.14)
Family Idiacanthidae—deepsea stalkeyefishes
Family Malacosteidae—deepsea loosejaws
Family Melanostomiatidae—scaleless dragonfishes
Family Stomiatidae—deepsea scaly dragonfishes
Family Chauliodontidae—viperfishes
Suborder Giganturoidei
Family Giganturidae—giganturids (Fig. 2.21)
Order Gonorynchiformes
Suborder Chanoidei

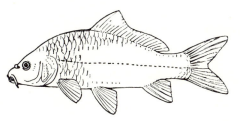

Fig. 2.23 Elephantfish, *Mormyrus caballus*. (Based on Boulenger, 1904).

Fig. 2.24 Carp, *Cyprinus carpio*. (Based on Hubbs and Lagler, 1941).

Fig. 2.25 American eel, *Anguilla rostrata*. (Based on Jordan and Evermann, 1900).

Fig. 2.26 Halosaurid eel, *Aldrovandia macrochir*. (Based on Jordan and Evermann, 1900).

Fig. 2.27 Deepsea spiny eel, *Notacanthus analis*. (Based on Jordan and Evermann, 1900).

Fig. 2.28 Atlantic saury, *Scomberesox saurus*. (Based on Jordan and Evermann, 1900).

Fig. 2.29 Atlantic cod, *Gadus morhua*. (Based on Jordan and Evermann, 1900).

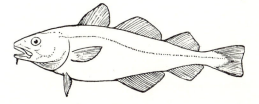

Family Chanidae—milkfishes
Family Kneriidae—kneriids
Family Phractolaemidae—phractolaemids
Suborder Gonorynchoidei
Family Gonorynchidae—gonorynchids
Order Cypriniformes (Ostariophysi)
Suborder Characoidei
Family Characidae—characins*
Family Gymnotidae—gymnotid eels (Fig. 6.2b)
Suborder Cyprinoidei
Family Cyprinidae—minnows and carps (Figs. 2.24, 6.4, 10.4 creek
    chub, and 10.6 bitterling)
Family Gyrinocheilidae—gyrinocheilids
Family Catostomidae—suckers (Fig. 5.3)
Family Homalopteridae—hillstream loaches
Family Cobitidae—loaches
Family Gastromyzonidae—suckerbelly loaches
ORDER SILURIFORMES
Suborder Siluroidei
Family Diplomystidae—diplomystid catfishes
Family Ictaluridae—North American freshwater catfishes (Figs. 3.1
    and 10.6 bullhead)
Family Bagridae—bagrid catfishes
Family Pimelodidae—pimelodid catfishes
Family Siluridae—Eurasian catfishes
Family Schilbeidae—schilbeid catfishes
Family Amblycipitidae—amblycipitids
Family Amphiliidae—amphiliids
Family Akysidae—akysids
Family Sisoridae—sisorids
Family Clariidae—airbreathing catfishes (Fig. 8.10b)
Family Heteropneustidae—stinging catfishes (Fig. 8.10c)
Family Chacidae—chacids
Family Olyridae—olyrids
Family Malapteruridae—electric catfishes
Family Mochokidae—upsidedown catfishes
Family Ariidae—sea catfishes
Family Doradidae—doradid armored catfishes
Family Aspredinidae—banjo or obstetrical catfishes (Fig. 10.6)
Family Plotosidae—plotosid sea catfishes
Family Ageneiosidae—ageneiosids

* The bases for recognition of characoid families are not sufficiently grounded to warrant
more than those recognized here.

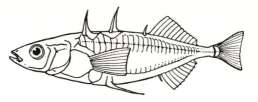

Fig. 2.30 Threespine stickleback, *Gasterosteus aculeatus*. (Based on Jordan and Evermann, 1900).

Fig. 2.31 Pipefish, *Pseudophallus starksi*. (Based on Jordan and Evermann, 1900).

Fig. 2.32 Oarfish, *Regalecus glesne*. (Based on Jordan and Evermann, 1900).

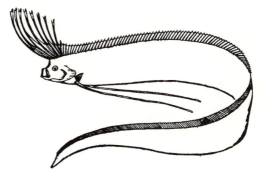

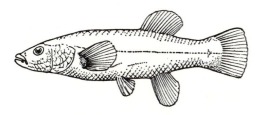

Fig. 2.33 Striped killifish, *Fundulus majalis*. (Based on Jordan and Everman, 1900).

Fig. 2.34 Priapiumfish, *Noestethus amaricola*. (Based on Berg, 1940).

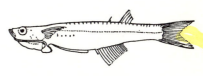

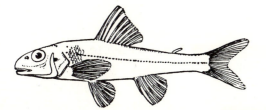

Fig. 2.35 Trout-perch, *Percopsis omiscomaycus*. (Based on Jordan and Evermann, 1900).

Family Hypophthalmidae—hypophthalmids
Family Helogeneidae—helogeneids
Family Cetopsidae—cetopsids
Family Trichomycteridae—parasitic catfishes
Family Callichthyidae—callichthyid armored catfishes
Family Loricariidae—armored catfishes
Order Myctophiformes
Family Synodontidae—lizardfishes (Fig. 2.19)
Family Harpadontidae—bombay duck
Family Alepisauridae—lancetfishes
Family Scopelarchidae—pearleyes
Family Myctophidae—lanternfishes (Fig. 4.6e)
Order Cetomimiformes
Suborder Ateleopoidei
Family Ateleopidae—deepsea ateleopids (Fig. 2.20)
Suborder Cetomimoidei
Family Cetomimidae—cetomimids
Family Mirapinnidae—mirapinnids
Family Eutaeniophoridae—eutaeniophorids
Family Megalomycteridae—megalomycterids
Suborder Rondeletioidei
Family Gibberichthyidae—gibberfishes
Family Barbourisiidae—barbourisiids
Family Rondeletiidae—rondeletiids
Order Beloniformes
Suborder Scomberesocoidei
Family Scomberesocidae—sauries (Fig. 2.28)
Family Belonidae—needlefishes (Figs. 3.1 and 5.5b)
Suborder Exocoetoidei
Family Exocoetidae—flyingfishes
Family Hemiramphidae—halfbeaks (Fig. 5.5c)
Order Cyprinodontiformes
Suborder Adrianichthyoidei
Family Oryziatidae—ricefishes
Family Adrianichthyidae—adrianichthyids
Family Horaichthyidae—horaichthyids
Suborder Cyprinodontoidei
Family Cyprinodontidae—killifishes (Figs. 2.33 and 10.7)
Family Goodeidae—Mexican livebearers
Family Jenynsiidae—jenynsiids
Family Anablepidae—foureyefishes
Family Poeciliidae—livebearers (Figs. 12.1 and 12.2)

Fig. 2.36 Stephanoberycid, *Steph-anoberyx monae*. (Based on Jordan and Evermann, 1900).

Fig. 2.37 Berycid, *Beryx splen-dens*. (Based on Jordan and Ever-mann, 1900).

Fig. 2.38 American John Dory, *Zenopsis ocellata*. (Based on Jor-dan and Evermann, 1900).

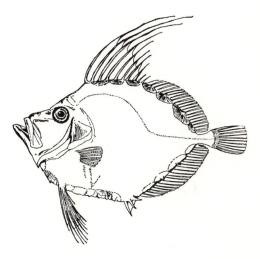

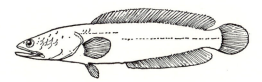

Fig. 2.39 Snakehead, *Channa striata*. (Based on Norman, 1931).

Order Gasterosteiformes
  Suborder Gasterosteoidei
    Family Gasterosteidae—sticklebacks (Figs. 2.30 and 10.6)
    Family Aulorhynchidae—tube-snouts
  Suborder Aulostomoidei
    Family Aulostomidae—trumpetfishes
    Family Fistulariidae—cornetfishes
    Family Macrorhamphosidae—snipefishes
    Family Centriscidae—shrimpfishes
  Suborder Syngnathoidei
    Family Syngnathidae—pipefishes and seahorses (Figs. 2.31, 4.6, and 10.6f)
Order Lampridiformes
  Suborder Lampridoidei
    Family Lamprididae—opahs
  Suborder Trachipteroidei
    Family Radiicephalidae—radiicephalids
    Family Lophotidae—crestfishes
    Family Trachipteridae—ribbonfishes
    Family Regalecidae—oarfishes (Fig. 2.32)
  Suborder Stylephoroidei
    Family Stylephoridae—tube-eyes
Order Beryciformes
  Suborder Stephanoberycoidei
    Family Stephanoberycidae—deepsea pricklefishes (Fig. 2.36)
    Family Melamphaeidae—deepsea bigscale fishes
  Suborder Polymixioidei
    Family Polymixiidae—beardfishes
  Suborder Berycoidei
    Family Diretmidae—diretmids
    Family Trachichthyidae—trachichthyids
    Family Berycidae—berycids (Fig. 2.37)
    Family Monocentridae—pinecone fishes
    Family Anomalopidae—lanterneye fishes
    Family Holocentridae—squirrelfishes
Order Zeiformes
    Family Zeidae—dories (Fig. 2.38)
    Family Grammicolepidae—grammicolepids
    Family Caproidae—boarfishes
    Family Antigoniidae—antigoniids
Order Mugiliformes
  Suborder Mugiloidei

Fig. 2.40 Cuchia, *Amphipnops cuchia*. (Based on Norman, 1931).

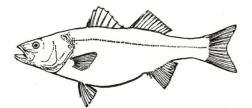

Fig. 2.41 Striped bass, *Morone saxatilis*. (Based on Jordan and Evermann, 1900).

Fig. 2.42 Lookdown, *Selene vomer*. (Based on Jordan and Evermann, 1900).

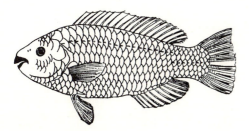

Fig. 2.43 Rainbow parrotfish, *Scarus guacamaia*. (Based on Jordan and Evermann, 1900).

Family Mugilidae—mullets
Suborder Sphyraenoidei
   Family Sphyraenidae—barracudas
Suborder Polynemoidei
   Family Polynemidae—threadfins
Suborder Atherinoidei
   Family Atherinidae—silversides
   Family Melanotaeniidae—(Australian) rainbowfishes
   Family Isonidae—isonids
   Family Neostethidae—neostethid priapiumfishes
   Family Phallostethidae—phallostethid priapiumfishes (Fig. 2.34)
Order Perciformes
Suborder Percoidei
   Family Centropomidae—snooks
   Family Percichthyidae—temperate basses (Fig. 2.41)
   Family Theraponidae—tigerfishes
   Family Serranidae—sea basses (Fig. $4.6c_1$)
   Family Grammistidae—soapfishes
   Family Kuhliidae—aholeholes
   Family Centrarchidae—sunfishes (Figs. 3.1, 3.15c, 6.3, and 10.6)
   Family Percidae—perches (Figs. 2.1, 3.1, 8.3, and 10.6 darter)
   Family Priacanthidae—bigeyes
   Family Apogonidae—cardinalfishes
   Family Branchiostegidae—tilefishes
   Family Pomatomidae—bluefishes
   Family Rachycentridae—cobias
   Family Carangidae—jacks and pompanos (Figs. 2.42 and 10.11)
   Family Coryphaenidae—dolphins (Fig. 10.4b)
   Family Bramidae—pomfrets
   Family Lobotidae—tripletails
   Family Lutjanidae—snappers
   Family Leiognathidae—slipmouths
   Family Gerridae—mojarras
   Family Pomadasyidae—grunts
   Family Sparidae—porgies
   Family Sciaenidae—drums (Fig. $4.6c_2$)
   Family Pempheridae—sweepers
   Family Ephippidae—spadefishes
   Family Kyphosidae—rudderfishes
   Family Mullidae—goatfishes
   Family Monodactylidae—fingerfishes
   Family Bathyclupeidae—deepsea herrings (Fig. 2.17)

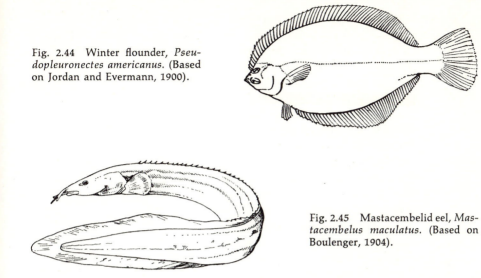

Fig. 2.44 Winter flounder, *Pseudopleuronectes americanus*. (Based on Jordan and Evermann, 1900).

Fig. 2.45 Mastacembelid eel, *Mastacembelus maculatus*. (Based on Boulenger, 1904).

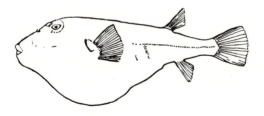

Fig. 2.46 Sharksucker, *Echeneis naucrates*. (Based on Jordan and Evermann, 1900).

Fig. 2.47 Northern puffer, *Sphoeroides maculatus*. (Based on Jordan and Evermann, 1900).

Fig. 2.48 Clingfish, *Gobiesox maeandricus*. (Based on Jordan and Evermann, 1900).

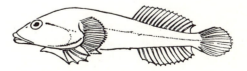

Family Toxotidae—archerfishes
Family Scatophagidae—scats
Family Chaetodontidae—butterflyfishes (Fig. 4.6d and 5.5f)
Family Nandidae—leaffishes (Fig. 4.6g)
Family Badidae—badids
Family Cichlidae—cichlids (Fig. 5.5d)
Family Embiotocidae—surfperches (Fig. 10.6)
Family Pomacentridae—damselfishes
Family Cirrhitidae—hawkfishes
Family Latridae—trumpeters
Family Labridae—wrasses
Family Scaridae—parrotfishes (Fig. 2.43)
Family Trichodontidae—sandfishes
Family Opisthognathidae—jawfishes
Family Bathymasteridae—ronquils
Family Zoarcidae—eelpouts
Family Pholidae—gunnels
Family Mugiloididae—sandperches
Family Trachinidae—weevils
Family Percophididae—flatheads
Family Trichonotidae—sanddivers
Family Dactyloscopidae—sand stargazers
Family Uranoscopidae—stargazers
Family Chiasmodontidae—deepsea swallowers
Suborder Echeneioidei
Family Echeneidae—remoras (sharksuckers; Fig. 2.46)
Suborder Ophidioidei
Family Gadopsidae—river blackfishes
Family Brotulidae—brotulas
Family Ophidiidae—cusk-eels
Family Carapidae—pearlfishes
Suborder Notothenioidei
Family Nototheniidae—Antarctic blennies
Family Bathydraconidae—dragonfishes
Family Channichthyidae—icefishes
Suborder Blennioidei
Family Blenniidae—combtooth blennies
Family Anarhichadidae—wolffishes
Family Clinidae—clinids
Family Stichaeidae—pricklebacks
Family Ptilichthyidae—quillfishes
Family Scytalinidae—graveldivers

Fig. 2.49 Leopard toadfish, *Opsanus pardus*. (Based on Jordan and Evermann, 1900).

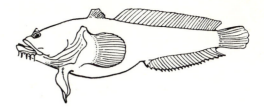

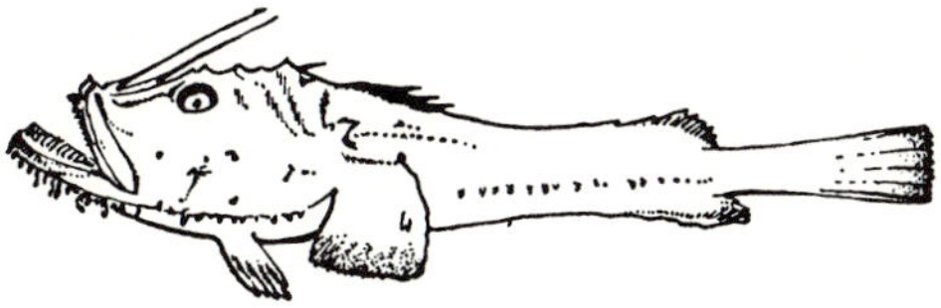

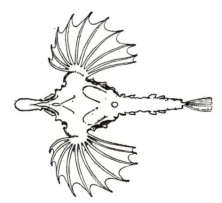

Fig. 2.50 Goosefish, *Lophius americanus*. (Based on Goode and Bean, 1895).

Fig. 2.51 Sea moth, *Pegasus umitengu*. (Based on Jordan, 1905).

Family Zaproridae—prowfishes
Suborder Icosteoidei
    Family Icosteidae—ragfishes
Suborder Schindlerioidei
    Family Schindleriidae—schindleriids
Suborder Ammodytoidei
    Family Ammodytidae—sand lances
Suborder Gobioidei
    Family Eleotridae—sleepers
    Family Gobiidae—gobies
    Family Microdesmidae—wormfishes
Suborder Kurtoidei
    Family Kurtidae—forehead brooders (Fig. 10.6)
Suborder Acanthuroidei
    Family Acanthuridae—surgeonfishes
    Family Siganidae—rabbitfishes
Suborder Scombroidei
    Family Gempylidae—snake mackerels
    Family Trichiuridae—cutlassfishes
    Family Scombridae—mackerels and tunas (Fig. 3.1)
Suborder Xiphioidei
    Family Xiphiidae—swordfishes
    Family Luvaridae—louvars
    Family Istiophoridae—billfishes (marlins, sailfishes, spearfishes)
Suborder Stromateoidei
    Family Stromateidae—butterfishes (Fig. 3.1 harvestfish)
    Family Nomeidae—shepherdfishes
    Family Tetragonuridae—squaretails
Suborder Anabantoidei
    Family Channidae—snakeheads (Figs. 2.39 and 8.10a)
    Family Anabantidae—climbing perches (Fig. 10.6 fightingfish)
    Family Luciocephalidae—pikeheads
Order Gobiesociformes
Suborder Callionymoidei
    Family Callionymidae—dragonets
Suborder Gobiesocoidei
    Family Gobiesocidae—clingfishes (Fig. 2.48)
Order Tetraodontiformes (Plectognathi)
Suborder Balistoidei
    Family Triacanthodidae—spikefishes
    Family Triacanthidae—triplespines
    Family Balistidae—triggerfishes

Family Monacanthidae—filefishes (Figs. 3.1 and 6.2c)
Family Aracanidae—keeled boxfishes
Family Ostraciidae—boxfishes
Suborder Tetraodontoidei
Family Triodontidae—pursefishes
Family Tetraodontidae—puffers (Fig. 2.47)
Family Canthigasteridae—sharpnose puffers
Family Diodontidae—porcupinefishes
Family Molidae—ocean sunfishes (Fig. 3.1)
Order Pleuronectiformes
Suborder Pleuronectoidei
Family Bothidae—lefteye flounders
Family Pleuronectidae—righteye flounders (Fig. 2.44)
Suborder Soleoidei
Family Soleidae—soles
Family Cynoglossidae—tonguefishes
Order Scorpaeniformes
Suborder Scorpaenoidei
Family Scorpaenidae—scorpionfishes and rockfishes
Family Triglidae—searobins
Suborder Platycephaloidei
Family Platycephalidae—flatheads
Suborder Hexagrammoidei
Family Hexagrammidae—greenlings
Family Anoplopomatidae—sablefishes
Family Zaniolepidae—combfishes
Suborder Cottoidei
Family Cottidae—sculpins
Family Agonidae—poachers
Family Cyclopteridae—lumpfishes and snailfishes
Order Mastacembeliformes
Suborder Mastacembeloidei
Family Mastacembelidae—mastacembelid eels (Fig. 2.45)
Suborder Chaudhurioidei
Family Chaudhuriidae—chadhuriids
Order Synbranchiformes
Family Synbranchidae—swamp eels
Family Amphipnoidae—cuchia (Fig. 2.40)
Order Dactylopteriformes
Family Dactylopteridae—flying gurnards
Order Pegasiformes
Family Pegasidae—seamoths (Fig. 2.51)

Order Percopsiformes
  Suborder Amblyopsoidei
    Family Amblyopsidae—cavefishes
  Suborder Percopsoidei
    Family Percopsidae—trout-perches (Fig. 2.35)
  Suborder Aphredoderoidei
    Family Aphredoderidae—pirate perches
Order Gadiformes
  Suborder Gadoidei
    Family Gadidae—codfishes (Fig. 2.29)
    Family Merluciidae—hakes
  Suborder Melanonoidei
    Family Melanonidae—grenadiers (or rattails; formerly Macruridae)
Order Lophiiformes
  Suborder Batrachoidei
    Family Batrachoididae—toadfishes (Fig. 2.49)
  Suborder Lophioidei
    Family Lophiidae—anglerfishes (Figs. 2.50 and 3.1)
  Suborder Antennarioidei
    Family Antenariidae—frogfishes
    Family Ogcocephalidae—batfishes
  Suborder Ceratioidei
    Family Ceratiidae—seadevils (Fig. 10.4 deepsea anglerfish)

SPECIAL REFERENCES ON MAJOR CLASSIFICATION OF FISHES

Bailey, R. M., and T. M. Cavender, 1971. (Fishes). In McGraw-Hill Encyclopedia of Science and Technology, 3rd ed., McGraw-Hill Book Company, New York.

Barlow, G. W., K. F. Liem, and W. Wickler. 1968. Badidae, a new fish family—behavioural, osteological, and developmental evidence. Jour. Zool., London, 156: 415–447.

Berg, L. S. 1940. Classification of fishes both recent and fossil. Trav. Inst. Zool. Acad. Sci. URSS, 5: 87–517. Reprint, 1947. Edwards Brothers, Ann Arbor, Mich.

Birdsong, R. S. 1975. The osteology of Microgobius signatus Poey (Pisces: Gobiidae) with comments on other gobiid fishes. Bull. Florida State Mus., Biol., Sci., 19 (3): 135–187.

Böhlke, J. E., and C. H. Robins. 1974. Description of a new genus and species of clinid fish from the western Caribbean, with comments on the families of the Blennioidea. Proc. Acad. Nat. Sci. Phila., 126 (1): 1–8.

Cohen, D. M. 1964. Suborder Argentinoidei. In Bigelow, H. B., ed., Fishes of the Western North Atlantic. Mem. Sers Found. Mar. Res. No. 1 (4): 1–70.

Colbert, E. H. 1955. Evolution of the vertebrates. John Wiley and Sons, New York.

Eaton, T. H., Jr. 1970. The stem-tail problem and the ancestry of chordates. Jour. Paleo., 44: 969–979.

Gosline, W. A. 1966a. The limits of the fish family Serranidae, with notes on other lower percoids. Proc. Calif. Acad. Sci., ser. 4, 33 (6): 91–112.

——. 1966b. Comments on the classification of the percoid fishes. *Pac. Sci.*, 22 (4): 409–418.

——. 1968. The suborders of perciform fishes. Proc. U.S. Nat. Mus., 124: 1–78.

——. 1971. Functional morphology and classification of teleostean fishes. Univ. Press of Hawaii, Honolulu.

Greenwood, P. H. 1968. The osteology and relationships of the Denticipitidae, a family of clupeomorph fishes. Bull. Brit. Mus. (Nat. Hist.), Zool., 16: 215–273.

Greenwood, P. H., R. S Miles, and C. Patterson, eds. 1974. Interrelationships of fishes. Published for the Linnaean Soc. of London by Academic Press, New York and London.

Greenwood, P. H., D. E. Rosen, S. H. Weitzman, and G. S. Myers. 1966. Phyletic studies of teleostean fishes with a provisional classification of living forms. Bull. Am. Mus. Nat. Hist., 131: 339–456.

Jarvik, E. 1968. The systematic position of the Dipnoi. Nobel Symp., 4: 223–245.

Jefferies, R. P. S. 1968. The subphylum Calcichordata (Jefferies 1967) primitive fossil chordates with echinoderm affinities. Bull. Brit. Mus. (Nat. Hist.), Geol., 16 (6): 241–339.

Jordan, D. S. 1923. A classification of fishes including families and genera as far as known. Stanford Univ. Publ. Biol. Sci., 3 (2): 79–243.

Lehtola, K. A. 1973. Ordovician vertebrates from Ontario. Contrib. Univ. Mich. Mus. Paleo., 24 (4): 23–30.

Lindberg, G. U. 1971. (1974 transl.) Fishes of the world, a key to families and a checklist. John Wiley and Sons (Israel Program for Scientific Translations Ltd.).

McAllister, D. E. 1968. The evolution of branchiostegals and the classification of teleostome fishes, living and fossil. Bull. Nat. Mus. Canada, 221: 1–239.

——. 1971. Old Fourlegs. A "Living Fossil." National Museums of Canada, Nat. Mus. Nat. Soc., Zool. Div., Odyssey Ser. 1: 1–25.

Miles, R. S. 1974. Relationships of acanthodians, pp. 63–103. *In* Interrelationships of fishes, P. H. Greenwood, R. S. Miles, C. Patterson, eds. Published for the Linnaean Society of London by Academic Press, New York.

Monod, T. 1968. Le complexe urophore des poissons téléostéens. Mém. Inst. Fondamental Afrique noire, no. 81, 705 pp.

Moss, M. L. 1964. The phylogeny of mineralized tissues. *Int. Rev. Gen. and Exptl. Zool.*, 1: 298–332.

Moy-Thomas, J. A., and R. S. Miles. 1971. Paleozoic fishes. 2nd ed. Chapman and Hall, London. W. B. Saunders Company, Philadelphia and Toronto.

Nelson, G. J. 1969. Gill arches and the phylogeny of fishes, with notes on the classification of vertebrates. Bull. Am. Mus. Nat. Hist., 141: 475–552.

Nybelin, O. 1971. On the caudal skeleton in *Elops* with remarks on other teleostean fishes. Acta Reg. Soc. Sc. Lit. Gothob. (Zool.), 7: 1–52.

Orvig, T., ed. 1968. Current problems of lower vertebrate phylogeny. Almquist and Wiksell, Stockholm, 539 pp.

Patterson, C. 1975. The braincase of pholidophorid and leptolepid fishes, with a review of the actinopterygian braincase. Philos. Trans. Roy. Soc. London B. Geol. Sci., 269 (899): 275–579.

Roberts, T. 1974. Interrelationships of ostariophysans, pp. 373–395. *In* Interrelationships of fishes, P. H. Greenwood, R. S. Miles, C. Patterson, eds. Published for the Linnaean Society of London by Academic Press, New York.

Romer, A. S. 1966. Vertebrate paleontology. 3rd ed. Univ. Chicago Press, Chicago, 468 pp.

——. 1968. Notes and comments on "Vertebrate Paleontology." Univ. Chicago Press, Chicago, 304 pp.

Schaeffer, B. 1967. Comments on elasmobranch evolution, pp. 3–35. *In* Sharks, skates, and rays. Gilbert, P. W., R. F. Mathewson, and D. P. Rall, eds. Johns Hopkins Press, Baltimore, Maryland.

Stensiö, E. A. 1969. Elasmobranchiomorphi Placodermata Arthrodires, pp. 71–550. *In* Traité de Paléontologie, 4 (2), J. Piveteau, ed. Masson et Cie, Paris.

Thomson, K. S. 1969. The biology of the lobe-finned fishes. *Biol. Rev.*, 44: 91–154.

Tyler, J. C. 1974. Tetraodontiformes. Encycl. Britannica, 15th ed., pp. 162–164.

Weitzman, S. H. 1974. Osteology and evolutionary relationships of the Sternoptychidae, with a new classification of stomiatoid families. Bull. Amer. Mus. Nat. Hist., 153: 327–478.

Zangerl, R. 1974. Interrelationships of early Chondrichthyans, pp. 1–14. *In* Interrelationships of fishes. P. H. Greenwood, R. S. Miles, C. Patterson, eds. Published for the Linnaean Society of London by Academic Press, New York.

3

Basic Fish Anatomy

Biologically speaking, a fish is composed of ten systems of bodily organs that work together to make up the whole individual. These ten systems cover the fish, handle its food, and carry away wastes; they integrate the life processes of the fish and relate it to conditions in the environment; they provide for breathing and for protection against injury; they support the body and enable movement; and finally, they work to perpetuate fish as species and, through evolution, as a major group of animals.

GROSS EXTERNAL ANATOMY (FIGS. 3.1 AND 3.2)

Form (Fig. 3.1; Chapter 6)

Commonly the fish body is torpedo-shaped (fusiform), and most often slightly to strongly ovoid in cross section. In free-swimming species the body approximates the theoretically perfect streamlined form, in which the greatest cross section is located close to 36 percent of the length back from the anterior tip (the entering wedge), and the contours sweep back gently in the tail race. Many fishes depart moderately to completely from the foregoing generalized shape. These departures range from globe shapes (globiform—puffers, Tetraodontidae) through serpentine (anguilliform—eels, Anguillidae), to threadlike in outline (filiform—snipe eels, Nemichthyidae). Some are strongly flattened from side to side (compressed—butterflyfishes, Chaetodontidae, and flounders, Pleuronectidae), others, flattened but greatly elongated (trachip-

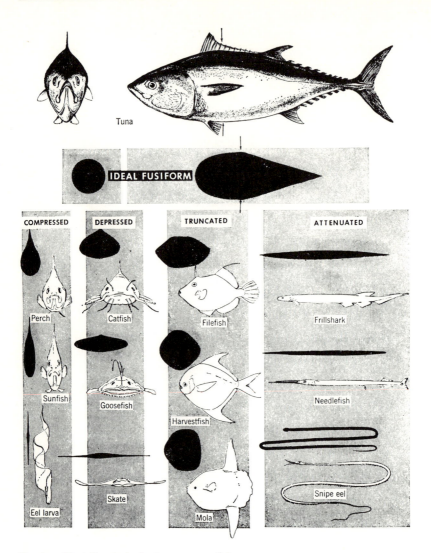

Fig. 3.1   Variation in body form among fishes.

teriform, ribbonfishes, Trachipteridae), and still others, flattened from top to bottom (depressed—the skates, Rajidae, and the batfishes, Ogcocephalidae, etc.).

In spite of the many variations in shape, the ground plan of body organiza-tion in fishes is bilateral symmetry, as for the vertebrates generally. The left and right halves of the body are basically mirror images of one another. Fur-thermore, there is strong cephalization in fishes, along with compactness of body that flows smoothly into the tail (urosome). The tail is consistently an integral part of the body rather than an appendage; only the fins are distinctly peripheral anatomical parts in most fishes.

Body Covering (Chapter 4)

An ordinary fish is covered with a relatively tough skin—at least tough enough in most species so that specimens can be readily skinned for the table or for the study of the muscles and other organs. The skin is continuous with the lining of all the body openings and is transparent as it runs over the surface of the eye. Much of the diverse coloration of fishes is due to its color cells, and the slimy coating is due to its mucous cells.

The skin in many fishes is devoid of scales, but in others it is armored by scales that develop in it. Scales range in size from microscopic to large, in thickness from tissue-thin to plate-thick, in ornamentation from simple to complex, in extent of body coverage from partial to complete, in structure from non-bony to quite bony, and in fastness from loosely deciduous (caducous) to very firmly attached. The types of scales may characterize major fish groups. Scale morphology and the numbers of scale rows along or around the body frequently serve as specific and generic characters. Living agnaths are scaleless; sharks and their relatives have dentinal placoid scales known as denticles; bony fishes have various types of bony scales.

Appendages (Chapter 5)

Appendages of fishes comprise the fins and the cirrhi (flaps of flesh) which attain extreme development in the sargassumfish (*Pterophryne*) and the leafy sea dragon (*Phyllopteryx*, Fig. 4.6f). The fins are classified as median or paired.

Median Fins (Fig. 3.2).   Rayed fins in line with the median axis of a typical fish are those of the back (dorsal fin or fins), the tail (caudal fin), and the lower edge of the body just behind the vent (anal fin). Although most commonly all of the foregoing are present, each is absent in some kinds of fishes. Also developed in the median axis may be a rayless, fatty adipose fin (as in the trouts, Salmonidae) or fins reduced to a few disconnected spines (as in the sticklebacks, Gasterosteidae). The anal fin may be modified into an intromittent organ, the gonopodium, for use in copulation (as in the livebearers, Poeciliidae, Fig. 10.5b).

Paired Fins (Fig. 3.2).   Paired fins are the pectorals and the pelvics (ventrals). The pectorals are supported by the pectoral girdle that joins the skull and, in spiny-rayed fishes, are situated higher on the sides of the fish than the pelvics. Although ordinarily present in fishes, the pectorals are wanting in such kinds

FISH TYPES

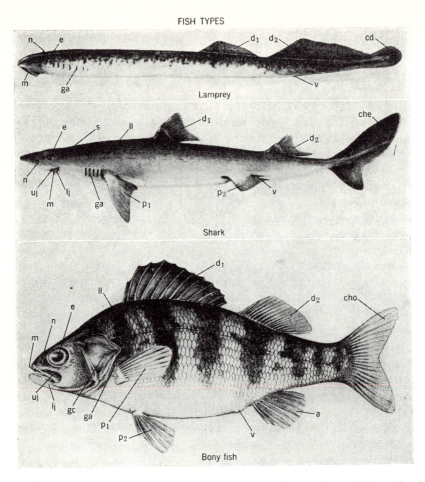

Fig. 3.2 Comparative external anatomy of major groups of living fishes. The following is an index to the labeling. *Appendages:* d_1—first dorsal fin, d_2—second dorsal fin, c—caudal fin (cd, diphycercal caudal; chc—heterocercal caudal; cho—homocercal caudal), a—anal fin, p_1—pectoral fin, p_2—pelvic fin; *Sense organs:* N—naris, e—eye, ll—lateral line; *Skeleton:* lj—lower jaw, uj—upper jaw, gc—gill cover; *Openings:* m—mouth, s—spiracle, ga—gill aperture, v—anus or vent).

as the lampreys (Petromyzonidae) and the hagfishes (Myxinidae). They are greatly enlarged in the soaring flyingfishes (Exocoetidae), the flying gurnards (Dactylopteridae), and the flying characins (Gasteropelecinae). In at least one topminnow (Cyprinodontiformes), one pectoral is modified into a clasper (10.5) and in some clingfishes (Gobiesocidae) and Asiatic suckerbelly loaches (*Gastromyzon*), the pectorals serve as a part of the ventral holdfast organ.

The pelvic fins vary substantially in position and in adaptive modification; typically their support is by a pelvic girdle anchored in the belly musculature.

In most soft-rayed fishes (malacopterygians) the pelvics are *abdomnial* in position (e.g., clupeoids, salmonoids, cyprinoids) but they may also be situated anteriorly, just below the pectorals, in a *thoracic* position (as in many spiny-rayed species, the acanthopterygians), or even under the throat, in a *jugular* position, as in blennies. The pectoral and pelvic girdles are connected by ligament when the pelvics are thoracic or jugular. The pelvics are lacking in the lampreys and hagfishes (the living cephalaspidomorphs) and have become lost in various other fishes, notably the eels (Anguillidae, etc.). In the sharks and some relatives they are modified into claspers (myxopterygia) (10.5), which serve as intromittent organs in copulation. In the clingfishes (Gobiesocidae), suckerbelly loaches (*Gastromyzon*), and certain other fishes, the pelvics form part of a holdfast organ that resembles a suction cup on the belly.

Openings (Fig. 3.2)

The mouth, the gill apertures (including the spiracle in some primitive forms), and the vent (anus) are the principal gross openings connected with the alimentary tract in the fish body. Openings for the sense organs include the narial apertures or nares (Fig. 3.21) and variously distributed small sensory pores.

Mouth (Fig. 3.2).   The fish mouth is situated anteriorly on the head, in *terminal* position, but adaptively its position may be *superior* (opening dorsally) or *inferior* (slightly to prominently overhung by the snout). In relative size it ranges from small (as in the tube-snout *Aulorhynchus* or in small-mouthed cyprinids such as *Phoxinus*) to huge (as in the deepsea swallowers and gulpers, Saccopharyngidae and Eurypharyngidae, respectively). The mouth is bordered by lips which may form a cartilaginous plate (stoneroller, a minnow, *Campostoma*) or may be membranous (as in most fishes) or variously fleshy and often quite papillose (suckers, Catostomidae). In most fishes the lips are scaleless and beset with minute sensory organs, but in the sharks and many of their relatives the placoid scales of the body covering are continuous over the lips to become abruptly enlarged as teeth on the jaws.

Gill Apertures (Fig. 3.2 and 3.13).   In the many fishes with gill covers (the operculate fishes—the bony fishes in general) there is a single opening on each side of the head. Normally this opening is in front of the pectoral fin bases, but in the batfishes (Ogcocephalidae), curiously, it is behind them. The openings are a single pair in the chimaeras (Holocephali) and in some hagfishes (as *Myxine*), neither of which, however, has a bony opercular apparatus. In the other hagfishes, the openings vary from 5 to 14 on each side. All lamprey (Petromyzonidae) have 7. Nearly all sharks (Squaliformes) have 5, only a few, 6 or 7; all rays (Rajiformes and Torpediniformes) have 5.

Principally in the shark group (Chondrichthyes), there is an opening that supplements the gill apertures and extends from the outside into the pharyngeal cavity. This is the spiracle that is located between the anteriormost gill slit and the eye. It lies between the hyoid arch and the cranium and has been cited as evidence of the evolution of function of the hyoid arch from one of gill support to one of jaw, tongue, and throat support. Inside the spiracle a small tuft of gill filaments (the hyoidean pseudobranch) persists. This structure is retained in higher bony fishes although they have no spiracle.

Anus (Figs. 3.2 and 3.12). The anus or vent of a fish is on the mid-ventral line of the body. Most commonly it is in the second half of the over-all length of the individual, behind the bases of the pelvic fins and just in front of the anal fin. Only rarely is it located anteriorly as in the adult pirate perch (*Aphredoderus*), where it is jugular in position. In the sharks and their relatives and in the lungfishes, the anus opens into a depression on the ventral body surface, the cloaca, which contains the exits of the urinary and genital ducts, as well as that of the intestine. In most fishes, however, the openings of the urogenital ducts are at the surface, behind the anus.

Abdominal Pores. Abdominal pores open anteriorly in the vent of some fishes including lampreys (Petromyzonidae) and sharks (Squaliformes, etc). These openings may not be identical in their origins in these two groups, but in both they afford a communication between the body cavity and the exterior. Normally they are paired, one on each side of the midline, although in some individuals only one is present. They may represent vestiges of exits for eggs and/or sperms in ancestral forms. They appear to have been lost entirely by higher fish groups. In certain herrings (Clupeidae) and other teleosts, however, the gas bladder opens posteriorly through a pore near the anus.

Sensory Organs

Nares (Figs. 3.2 and 3.21). One or two nares (nostrils) on each side of the snout (dirhinous condition) leading to a blind sac represent the organs of smell externally among fishes. However, in the lampreys and hagfishes (cyclostome Cephalaspidomorphi) the nostril is single and median (monorhinous condition). Most fishes have the narial openings at the top and sides (dorsolaterally) of the snout. In others, such as the sharks, rays, and skates (Elasmobranchii) the nares are on the ventral surface of the snout. There are differences in the minute anatomy of the narial apertures—ranging from a broad, groove-like opening in the dogfish shark (*Squalus*), through partial division of the aperture by a flap of tissue into incurrent (anterior) and excurrent (posterior) pores to complete separation of the incoming and outgoing water through the development of a separate pore for each current, in most teleosts (Fig. 3.21). There is often value for classification in the fleshy, valvular tissue of the nostrils.

In the living lungfishes (Dipnoi) the external, incurrent nares communicate by passages with the excurrent nares that have come to lie in the mouth cavity; such a communication also appears very rarely among certain of the higher bony fishes (Actinopterygii).

Eyes (Fig. 3.22). The essentially lidless eyes that cannot be closed are situated in orbits, one on each side of the midline of the fish head. Most often the eyes are lateral, with partially independent fields of vision and movement. In many bottom dwellers, including the skates (Rajidae), most sculpins (Cottidae), and the goosefishes (Lophiidae), the eyes are dorsal. In adults of the flounders and their relatives (Pleuronectiformes) both eyes are on one side of the head. The eyes are variously reduced or absent in cavefishes (Amblyopsidae, and the blind cave characin, *Anoptichthys*, among others).

Visible through the transparent skin (conjunctiva) that covers the eye and through the transparent cornea of the eyeball are: (*a*) the opening of the pupil of the eye and, through it, the spherical crystalline lens inside the eyeball; and (*b*) the colored, washer-shaped iris surrounding the pupil.

Skin Organs. Numerous microscopically small openings of skin sensory organs are developed on the surface of the fish body. In most fishes a series of these pores, extending along each side in a single row from the head to the caudal fin, comprises the lateral line (Fig. 3.2). The lateral-line system forms branches about the head, including one above and one below the orbit of each eye (Fig. 11.24). In some fishes (including the northern pike, *Esox lucius*, and the silversides, Atherinidae) the pores and their sensory organs are not linearly concentrated on the body but are rather widely scattered. In sharks and their relatives (Elasmobranchii), there are extra and specialized parts of the lateral-line system, especially in the snout region. In a few fishes taste buds occur in the skin and integumentary tactile sensory structures have also been described in it; neither of these are grossly visible.

SKELETON (CHAPTER 5)

In the skeleton of fishes are included notochord, connective tissues, bone, cartilage, non-bony scale and tooth components such as enamel and dentine, supporting cells of the nervous system (neuroglia), and fin rays (horny in sharks, ceratotrichia, and scaly and spiny in higher fishes, lepidotrichia and actinotrichia, respectively).

The many kinds of skeletal materials are organized into external skeletal features and internal ones, both with soft and with hard parts.

External Skeletal Features

A part of the fish skeleton is, as already described, evident on gross external inspection. Included are such features of the integumentary skeleton as scales or bony scale-plates in the skin, fin rays, and connective tissues that toughen the skin and join it to underlying musculature, bone, and cartilage. Evident too are parts of the internal skeleton (endoskeleton), principally the superficial bones of the head and the shoulder (pectoral) girdle. These and other parts of the deep skeleton are described briefly below.

Membranous Skeleton (Figs. 3.7 and 3.18)

A connective tissue envelope joins the skin and its appendages to the underlying musculature and firm skeletal elements. This envelope is continuous at the mid-dorsal and mid-ventral body lines with the median skeletogenous septum. In the tail region this median septum divides the fish into two lateral halves and covers both sides of median structures, such as the caudal vertebrae. At the level of the body cavity, the median or axial septum, after partitioning the body into two lateral halves along the back, divides itself to surround the vertebrae and then joins the parietal peritoneum to pass ventrally along the inside of the body wall. Mid-ventrally, it becomes visible as a white line of connective tissue, the *linea alba*.

From near the middle of each side, the horizontal skeletogenous septum runs inward from the superficial envelope to the median septum. The result of median and horizontal partitioning of the tail is an appearance of four quadrants of muscle mass. Separation of muscles into strongly joined bundles is accomplished by the myosepta. These septa are readily seen between the myomeres—the regular blocks of muscle on the side of a skinned fish. Descriptive terms that have been assigned to further subdivisions of the membranous skeleton include the following:

a. *Perineural sheath,* surrounds the central nervous system—its outer layer is the *dura mater* and that most closely affixed to the brain and spinal chord is the *pia mater.*
b. *Pereneurium,* envelops nerves.
c. *Perichondrium,* wraps cartilage.
d. *Periosteum,* invests bone.
e. *Perimysium,* covers muscles.
f. *Peritoneum,* covers organs of the body cavity (visceral peritoneum) and lines the visceral cavity (parietal peritoneum).

g.   *Pericardium,* covers the heart and lines the pericardial cavity.

h.   *Tendons* and *ligaments,* attach certain muscles to one another or to the firm skeleton.

i.   *Mesenteries,* support organs in the body cavity.

The membranous skeleton behind the head at once partitions the fish into numerous "compartments" and serves to anchor these compartments to one another. It thus serves to join muscles to one another and to the firm skeleton. Obviously, many divisions of the membranous skeleton named above are joined by connective tissue to one or more other divisions; thus the divisions are more or less continous with one another.

Notochord

The notochord is neither essentially membranous nor firm in consistency. It appears early in embryonic development of fishes (and all other chordates) as an elongated rod of tissue that lies in the midline and axis of the body. When fully developed it is composed of a relatively substantial perichordal sheath (like the very stout skin of a sausage), which is filled with cells that are turgid with viscid contents. The notochord may be rod-like throughout its length (lungfishes and chondrosteans). It may also be relatively very large (an adult, 4-foot sturgeon, *Acipenser,* has one about 3.5 feet long and more than half an inch in diameter). Or, it may be reduced to a series of small dots of material, one between each two adjacent vertebrae (in many higher bony fishes). In some fishes including the salmonoids, it is moniliform (resembling a string of beads); in the sharks and relatives (Chondrichthyes) this condition results from the invasion of the perichordal sheath by cartilage. In gnathostome fishes generally, vertebral parts arise embryonically and variously in the perichordal sheath.

Axial Firm Skeleton (Fig. 3.3)

The axial firm skeleton of a fish is composed of the skull, the vertebral column, the ribs, and the intermuscular bones.

Skull (Fig. 3.4).   The living cephalaspidomorphs (lampreys and hagfishes) have firm axial skeletons that are relatively primitive in some respects and highly specialized in others. The axial skeleton is cartilaginous and includes the skull, vertebral elements, and fin rays. The lamprey skull (Petromyzonidae) is composed of a brain case (neurocranium) and sense capsules for the organs of smell, sight, and hearing-and-balance (respectively the olfactory, optic, and otic capsules). A branchial basket supports the pharynx and gill pouches, and other cartilages support the face, the buccal funnel around the

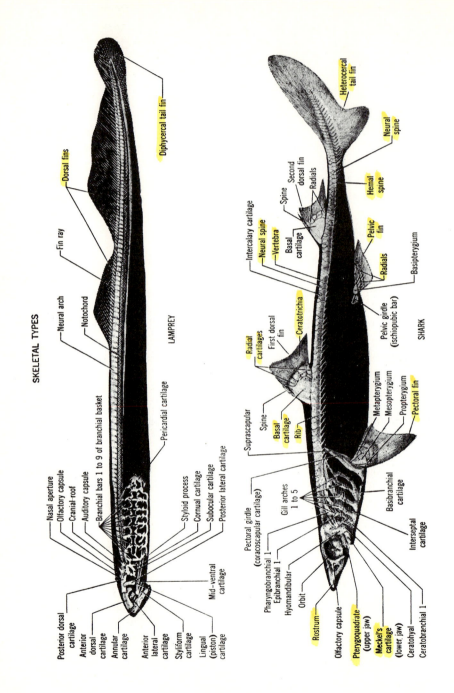

SKELETAL TYPES

LAMPREY

- Diphycercal tail fin
- Dorsal fins
- Fin ray
- Neural arch
- Notochord
- Branchial bars 1 to 9 of branchial basket
- Pericardial cartilage
- Nasal aperture
- Olfactory capsule
- Cranial-roof
- Auditory capsule
- Styloid process
- Cornual cartilage
- Subocular cartilage
- Posterior lateral cartilage
- Posterior dorsal cartilage
- Anterior dorsal cartilage
- Annular cartilage
- Anterior lateral cartilage
- Styliform cartilage
- Lingual (piston) cartilage
- Mid-ventral cartilage

SHARK

- Heterocercal tail fin
- Neural spine
- Hemal spine
- Radials
- Second dorsal fin
- Spine
- Pelvic fin
- Intercalary cartilage
- Neural spine
- Vertebra
- Basal cartilage
- Radials
- Basipterygium
- First dorsal fin
- Ceratotrichia
- Radial cartilages
- Suprascapular
- Spine
- Basal cartilage
- Rib
- Metapterygium
- Mesopterygium
- Propterygium
- Pectoral fin
- Pelvic girdle (ischiopubic bar)
- Pectoral girdle (coracoscapular cartilage)
- Pharyngobranchial 1
- Epibranchial 1
- Hyomandibular
- Orbit
- Gill arches 1 to 5
- Basibranchial cartilage
- Interseptal cartilage
- Rostrum
- Olfactory capsule
- Pterygoquadrate (upper jaw)
- Meckel's cartilage (lower jaw)
- Ceratohyal
- Ceratobranchial 1

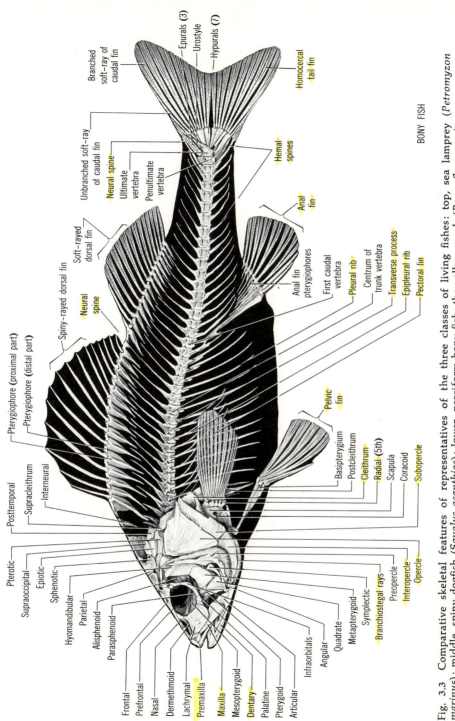

Epurals (3)
Urostyle
Hypurals (7)

Branched
soft-ray of
caudal fin

Homocercal
tail fin

BONY FISH

Unbranched soft-ray
of caudal fin
Neural spine
Ultimate
vertebra
Penultimate
vertebra

Hemal
spines

Anal
fin

Soft-rayed
dorsal fin

Spiny-rayed dorsal fin

Neural
spine

Anal fin
pterygiophores

First caudal
vertebra

Pleural rib

Centrum of
trunk vertebra

Transverse process

Epipleural rib

Pectoral fin

Pterygiophore (proximal part)

Pterygiophore (distal part)

Pelvic
fin

Basipterygium
Postcleithrum
Cleithrum
Radial (5th)
Scapula
Coracoid
Subopercle

Posttemporal
Supracleithrum
Interneural

Pterotic
Supraoccipital
Epiotic
Sphenotic
Hyomandibular
Parietal
Alisphenoid
Parasphenoid

Infraorbitals
Angular
Quadrate
Metapterygoid
Symplectic
Branchiostegal rays
Preopercle
Interopercle
Opercle

Frontal
Prefrontal
Nasal
Dermethmoid
Lachrymal
Premaxilla
Maxilla
Mesopterygoid
Dentary
Palatine
Pterygoid
Articular

Fig. 3.3 Comparative skeletal features of representatives of the three classes of living fishes: top, sea lamprey (*Petromyzon marinus*); middle, spiny dogfish (*Squalus acanthias*); lower, perciform bony fish, the yellow perch (*Perca flavescens*).

63

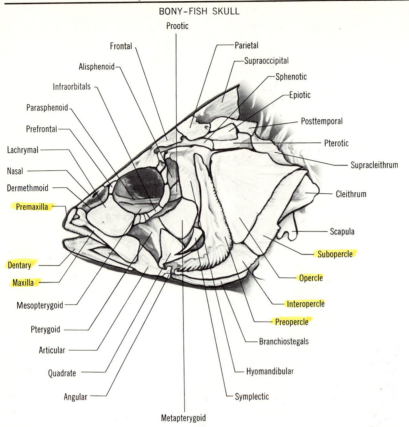

BONY-FISH SKULL

Prootic

Frontal

Alisphenoid

Infraorbitals

Parasphenoid

Prefrontal

Lachrymal

Nasal

Dermethmoid

Premaxilla

Dentary

Maxilla

Mesopterygoid

Pterygoid

Articular

Quadrate

Angular

Metapterygoid

Parietal

Supraoccipital

Sphenotic

Epiotic

Posttemporal

Pterotic

Supracleithrum

Cleithrum

Scapula

Subopercle

Opercle

Interopercle

Preopercle

Branchiostegals

Hyomandibular

Symplectic

Fig. 3.4 Superficial head bones of a spiny-rayed, perciform bony fish, the yellow perch (*Perca flavescens*).

mouth, and the tongue. The skull in hagfishes (Myxinidae) is not as complete as that of the lampreys; the "arches" that give rise to the complex branchial basket of the lampreys are reduced to a simple ring-cartilage that surrounds the duct leading from esophagus to the exterior in the hagfishes.

The shark-type (elasmobranch) skull is composed of a single-unit cartilaginous cranium (chondrocranium) and the visceral arches (branchiocranium) and their derivatives. The chondrocranium has an incompletely roofed brain case (neurocranium) and two pairs of sense-organ capsules, those for the ears and eyes. Because of the large opening (anterior fontanelle) that renders the neurocranial roof in these fishes incomplete, it may be said of them truly that they have holes in their heads. The visceral arches are thought to have numbered eight primitively. The first became lost, or perhaps is represented by the labial cartilages of those sharks that have them. The second arch is modified into the upper jaw (pterygoquadrate or maxillary cartilages) and the lower jaw (Meckel's or mandibular cartilages). The third arch in this hypothetical series is the hyoid which suspends the jaws and supports the tongue. The fourth through the eighth visceral arches are the first through the fifth gill arches, respectively.

The bony-fish skull (Fig. 3.4) is also composed of two distinctive parts, the neurocranium and the branchiocranium. The neurocranium in turn has two major parts: (a) a series of inner (endosteal) bony elements that provide a floor to the brain case and surround and protect the olfactory, optic, and otic capsules and the anterior part of the notochord; (b) a series of outer (ectosteal) dermal bones that roof the brain case, and give form to the face. The branchiocranium has three regions: (1) jaw or mandibular; (2) hyal (the jaw-supporting hyoid arch and the bones of the gill-covering opercular series); and (3) branchial (the gill arches).

The bones of the neurocranium may be grouped by location into the following regions: olfactory (nasal area); orbital (about the eye); otic (about the ear); and basicranial. Present in each of these regions are both cartilage- and dermal-bones. Cartilage-bone, also termed replacement-bone, is usually deeper in location than dermal-bone and is first laid down as cartilage and then replaced by bone. Dermal-bone is generally superficial in position and originates in inner layers of the skin. Bones of both these types are mostly bilaterally paired; only a few are median and thus unpaired.

The olfactory region of the neurocranium remains partly cartilaginous in the adult. However, cartilage-bone in it includes the paired lateral ethmoids or parethmoids, the paired preethmoids, and the median ethmoid. The parethmoids are covered by the prefrontals dorsolaterally; the ethmoid and preethmoids are covered ventrally by the vomer (prevomer) and dorsolaterally by the nasals. Dermal bones of this region are the paired prefrontals and nasals, and the unpaired vomer which bears teeth in many fishes.

The orbital region has, as cartilage-bones, the median orbitosphenoid and the paired alisphenoids. The frontals cover all three of these bones dorsally. The small dermal-bones surrounding the orbital region are collectively termed the circumorbitals or sclerotic bones. All are paired, including the frontal bones that dorsally make up part of the bony socket of the eye. Present are supraorbitals and several infraorbitals. The infraorbitals have names but are also conventionally numbered beginning with the anteriormost: no. 1, lachrymal or preorbital; no. 2, jugal; no. 3, true infraorbital; nos. 4, 5, and 6 (dermosphenotics) bear the infraorbital branch of the lateral line sense organ (Fig. 11.24).

The otic region has several cartilage-bones, often difficult to identify even when the neurocranium is carefully disarticulated. Present are the paired sphenotics (autosphenotics), pterotics (autopterotics), prootics, epiotics, opisthotics, exoccipitals, and the unpaired supra-occipital situated dorsomesially and extending caudad. Associated bones of dermal origin are the paired parietals, posttemporals, and supracleithra. In this complex, the sphenotic bone is covered dorsolaterally by dermosphenotics and the posttemporal connects the pectoral girdle with the skull.

In the basicranial region, the sole cartilage-bone is the unpaired basioccipital, with a concave posterior surface that articulates with the first vertebra. The single dermal-bone is the median parasphenoid, shaped somewhat like a crucifix. This long bone extends from the olfactory region to the basioccipital.

The mandibular region of the branchiocranium is composed of the oromandibular (upper and lower jaws) and associated bones, all of which are paired elements. Cartilage-bone constituents of the upper-jaw mechanism are the palatines, quadrates, and metapterygoids. The palatines are toothed bones, typically. The quadrates are the "keystones" of the junction of the pterygoid, upper hyoid bones and the lower jaw. Dermal-bones of the upper jaw are the premaxillae, maxillae, pterygoids, and mesopterygoids. The premaxillae bear teeth in most toothed fishes but the maxillae are typically excluded from the gape of the mouth in spiny-rayed fishes and consequently have no teeth on them (edentulous condition). In many fishes there is a small bone of dermal origin, the supramaxilla, attached to the maxilla posterodorsally. In the lower jaw the sole cartilage-bones are the articulars which articulate posteriorly with the quadrates. The posterior end of each articular is somewhat rounded and broad; the anterior end narrows to fit the inclined, V-shaped notch of the dentary. Dermal-bones of the lower jaw are the paired angulars and the dentaries. Each angular is a small bone attached to the posteroventral part of the articular. The dentaries are the principal components of the lower jaw and are subtriangular in shape. The dentaries bear teeth in many fishes.

The hyoid region is composed of both paired and unpaired bones of both cartilage and dermal types. Paired cartilage-bones include the hyomandibulars, symplectics, interhyals, epihyals, ceratohyals, and hypohyals. The basihyal (glossohyal, entoglossal or tongue bone) is a median cartilage-bone. Dermal-bones of the region are the paired preopercles, opercles, subopercles, interopercles and branchiostegals, and the unpaired urohyal.

The hyomandibular articulates dorsally with the sphenotic, prootic and pterotic; anterodorsally it joins the metapterygoid. The hyomandibular is connected to the interhyal and the symplectic by connective tissue and cartilage, and also articulates with the opercle. The symplectic is small and fits into a narrow groove in the quadrate. The interhyal connects the epihyal to the hyomandibular and the symplectic. The epihyal is a triangular bone between the interhyal and the ceratohyal. The ceratohyal and epihyal have bony processes that interdigitate to make a strong connection between these two bones. The hypohyal has two parts and articulates with the ceratohyal.

On each side of the skull, the series of opercular bones that make up the gill cover and frame the cheek posteriorly is composed of the preopercle, opercle, interopercle, and subopercle. The opercle is typically the largest bone in the series and extends farthest posteriorly. The branchiostegal rays support the branchiostegal membrane. They are attached anteriorly to the hyoid bones, principally the ceratohyals. Their number and attachment differ in the different groups of fishes (Table 2.3). The urohyal is a substantial bone with flat horizontal and vertical components that give it the aspect of an inverted T in cross-section. It lies in the median septum of the throat in the heavy hypobranchial muscles between the rami of the lower jaw.

The branchial region of the firm skeleton is entirely made up of cartilage-bone. Paired components proceeding ventrally in a typical gill arch are the pharyngobranchials, epibranchials, ceratobranchials, and hypobranchials, and the unpaired basibranchial. Typically in bony fishes there are three basibranchials, three pairs of hypobranchials, and four pairs each of the ceratobranchials, epibranchials, and pharyngobranchials. In many perciform fishes, there are teeth on the pharyngeal surfaces of various of these branchial bones. Dorsal pharyngeal teeth are suprapharyngeals, and ventral, infrapharyngeals. The throat (ventral pharyngeal) teeth that characterize minnows (Cyprinidae) and suckers (Catostomidae) are on the fifth pair of ceratobranchials.

Vertebral Column and Ribs (Fig. 3.5).   The backbone of a fish is composed of a series of segments, the vertebrae. Grossly, there is one vertebra per body segment but two may occur (diplospondyli, as in the tail of some sharks, Elasmobranchii). Although in any species these vertebrae are more or less alike, they are modified generally according to body region. Anteriorly one or two (the atlas and axis) are altered for joining the column to the cranium. Throughout the length of the trunk, the bodies of the vertebrae (vertebral centra) often have lateral processes that bear ribs. Throughout the length of the column, the vertebrae also form, above the centra, a series of arches that protect the spinal cord. Below each centrum in the tail there is an arch that partly encases main axial blood vessels. The rays of the caudal fin are supported by altered vertebral elements (penultimate, hypurals, epurals, urostyle; Fig. 3.3).

In the living lampreys (Petromyzonidae) the vertebrae are primitive, have become cartilaginous, lack a centrum, and are degenerate and also otherwise less complete than in extinct jawless fishes. The living hagfishes (Myxinidae) lack even such incomplete vertebrae. The remaining trace of the vertebral column is a cartilaginous sheath around the notochord. In the shark group (Elasmobranchii), the centra of the cartilaginous trunk vertebrae bear dorsal (neural) arches and transverse processes (basapophyses), to which short ribs are attached. The centra of the caudal vertebrae in the group have both

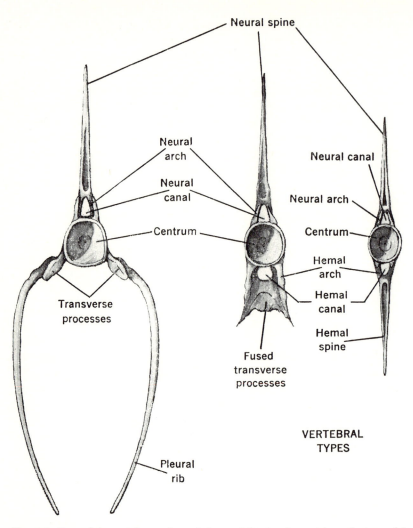

Fig. 3.5 Bony-fish vertebrae; from left to right trunk vertebra, first caudal vertebra, caudal vertebra.

neural and ventral (hemal) arches. The shark-related holocephalans, the chimaeras, have incomplete centra that are rings of calcified cartilage but do not have ribs (hence no rib-supporting processes on the reduced centra). The vertebrae of bony fishes (Osteichthyes) are typically bony.

Structurally, the vertebrae of fishes are less complex than those of the land vertebrates. Typically articulation is effected by simple apposition of the centra, rather than through complicated interdigitating processes. This is due to the support that water offers to the body mass of fishes in contrast to the absence of this support in land animals.

Ribs of fishes, according to location, are of two types, dorsal and ventral. The ventral ribs are reportedly limited to bony fishes (Osteichthyes) and are borne by the parapophyses of the trunk vertebrae. They develop within the connective tissue partitions (myosepta) of the lateral muscle bundles (myomeres) at the peritoneum. The dorsal ribs are also in the trunk region but lie in the horizontal skeletogenous septum at junctions of this septum with the myosepta; hence they are often called "intermuscular bones."

With the exception of the cephalaspidomorphs and the holocephalans, other living fish groups (Elasmobranchii and Osteichthyes) all have dorsal ribs. In addition many species have ventral ribs as well. Most of the higher bony fishes also have segmental intermuscular bones.

Intermuscular Bones. Many bony fishes have small, splint bones of assorted shapes in the myosepta; groups represented include the herrings (Clupeidae), pikes (Esocidae), and some salmons and relatives (Salmonidae), suckers (Catostomidae), and carps (Cyprinidae), to name only a few. In the trout genus *Salmo*, the bones are quite straight and run laterad and caudad from the vertebral column with which each intermuscular bone has a ligamentous connection. In the herrings (such as *Alosa* and *Dorosoma*) bones of this kind are often C-shaped and run laterad in the myosepta from a tendinous connection with a vertebral neural spine. In the pike family (Esocidae) and in some of the suckers (Catostomidae) the intermuscular bones are forked or Y-shaped. One arm of each Y is connected by ligament to a vertebral neural spine, from which it passes outward. The other arm and the base of the Y more or less parallel the vertebral column in an epaxial myoseptum. The presence of Y-shaped intermuscular bones in a species does not preclude the simultaneous occurrence of other shapes. Intermuscular bones can be most annoying in food fishes and sometimes cause real difficulty to man when they become lodged in the throat during eating.

Appendicular Firm Skeleton

The skeletal support of the median and paired fins differs fundamentally in that the pectoral and pelvic fins are supported by girdles whereas the unpaired fins are not.

Dorsal and Anal Fin Supports (Fig. 3.3). The internal skeletal supports of the common type of dorsal and anal fin in bony fishes (Osteichthyes) are a three-bone series. Inwardly, lying in the median skeletogenous septum between two adjacent vertebral spines, is a proximal pterygiophore (axonost) derived from an interneural, dorsally, or an interhemal, ventrally. Intermediately, enroute to the fin ray to be supported, is a pterygiophore. Outermost in the series, articulating with the fin ray, is a distal pterygiophore (baseost). In the sharks and relatives (Chondrichthyes), dorsal fins are supported internally by med-ially situated basal cartilages that rest on neural spines of the contiguous vertebrae; cartilaginous radial cartilages extend outward to support the numerous fin rays, although the fin spine (as in the spiny dogfish, *Squalus acanthias*) may join the basal cartilage directly. In the lampreys (Petro-myzonidae), the cartilaginous fin rays have only membranous support. In some fishes the proximal pterygiophore articulates with the fin ray nearest in line with it and the distal pterygiophore articulates with the next most caudad fin ray; the intermediate element is missing.

Caudal Fin Supports.   There is sufficient distinctiveness in the major types of skeletal support of caudal fins to have challenged ichthyologists both into classifying them and into theorizing as to their sequence in evolution. Ar-ranged by named types, major kinds follow.

a.   Proterocercal (diphycercal)—probable original type with body axis con-tinued to support caudal fin as if it were divided equally into dorsal and ventral parts. Examples are the lampreys and hagfishes (Cephalaspido-morphi, Fig. 3.3).

b.   Heterocercal—body axis upturned caudally, supporting caudal fin on ventral elements, and present in the archaic ostracoderms and placo-derms, sharks and relatives (Chondrichthyes) (Fig. 3.3) and primitive bony fishes (Chondrostei and Holostei) (in the latter in abbreviated form). The proterocercal and heterocercal types are transitory in the embryonic development of many fishes that have the next type.

c.   Homocercal—body axis terminating in a penultimate vertebra followed by a urostyle (the fusion product of several vertebral elements) and typically supporting the caudal fin rays along a vertical, slightly rounded line that terminates the urostyle and its related bony elements (a combination often loosely called the hypural plate, which is in fact only rarely, as in Gasterosteidae, a single element). The nearly or quite homocercal caudal fin is essentially prevalent in the highest bony fishes (Teleostei; Fig. 3.3).

Pectoral Fin Supports (Fig. 3.6).   The lampreys and hagfishes (Cephalaspido-morphi) lack not only paired fins, but also the girdles that support them. In the sharks and relatives (Chondrichthyes), however, the pectoral girdle is composed basically of a strong coracoscapular cartilage that is broadly U-Shaped. Paired coracoid elements make up the ventral part of the U, with fin articulations at its corners, and the upper extremities are the scapular

PAIRED–FIN GIRDLES

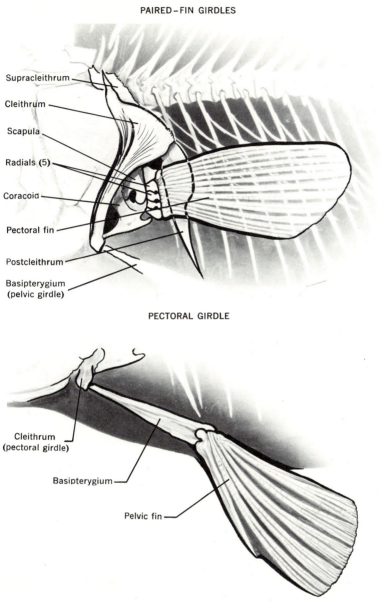

Supracleithrum

Cleithrum

Scapula

Radials (5)

Coracoid

Pectoral fin

Postcleithrum

Basipterygium
(pelvic girdle)

PECTORAL GIRDLE

Cleithrum
(pectoral girdle)

Basipterygium

Pelvic fin

PELVIC GIRDLE

Fig. 3.6   Pectoral and pelvic girdles of a spiny-rayed bony fish, the yellow perch (*Perca flavescens*).

parts. In the rays, skates, and chimaeras, the scapular cartilages attach by their dorsal ends to vertebrae.

In the bony fishes (Osteichthyes), the pectoral girdle is composed of both cartilage-bone and dermal-bone. The elements present in most bony fishes as cartilage-bones are the paired coracoids and scapulae and four pairs of radials and, as paired dermal-bones, the posttemporals, supracleithra, cleithra, and the postcleithra. Starting at the dorsal articulation of the pectoral girdle with the neurocranium and proceeding ventrally, the posttemporal is a forked bone with the two prongs of its fork directed anteriorly. One prong articulates with the epiotic and the other with the opisthotic. Posteroventrally the posttemporal attaches to the supracleithrum. The supracleithrum leads to the large cleithrum. The postcleithrum attaches ventrally to the cleithrum and extends into lateral trunk musculature; it often has two parts. The coracoid is a roughly triangular bone that is attached to the scapula above and to the cleithrum by connective tissue. The scapula is a quadrangular bone with a large opening or foramen in it. Most of the fin-ray supporting radials and, in certain fishes, some fin rays, articulate with the ventral edge of the scapula. The pectoral girdle is reduced or wanting among the puffers (Tetraodontidae).

The pectoral girdle is connected to the rays of the pectoral fin by various intermediate skeletal elements. In the sharks and relatives these are in outward sequence on each side of the fish, a propterygium, mesopterygium, and metapterygium, and numerous radials, which in turn bear the fin rays. In the lobefins (Crossopterygii), the mesopterygium persists and elongates to form more or less of an axial support for the radials, thus creating the lobate paired-fin bases for which these fishes are named. These elements are generally reduced in the higher bony fishes (Actinopterygii), with the exception of the primitive bichir (*Polypterus*) in which a distinctive lobate condition exists. The typical bony fish has only four small hourglass-shaped radials on each side.

Pelvic Fin Supports (Fig. 3.6).   The pelvic girdle in the sharks and relatives (Elasmobranchii) is a simple cartilaginous bar, termed the ischiopubic, that bears the fin-ray supporting radials. In the chimaeras (Holocephali) this bar is paired as is its bony derivative in the pelvic girdle of primitive lobefins (Crossopterygii) and lungfishes (Dipnoi). In the bony fishes (Osteichthyes) the pelvic girdle is a pair of cartilage-bones, the basipterygia, separated or variously fused. Attached posteriorly to each basipterygium are the radials which support the pelvic fin rays in lower bony fishes (Holostei). In the highest of the bony fishes (Teleostei) the pelvic radials disappear and the fin rays articulate directly with the basipterygia. Primitively abdominal in position, this girdle migrated forward during the evolution of the bony fishes far enough to contact the pectoral fins in the higher bony fishes. In some such fishes, as the cods (Gadidae), the pelvic fins are actually inserted in front of the pectorals.

MUSCLES (FIGS. 3.7 THROUGH 3.12)

Fishes, like all other vertebrates, have three major kinds of muscles. These are smooth (largely the involuntary muscles of the gut), cardiac (heart muscle), and skeletal (striated muscle, the bulk of a fish other than its skeleton). A simple arrangement of musculature is found in the lampreys and hagfishes (Cyclostomata) not only because of their primitive evolutionary position but also because they do not have the specialized muscles required by paired appendages. In these fishes the skeletal muscle behind the head is rather uniformly segmental (isomerous). The myosepta join the blocks of muscle, the myotomes, which constitute the great bulk of the body. Segmental muscle is also predominant in all the jawed fishes. In both agnaths and gnathostomes, axial muscle is the principal means of locomotion and contributes importantly to form. Derivations of segmental muscles in the jawed fishes supply the paired fins. The musculature of fishes is far less complicated than that of the higher vertebrates.

In fishes, as in other vertebrates, the three main types of muscle—striated, smooth and cardiac—may readily be distinguished histologically. From the point of view of attachment, there are two kinds—skeletal (striated) and non-skeletal (smooth and cardiac). Functionally, there are also two types— voluntary (skeletal or striated) and involuntary (smooth and cardiac). The lateral skeletal musculature in fishes has also been classified according to architectural type as cyclostomine (living agnaths) and piscine (living Chondrichthyes and Osteichthyes) (Fig. 3.7). Externally the cyclostomine myomere is

} -shaped, with one forward and two backward flexures, all of which are

broadly rounded. Internally the slope of the entire myomere is craniad. The

typical piscine myomere is ⟩ -shaped, with sharp flexures. Internally the

flexures of successive myomeres form nests of cones. The open ends of the cones of the flexures point backward and their apices are directed craniad. The open ends of the flexures face forward but point caudad. For study in gross anatomy, however, classification according to location is most apt, although for comparative purposes, regard must also be given to embryonic origin. The following treatment is based on the assortment of kinds by location.

It is well to be reminded that individual muscles, especially skeletal ones,

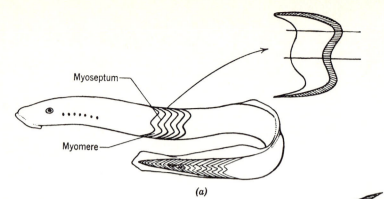

Myoseptum

Myomere

(a)

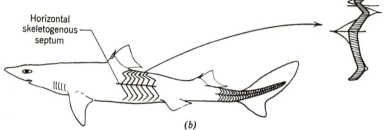

Horizontal
skeletogenous
septum

(b)

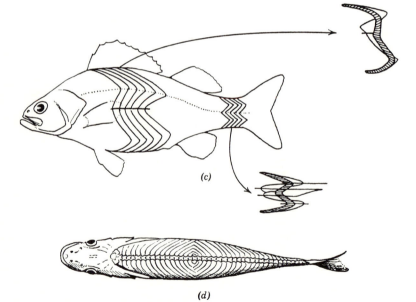

(c)

(d)

Fig. 3.7 Schematized myomere patterns of lateral musculature in the major living fish groups (Source: Nursall, 1956): (a) cyclostomine as in lateral view of a lamprey (*Petromyzon*); (b) piscine, shark (*Squalus*); (c) piscine, bony-fish, perch (*Perca*); (a-c) with details of lateral view of a single myomere. (d) Nearly frontal section through a perch (*Perca*) to show the arch-like appearance of the myomeres.

are mostly named for their action, location, and/or origin and insertion.

Skeletal Musculature of the Trunk (Figs. 3.7 and 3.8)

The trunk muscles are effectively attached to the firm skeleton, and are of primary use in movement of skeletal parts and in locomotion. Prominent blocks of lateral trunk muscle (myotomes) are visible as the meat of a fish when it is skinned or when it is before one as a steak (cross section). The thin partitions that join the myotomes have already been described as myosepta (parts of the membranous skeleton). Myotomes arise segmentally in embryos as myomeres. Also derived from these myomeres at appropriate positions in the body are:

a. Oculomotor muscles—three pairs for each eye, (Fig. 3.23).
b. Hypobranchial muscles—floor of pharynx generally, jaws, hyoid, and gill arches (important as extensors; Fig. 3.8).
c. Branchiomeric muscles—face, jaws, and gill arches (important as constrictors; Figs. 3.8 and 3.9).
d. Appendicular muscles—radial muscles at the bases of fin rays and protractors, retractors, levators, depressors, adductors, and abductors at the paired fins, etc. (Figs. 3.10 through 3.12).

In an adult fish, the segments of lateral trunk musculature extend from the skin deeply to the body axis and are limited at the midline by the median skeletogenous septum. Lengthwise, they extend from the occiput (at the rear of the cranium), the pectoral girdle, and the throat to the base of the caudal fin. In passing inward to the midline, the blocks of lateral muscle typically do not extend at right angles to the sagittal plane of the body, but rather slope variously to intercept the plane at sharp angles (Fig. 3.7). The result is that when a dissecting needle is thrust into the tail at right angles to the sagittal plane, it will enter one myotome, pass through it, and then pass through parts of one or more additional ones before attaining the median skeletogenous septum. As a result of the sloping nature of the myotomes that causes them to overlap, the cross section of a fish always shows the cut surfaces of multiple myotomes.

The two major lateral muscle masses of each side are separated by the horizontal skeletogenous septum (Fig. 3.18). The upper (epaxial) mass on each side of the sagittal plane lies above the horizontal axial septum and the ventral (hypaxial) mass below this septum. A third mass of relatively dark muscle is situated generally as a flattened wedge at, about, or extending well above and below the intercept between the horizontal skeletogenous septum and the connective tissue capsule of the body under the skin. This mass has

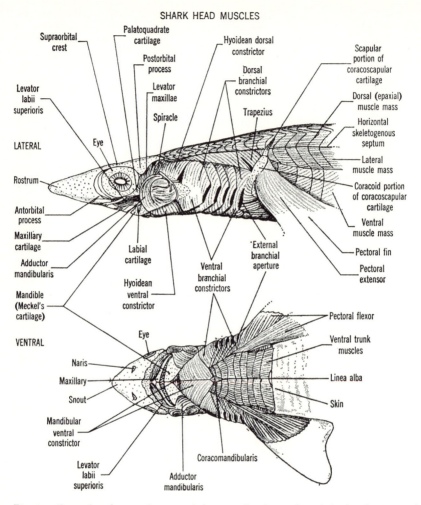

Fig. 3.8   Lateral and ventral views of the superficial muscles of the head region of a spiny dogfish (*Squalus acanthias*). (Source: Stockard, 1949).

been variously called the "dark meat" or "red muscle," and, often, in bottom feeders such as the carp (*Cyprinus carpio*), the "mud streak" or "mud stripe." This dark tissue is prominent in the tail of many bony fishes and typically ends anteriorly before attaining the pectoral girdle. The technical name, *M. lateralis superficialis*, has been given to it; it is often particularly fatty (for example, in the chinook salmon, *Oncorhynchus tshawytscha*). The fibers differ biochemically from other muscles and have different contractile properties.

Deep trunk muscles, of minor mass in proportion to the lateral ones, are the supracarinales and the infracarinales. The supracarinales extend from the scapular region of the pectoral girdle backward to the caudal fin, more or less in a median axial position (for flexing the body dorsally). The infracarinales are ventral longitudinal muscles that lie on either side of the midventral line from the throat to the tail fin (for flexing the body ventrally and for moving the pelvic girdle and anal fin).

Skeletal Musculature of the Head (Figs. 3.8 and 3.9)

On the head of a fish the musculature is associated primarily either with the jaws or with the gill arches. For both, there are superficial and deep components that differ not only among major groups of fishes but even among related species. Only an introduction to the evident superficial muscles is given here for two fishes—one, a shark, in which the jaws are supported by the hyoid arch (the hyostylic condition) and the other, a bony fish, in which the jaws are not so supported (the autostylic condition).

The face of a shark, when skinned, shows several jaw muscles (Fig. 3.8) among which the biting muscle of the lower jaw (*M. adductor mandibularis*) is the most prominent. At the level of the gills, the external constrictors (dorsal and ventral) are superficially most evident. Similar jaw muscles are visible in a bony fish (Fig. 3.9), but special muscles have come into being to service the gill cover.

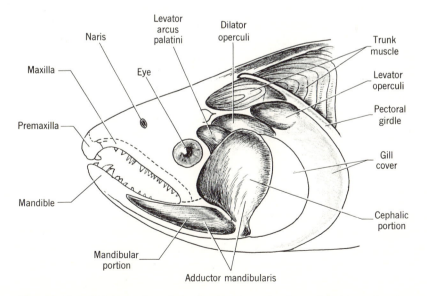

Fig. 3.9 Superficial muscles of the head region of a bony fish (chinook salmon, *Oncorhynchus tshawytscha*). (Source: Greene and Greene, 1914).

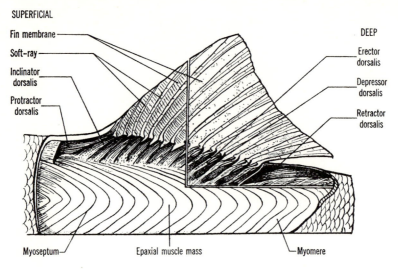

Fig. 3.10   Superficial and deep muscles of the median fin of a fish (dorsal fin of chinook salmon, *Oncorhynchus tshawytscha*). (Based on Greene and Greene, 1914).

Skeletal Muscles of the Median Fins (Figs. 3.10 and 3.11)

The muscles of the median fins function to move the fins for locomotion and for maneuvering the fish (while it is impelled by the action of other muscles). Superficially these muscles, for dorsal and anal fins, are organized as pairs: protractor (erecting) and retractor (depressing); lateral inclinators (bending) to each fin ray from each side, and, likewise to each fin ray nearest the median plane, an erector anteriorly and a depressor posteriorly (Fig. 3.10). The caudal fin has the lateral muscle masses anchored by tendons at its base. In addition, there arise internally from the superficial lateral musculature, slips of muscle that have become specialized for manipulating the caudal fin: for bending and raising it (dorsal flexors), and lowering it (ventral flexors), and for expanding and contracting it like a fan (flexors, interfilamentals between fin rays) (Fig. 3.11).

Skeletal Muscles of the Paired Fins (Figs. 3.8 and 3.12)

Of interest here are the relatively small but specialized muscle masses that arose as slips from embryonic axial myomeres to supply the paired fins (as did other slips for the median fins). Superficial bundles of appendicular muscle are visible about the bases of the fins when the overlying skin is removed. They account for movements of the appendage that are independent of the movements of the trunk.

Basically, the muscles of the paired fins are abductors (outward movement) and adductors (inward movement), with such supplements as thin sheets of muscle between the fin rays (for folding) and those that serve to hold or move the girdle (Figs. 3.8 and 3.12). Specializations of these muscles may be expected where the paired fins have functions other than maneuvering, such as for guiding sperms into the female (hugging or amplexus and intromission as in male sharks and relatives, Elasmobranchii).

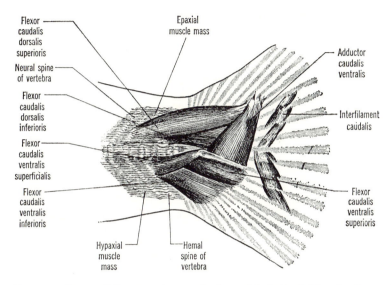

Fig. 3.11   Deep tail-fin musculature of a bony fish (tail of chinook salmon, *Oncorhynchus tshawytscha*). (Based on Greene and Greene, 1914).

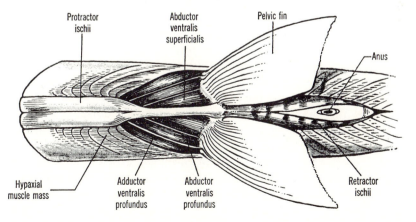

Fig. 3.12   Superficial muscles of the pelvic fins and girdle in a bony fish (pelvic region of chinook salmon, *Oncorhynchus tshawytscha*). (Based on Greene and Greene, 1914).

Heart Muscle

Muscle and connective tissue are the two principal components of the heart. Typically, cardiac musculature is dark red, in contrast to skeletal muscle that ordinarily ranges from white through pink (rarely to reddish brown) according to species. Its contraction is involuntary. The mass is thickest in the walls of the ventricle; the atrium is relatively thin-walled. The musculature of the heart (myocardium) is covered externally by epicardium and internally by endocardium, both membranous skeletal components of the pericardium.

Smooth Muscles

In fishes, as in vertebrates generally, involuntary smooth muscles are located in many different organs. Included are the following:

a.   Digestive tract—both longitudinal and circular fibers, accounting for movements of food in the tract (peristalsis); gas bladders also have these two kinds of fibers.
b.   Arteries—circular fibers, maintaining blood pressure.
c.   Reproductive and excretory ducts—moving products.
d.   Eye—accommodating vision by movement of the lens and regulating light intensity (dilation or constriction of iris) and differing in development and function among fish groups.

GILLS AND GAS BLADDERS (CHAPTER 8; FIGS. 3.13 THROUGH 3.15)

Gills in the agnathous lampreys and hagfishes (cyclostomes) are in pouches (as reflected in the term *marsipobranch*, meaning "pouch-gills," that is often used as a collective for these agnaths). Each pouch has an internal opening from the pharynx and one that leads to or toward the outside. Other parts of the water route in larval lampreys and in hagfishes include the oral vestibule, the mouth, and the pharynx. In the hagfishes (Myxinidae) a pituitary duct connects the exterior with the pharynx dorsally. In the adult lamprey (Petromyzonidae) the pharynx becomes divided from the dorsal food pathway anteriorly by a shelf of tissue, the velum. The ventral respiratory tube thus formed leads to the gill pouches via the internal branchial apertures. Lampreys have seven pairs of external gill openings. Some hagfishes have numerous gill pouches on a side, each with an external and an internal aperture (five to fourteen in *Eptatretus*). In other hagfishes (as *Myxine*), the six pairs of gill pouches open on each side into a common duct, which in turn opens to the outside near the level of the sixth pouch. Functionally this arrangement parallels the opercular gill covers in the chimaeras and the bony

GILL TYPES

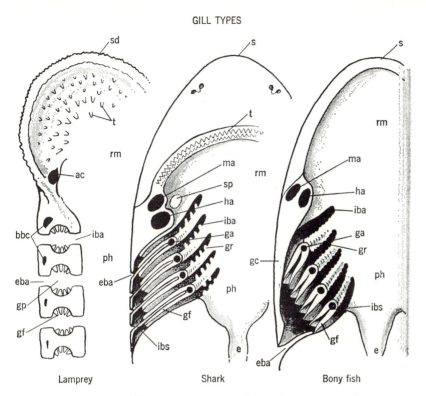

Fig. 3.13   Diagram of gill organization among fishes shown in frontal sections: ac, annular cartilage; bbc, branchial basket cartilage; e, esophagus; eba, external branchial aperture; ga, gill arch; gc, gill cover (operculum); gf, gill filament of a hemibranch; gp, gill pouch; gr, gill raker; ha, hyoid arch; iba, internal branchial aperture; ibs, interbranchial septum; ma, mandibular arch; ph, pharynx; rm, roof of mouth; s, snout; sd, sucking disc; sp, spiracle; t, teeth. (Shark and bony fish based on Kingsley, 1926).

fishes. Every hagfish, furthermore, has a tube (esophageocutaneous duct) aft of the gill pouches, on the left side only. This channel leads from the esophagus indirectly to the exterior through the left respiratory tube in the genera with a single gill aperture, or through the enlarged last gill aperture on the left side in the genera with several gill apertures.

In the gnathostomes, the gills continue as the chief organs of respiration, although supplementary specialized innovations have evolved in several groups (notably in the lungfishes that seasonally, at least, may use an adapted gas bladder for respiratory exchange with the atmosphere). The gills in gnathostomes are carried on branchial arches. In the sharks and relatives (Elasmobranchii), the gills are in chambers somewhat resembling those of the pouched-gill lampreys and hagfishes. The interbranchial septa (Fig. 3.13 and 8.1) that separate the chambers in the Chondrichthyes are variously

OPEN GAS-BLADDER TYPES

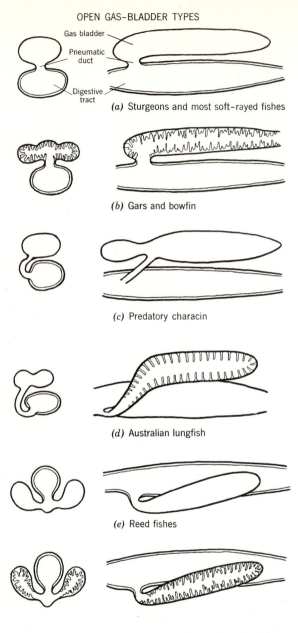

(a) Sturgeons and most soft-rayed fishes

(b) Gars and bowfin

(c) Predatory characin

(d) Australian lungfish

(e) Reed fishes

(f) African and South American lungfishes

Fig. 3.14 Diagrams of variations of gas bladder relationships to gut in physostomous fishes: (a) sturgeons (*Acipenser*); (b) gars (*Lepisosteus*) and bowfin (*Amia*) with roughened lining; (c) climbing perches (*Erythrinus*); (d) Australian lungfish (*Neoceratodus*) with sacculated lining; (e) bichirs (*Polypterus*) and reedfishes (*Calamoichthys*); and (f) African and South American lungfishes (respectively *Protopterus* and *Lepidosiren*). (Source: Dean, 1895).

reduced in the Osteichthyes where the gills of all arches lie in a more or less common chamber. The septa are thus either complete (Chondrichthyes) or variously partial (Osteichthyes) and each typically bears a hemibranch on each side. Since there is no separation of the last arch and its septum from the trunk, the last, caudally facing hemibranch is typically wanting.

Gill filaments in Chondrichthyes are ridge-like folds on the interbranchial septa (giving rise to the group name, "elasmobranch," which means "plate-gill"). Departing from this condition one may choose examples among the Osteichthyes to illustrate how the plate may have become free at its tip and how, correspondingly, the septum between two apposed filaments may have become shortened almost to the point of disappearance. When the septum is reduced, apposed filaments are free from one another throughout all or most of their lengths, depending on the extent of septal reduction (extreme in Fig. 8.1).

Gas bladders (Fig. 3.14) are unknown in the cyclostomes and chondrichthyans. In primitive bony fishes (paleopterygians of Regan) such vesicles (also termed swim bladders, air bladders, etc.) appeared presumably as accessory respiratory structures derived from the last pair of embryonic gill pouches. In the African bichirs (*Polypterus*), "living fossils," the gas bladder is indeed laterally bilobed and retains a pneumatic duct that opens into the floor of the esophagus. This paired condition of the sac is also present in the African and South American lungfishes (dipneumonans; *Protopterus* and *Lepidosiren*, respectively) but a single sac is found in the Australian lungfish (monopneumonan; *Neoceratodus*). In the lungfishes the walls of the organ have been strongly modified by muscularization, sacculation, and vascularization for gain in efficiency as an air breathing organ (Fig. 3.14). It is surmised that this bladder first developed in archaic fishes to serve as a lung and that the ordinary gas bladder was derived from such an organ.

In a higher bony fish (Neopterygii of Regan), a monopneumonan-like single gas bladder is found. This organ typically lies in the midline. It has been secondarily partitioned to produce a linear series of chambers from one developmental unit (the American redhorse suckers, *Moxostoma*, each have a three-chambered bladder). Secondarily, too, the bladder has been reduced or lost (as in most darters of the percid American subfamily Etheostomatinae). The embryonic connection between the gut and the gas bladder has been retained as a functional pneumatic duct in most soft-rayed fishes (thus collectively termed physostomous), but lost generally in spiny-rayed fishes (thus collectively, physoclistous) as mentioned in Chapter 2. Additionally, in several fishes (for example, Clupeidae, the herring family) the gas bladder communicates with the exterior by a pore near the anus.

Variations in the attachment of the pneumatic duct and in the degree of lateral pairing suggest the possibility that the gas bladder or lung evolved more than once. A cellular gas bladder of respiratory value has evolved in several holostean and teleostean fishes, including the gars (Lepisosteidae) and the tarpon (*Megalops*).

DIGESTIVE TRACT (CHAPTER 5; FIG. 3.15)

The ground plan of the digestive tract of vertebrates was established among fishes. The following constitute a list of features encountered by a morsel of food moving through the tract of a common spiny-rayed teleost such as one of the sea basses (Serranidae), perches (Percidae), sunfishes (Centrarchidae), or cichlids (Cichlidae).

a. *Mouth* bordered by toothed jaws, and rimmed internally by oral valves.
b. *Oral cavity* with vomerine and palatine teeth in roof, and tongue with tongue teeth in floor.
c. *Pharynx* with pharyngeal tooth pads on the throat sides of the gill arches, and gill rakers that guard the internal branchial openings.
d. *Esophagus* (gullet).
e. *Stomach.*
f. *Pylorus* (pyloric valve; followed by openings into pyloric caeca in many fishes).
g. *Small intestine*, passing the openings of the duct(s) that bring in bile and pancreatic secretions, and going through at least a major S-curve into the next section.
h. *Large intestine.*
i. *Anus.*

Among the bony fishes there are, however, many special adaptations of the digestive tract (Chapter 5).

The lampreys and hagfishes (cyclostomes) depart most notably from the fundamental plan of the digestive tract by the absence of jaws, of a well-defined stomach, and of curvatures in the intestine. In a parasitic lamprey, a pair of relatively large "salivary" glands opens by ducts below the tongue, the secretion of which retards coagulation of host blood and also dissolves tissue (histolytic effect). The most striking departure by sharks and their relatives is the development of a spiral valve (Fig. 5.8), rather than the curving of the intestine, to lengthen the food passage. A similar valve also appears in *Latimeria*, the sole living lobefin. A spiral valve is also present in some primitive teleosts, such as in the ladyfish ("10-pounder"), *Elops*. Its basic plan is visualized as a broad-flanged corkscrew (the valve) attached throughout its length to the walls of a cylinder (the gut). In the lampreys, a fold of the intestinal wall (typhlosole) extends into the lumen of the digestive tube and in effect may enhance both digestion and absorption.

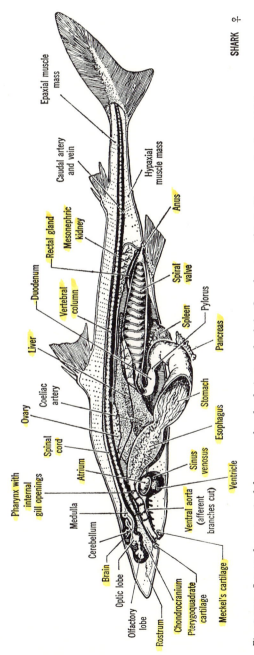

SHARK ♀

Epaxial muscle
mass

Caudal artery
and vein

Hypaxial
muscle mass

Rectal gland

Mesonephric
kidney

Anus

Duodenum

Spiral
valve

Vertebral
column

Liver

Spleen

Pylorus

Pancreas

Ovary

Coeliac
artery

Stomach

Spinal
cord

Esophagus

Pharynx with
internal
gill openings

Atrium

Sinus
venosus

Medulla

Ventricle

Cerebellum

Ventral aorta
(afferent
branches cut)

Brain

Optic lobe

Olfactory
lobe

Rostrum

Chondrocranium

Pterygoquadrate
cartilage

Meckel's cartilage

Fig. 3.15a Internal anatomical features of a shark (spiny dogfish, *Squalus acanthias*). (Source: Lagler, 1954).

85

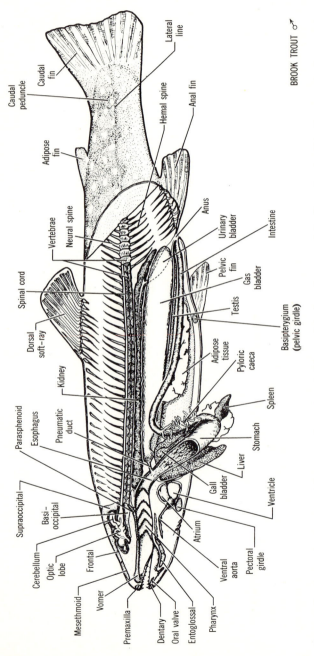

Caudal peduncle

Caudal fin

Lateral line

Adipose fin

Hemal spine

Anal fin

Vertebrae

Neural spine

Anus

Urinary bladder

Intestine

Spinal cord

Pelvic fin

Gas bladder

Dorsal soft-ray

Testis

Kidney

Basipterygium (pelvic girdle)

Parasphenoid

Esophagus

Pneumatic duct

Adipose tissue

Pyloric caeca

Supraoccipital

Spleen

Cerebellum

Basi-occipital

Frontal

Stomach

Liver

Optic lobe

Gall bladder

Ventricle

Vomer

Mesethmoid

Atrium

Premaxilla

Dentary

Oral valve

Ventral aorta

Entoglossal

Pharynx

Pectoral girdle

BROOK TROUT ♂

Fig. 3.15b Internal anatomical features of a soft-rayed bony fish (brook trout, *Salvelinus fontinalis*). (Source: Lagler, 1954).

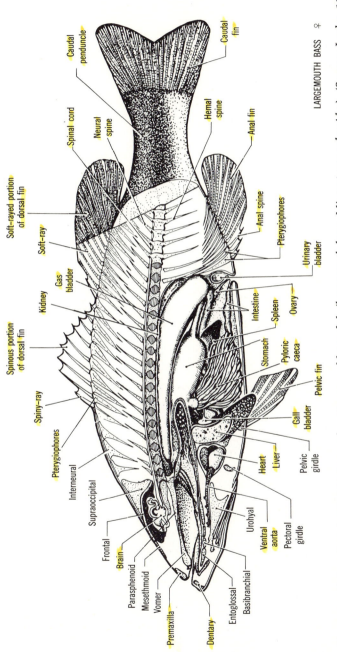

Caudal peduncle

Caudal fin

Neural spine

Spinal cord

Hemal spine

Anal fin

Soft-rayed portion of dorsal fin

Soft-ray

Anal spine

Pterygiophores

Kidney

Gas bladder

Urinary bladder

Spinous portion of dorsal fin

Intestine

Spleen

Ovary

Spiny-ray

Stomach

Pyloric caeca

Pterygiophores

Gall bladder

Pelvic fin

Interneural

Supraoccipital

Heart

Liver

Pelvic girdle

Frontal

Urohyal

Brain

Ventral aorta

Pectoral girdle

Parasphenoid

Mesethmoid

Vomer

Premaxilla

Entoglossal

Basibranchial

Dentary

LARGEMOUTH BASS ♀

Fig. 3.15c Internal anatomical features of a spiny-rayed bony fish (largemouth bass, *Micropterus salmoides*). (Source: Lagler, 1954).

87

CIRCULATORY SYSTEM (CHAPTER 7; FIGS. 3.16 AND 3.17)

The blood in fishes circulates by means of a more or less continuous tubular system of heart and vessels. The heart is a valved pump that forces blood forward toward the gills for aeration. Having traversed aortic arches in passing to and from the gills, arterial blood is ultimately dispersed into capillaries in the other tissues. Venous blood from the tissues returns to the heart, although that flowing through the kidneys and the liver is first again dispersed in capillaries respectively in the renal (except in cyclostomes) and the hepatic portal systems. Arteries carry blood away from the heart, the veins, toward it.

Blood

Fish blood, like the same tissue of other vertebrates, has two parts: one fluid, the other solid. The fluid part is the plasma in which the solid parts, the blood cells are transported. The cells are red (erythrocytes) and white (lymphocytes, leucocytes). Hemoglobin in red corpuscles greatly enhances the ability of blood to transport oxygen. All fishes, however, do not possess red cells and hemoglobin. Certain Antarctic fishes (Chaenichthyidae; "icefishes" or "white crocodile fishes") have colorless blood lacking erythrocytes. The blood of young eels (leptocephalus larvae) is also colorless. The blood pigment of the lamprey, *Petromyzon*, furthermore is not the typical hemoglobin of other vertebrates.

Heart (Fig. 3.16)

The fish heart is constructed fundamentally to afford a single circulation, rather than a completely double one as in the mammals. Typically, blood returning to the heart enters a sinus venosus, passes through the auricle or atrium, and is propelled toward the gills for aeration by the relatively thick-walled, muscular ventricle. The foregoing is the plan for the cyclostomes. The Chondrichthyes add a contractile, muscular valved base (the conus arteriosus) to the ventral aorta where it leaves the ventricle. In higher bony fishes the plan is like that of the shark and its relatives, but the first section of the ventral aorta is typically thin-walled and valved (the bulbus arteriosus), and not contractile but elastic, alternately enlarging and shrinking in response to changes in blood pressure from alternate ventricular contraction (systole) and rest (diastole). With many variations in the teleost series, the conus is reduced and replaced anteriorly by the noncontractile bulbus.

The ventral aorta in a fish is median in position, beneath the gills. From it branch the afferent branchial arteries to each gill pouch or arch. Within the gills, afferents break down into capillaries and collect again (after a drop in

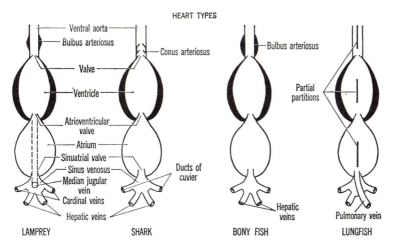

HEART TYPES

LAMPREY   SHARK   BONY FISH   LUNGFISH

Fig. 3.16   Comparative generalized schemes of the heart in major groups of living fishes.

blood pressure) into efferents that form the dorsal aorta, main vessel for distribution of blood to the body.

Striking difference from the previous account of heart structure is witnessed in the lungfishes. Both the atrium and the ventricle are partly divided into two chambers. The effect of this division is extended forward in the shortened ventral aorta, the truncus arteriosus, so that the blood from the right side of the heart is concentrated in its flow to the last two branchial arches. In the Australian lungfish (*Neoceratodus*), the lung is supplied by a substantial vessel from the fourth aortic arch on each side (pulmonary arteries), but, in each of the other two living lungfishes, the pulmonary artery arises from the dorsal aorta. From the lung, the blood returns by a pulmonary vein to the left side of the sinus venosus. Owing to partitioning of the atrium, such blood tends to concentrate on the left side and to be delivered anteriorly mostly into aortic arches one and two. Thus a partial separation of pulmonary and systemic circulations was effected in ancestral lungfishes (Dipnoi), perhaps to herald the somewhat more advanced condition in amphibians that led to the completely double circulation of higher reptiles, birds, and mammals.

Blood Vessels (Fig. 3.17)

Oxygenated blood leaves the gills to supply the head region principally through carotid arteries and to supply the body by branches of the dorsal aorta, including many that are segmentally arranged. In the tail the dorsal

CIRCULATION

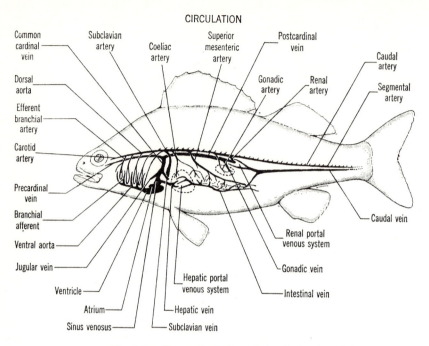

Fig. 3.17   Generalized plan of circulation in a fish.

aorta is termed the caudal artery and traverses the hemal arches of the caudal vertebrae. Blood from the tail collects principally in the caudal vein as it passes forward through the hemal arches just beneath the caudal artery. In fish groups above the cyclostomes, the caudal vein drains into the renal portal system of the kidneys and upon leaving the kidneys joins blood from the dorsal musculature in the posterior cardinal veins. The postcardinals course forward, and receive blood from anterior blood spaces such as the anterior cardinals. Cardinals join the ducts of Cuvier (common cardinal veins) on each side of the esophagus, near the level of the pectoral girdle. The ducts of Cuvier receive additional blood from beneath the head (via the jugular veins) and from the lateral body wall (via the lateral abdominal veins). In addition to receiving the ducts of Cuvier, the sinus venosus also collects blood from the liver (via the hepatic veins). The liver is supplied from the viscera through the hepatic portal system.

Lymph Vessels

The lymphatic system is much less well known in fishes than the arterial and venous systems. The lymph vessels are thin-walled sinuses and not readily discernible on gross dissection. (See Chapter 7 and Fig. 7.7).

KIDNEYS (CHAPTER 9; FIGS. 3.15, 3.18, 3.19 AND 9.2 THROUGH 9.5)

Although differing greatly in function, the renal excretory and the reproductive system in fishes (as in other vertebrates) are antomically interrelated.

The kidneys of fishes are paired, longitudinal structures that lie above the body cavity, ventral to the vertebral column and the dorsal aorta, just outside the peritoneum. Commonly they are reddish brown, pulpy, and bloody when broken open (as upon their removal when cleaning a fish for the table). Each kidney drains to the exterior through a duct, that may fuse caudally into a single duct or into an enlarged sinus (the urogenital sinus in sharks and relatives, Elasmobranchii, the urinary bladder in many bony fishes, Actinopterygii). However, the actinopterygian urinary bladder differs from that of the higher vertebrates in being of different embryonic origin—mesodermal instead of endodermal.

Two basic anatomical types of kidneys are recognized in fishes, pronephric and mesonephric (Fig. 3.18). In the pronephric type, anterior funnels lead

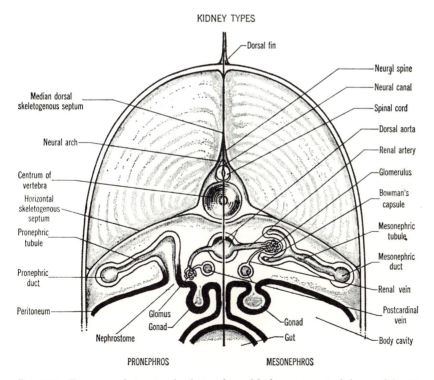

KIDNEY TYPES

Dorsal fin

Neural spine

Neural canal

Spinal cord

Dorsal aorta

Renal artery

Glomerulus

Bowman's capsule

Mesonephric tubule

Mesonephric duct

Renal vein

Postcardinal vein

Body cavity

Median dorsal skeletogenous septum

Neural arch

Centrum of vertebra

Horizontal skeletogenous septum

Pronephric tubule

Pronephric duct

Peritoneum

Glomus

Gonad

Nephrostome

Gonad

Gut

PRONEPHROS

MESONEPHROS

Fig. 3.18   Diagram of structural relationship of kidney types in fishes and location of the gonads as shown in cross-sections at two different levels of the body trunk. (Based on Kingsley, 1926).

directly from the body cavity to the pronephric duct by way of pronephric tubules. In the mesonephric type, funnels opening into the body cavity are absent but instead branches of the mesonephric duct, the mesonephric tubules, each possess an enlarged blind end. This blind end, termed a Bowman's capsule (Fig. 9.3), resembles in contour a rubber ball being pressed by a finger. In the corresponding indentation of a Bowman's capsule, a ball of capillaries (a glomerulus, from a renal artery branch) is located. Materials to be excreted pass from a glomerulus into the capsule and its draining tubule, thence they go via the kidney duct toward the exterior.

The pronephros is anterior to the mesonephros. In most fishes it is transitional in that its functions in early life are later taken over by the mesonephros. However, a pronephros may remain as a functional kidney (as in the hagfish *Myxine*) and grade posteriorly into a likewise functional mesonephros. The posterior part of the kidneys has been called an opisthonephros when the mesonephric tubules themselves become wrapped in capillaries to extend the functional part of the system beyond that of the simple mesonephric type, in which action is concentrated at the glomeruli and their capsules.

In some of the most highly evolved marine groups of bony fishes, there is extreme reduction in numbers of glomeruli in the kidney. The extreme is among blennioid fishes, toadfishes (Batrachoididae), goosefishes (Lophiidae), and frogfishes (Antennariidae), where the glomeruli have completely disappeared and left purely tubular (aglomerular) kidneys.

REPRODUCTIVE GLANDS—GONADS (CHAPTER 10; FIGS. 3.15 AND 3.18)

Female gonads are ovaries and male gonads, testes. Normally they occur in separate adult individuals. As abnormalities, combined male and female gonads (ovotestes) have been described in many species (giving evidence of common embryonic origin). Normal hermaphroditism occurs in some groups of fishes (as in the sea basses, Serranidae). Self-fertilization is, however, the real exception.

Testes

Testes in fishes are internal and longitudinal. They originate as paired structures, and remain so in most species. They are suspended by lengthwise mesenteries (mesorchia) in the upper section of the body cavity, and are found alongside or below the gas bladder when one is present. They are composed of follicles in which the spermatozoa develop. The size and color vary according to stage of sexual maturity and ripeness. The weight may equal 12 percent or more of that of the body. Most often the testes are creamy-white and smooth. The gross aspect of the contents is flocculent rather than granular. In ripe testes the spermatic tubules are large enough in some species to produce a somewhat granular appearance, but a little teasing or squashing of fragments under low magnification will distinguish its tubules from the ova or ovigerous tissue in ovaries.

In the sharks and relatives (Chondrichthyes), spermatozoa from a testis make their way toward the exterior in the anterior part of the organ first through small coiled tubules (vasa efferentia, modified mesonephric tubules), then in sequence caudally to the sperm duct (modified mesonephric duct), a temporary storage organ (seminal vesicle), urogenital sinus, urogenital papilla and urogenital pore. At the level of the anterior part of the testis, the sperm duct (Wolffian duct; modified mesonephric duct) is convoluted (epididymis); in midlength it is a vas deferens proper, and caudally it becomes a thick-walled seminal vesicle. The urogenital sinus carries a pair of side pockets, the sperm sacs.

Bony fishes (Osteichthyes) retain the basic shark plan; however, seminal vesicles and/or sperm sacs are present only in a few kinds (as in the toad-fishes, *Opsanus*). The relationship of the vas deferens to the mesonephric duct differ among groups of higher fishes; the evolutionary relationships of the ducts remain uncertain. Sperm ducts (vasa deferentia) are lacking in the salmons and trouts (Salmonidae); the spermatozoa rupture into the body cavity and leave it posteriorly via a spermatic opening behind the anus. In the sharks and relatives (Chondrichthyes), lungfishes (Dipnoi), sturgeons (Acipenseriformes), and gars (*Lepisosteus*) varying use is made of kidney tubules and the mesonephric (Wolffian) duct. In the bowfin (*Amia*) there is an intermediate condition in which the testicular tubules (vasa efferentia) skip the kidney and communicate directly with the Wolffian duct. The most advanced arrangement is that of the bichir (*Polypterus*) and highest bony fishes (Teleostei) in which (as in the bowfin) there is no passage through the kidney but in which the sperm duct is new and originates from the testes, rather than by usurpation of the mesonephric duct. The functional sperm duct of the higher bony fishes is thus not homologous with that of certain lower bony fishes and the sharks and their relatives.

Ovaries

Ovaries, like testes, are internal, usually longitudinal, and originate as paired structures, but are often variously fused and shortened. They are suspended from high on the sides of the body cavity by a pair of mesenteries (mesovaria), and are thus typically situated below the gas bladder when one is present. The size and extent of occupancy of the body cavity vary with the stage of sexual maturity of the female. When ripe, the ovaries may compose as much as 70 percent of the body weight. The color varies, most often ranging from whitish in the young through greenish when immature to golden yellow (yolk color) in ripe adults. Textures of the organs range from floccular (almost testis-like in young), through microscopically granular, in juveniles, to grossly granular, varying with size of individual eggs, in adults. Commonly, just before being laid, the eggs enlarge and become clearer.

In cyclostomes, eggs pass from the ovary into the body cavity and then to the exterior through abdominal pores leading to the urogenital sinus at the vent. In the chondrichthyans, an oviduct (Müllerian duct, not a modification of mesonephric duct), with a funnel entrance (ostium tubae abdominale) at its head end is situated anteriorly in the body cavity and conducts the eggs caudally to their cloacal exit via genital openings (the ovarian capsule is not continuous with the oviduct; gymnoarian condition). The sharks and their relatives are either egg-layers (oviparous) or live-bearers (ovoviviparous or viviparous). In egg-layers, oviducal tissue is modified anteriorly into a shell gland. Among live-bearers, the oviduct is enlarged posteriorly into a uterus for retention of young during their embryonic development. The chondrichthyan, gymnoarian condition and resultant system of egg travel also holds in other groups including the lungfishes (Dipnoi), sturgeons (Acipenseriformes), and a primitive bony fish, the bowfin (*Amia*). Most higher bony fishes have the ovarian capsule continuous with the oviduct (cystoarian condition). Eggs thus pass from inside the ovary directly into the egg canal and do not go through the body cavity as they do in the chondrichthyans. Oviducts are lacking, however, in the trouts and salmons and other salmonoids where (as in lampreys, Petromyzonidae) the eggs rupture into the body cavity and find exit through pores adjacent to the urinary and rectal openings. The pores open at the time of reproduction and may represent greatly shortened oviducts.

ENDOCRINE ORGANS (CHAPTER 11; FIGS. 3.15, 3.18, AND 3.19)

From points of view of both structure (morphology) and function (physiology), the ductless (endocrine) glands such as the pituitary and thyroid are relatively less well known than those with ducts and with obvious secretions (exocrine) such as the liver. However, most endocrine organs of the higher vertebrates as well as others have been identified in fishes. The following list names them (and the diagram in Fig. 3.19 suggests their location): pituitary (hypophysis), urohypophysis, thyroid, thymus, pancreatic islets (islands of Langerhans), chromaffin tissue, interrenal tissue, interstitial tissue in the gonads, ultimobranchial body (parathyroid gland?), intestinal tissue (secretin-producing), kidney tissue (renin-producing), pineal body, corpuscles of Stannius. There is considerable variation in the location and form of many of the foregoing secretory organs and tissues among the fish groups. Among species there is also variety in the organization of the parts of the gland, as for example in the pituitary (Fig. 11.8). The thyroid of sharks and their relatives and of some bony fishes is a capsulated discrete gland. In most bony fishes, however, thyroid follicles are variously diffused in the throat region. Details of morphology and physiology of the endocrine glands are given in Chapter 11.

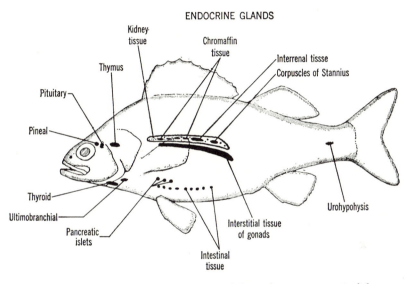

Fig. 3.19   Schematic location of the endocrine organs in fishes.

NERVOUS SYSTEM (CHAPTER 11; FIGS. 3.20 THROUGH 3.25)

The vertebrate nervous system, already well represented in even the most primitive living groups of fishes, is conveniently classified as follows on the basis of anatomy.

Cerebrospinal system
   Central division—brain and spinal cord
   Peripheral division—cranial and spinal nerves, special-sense organs
Autonomic system—ganglia and fibers, sympathetic and parasympathetic parts

The gross anatomy of the foregoing nervous components, excepting the microscopic receptors for taste, touch, and temperature, is described here. Details of anatomy and physiology of this system are given in Chapter 11.

Brain and Spinal Cord (Fig. 3.20)

The fish brain is an enlargement of the anterior end of the spinal cord. Its parts progress linearly from a forebrain region (the enlarged cerebral hemispheres and the connecting 'tween brain), through the midbrain with its swellings (the optic lobes), to the hind brain (cerebellum and medulla), and continue toward the caudal fin with the spinal cord.

The homologies of brain parts are summarized as follows:

Embryonic Parts		Definitive Brain Parts	Brain Cavities
Prosencephalon	Telencephalon	Forebrain (cerebral hemispheres)	Paired lateral ventricles
	Diencephalon	'Tween brain	Third ventricle
Mesencephalon	Mesencephalon	Midbrain (optic lobes)	
Rhombencephalon	Metencephalon	Cerebellum	Metacoel
	Myelencephalon	Medulla oblongata	Fourth ventricle

Both the brain and spinal cord are whitish and soft. The brain is housed in the cranium of the skull (already described). The spinal cord runs lengthwise of the fish in the neural canal of the vertebral column. Visible on the roof of the brain are mats of blood vessels, the choroid plexi (tela choroidea).

The cerebral hemispheres and the cerebellum are more prominent in sharks and relatives (Chondrichthyes) and bony fishes (Osteichthyes) than in the lampreys and hagfishes (cyclostomes). The midbrain, prominent in the cyclostome is also prominent in the chondrichthyans. However that of the higher bony fishes (Actinopterygii) is often very large. The cavities of the brain are continuous with that of the spinal cord.

Cranial Nerves (Fig. 3.20; Chapter 11)

Ten pairs of cranial nerves are associated with the fish brain (twelve in higher vertebrates). They are as follows: I, olfactory; II, optic; III, oculomotor; IV, trochlear; V, trigeminal; VI, abducens; VII, facial; VIII, acoustic; IX, glossopharyngeal; X, vagus. The functions of these nerves are described in Chapter 11.

Spinal Nerves

Paired spinal nerves arise segmentally from the cord behind the brain and carry sensory and motor fibers from and to the mass of the fish body. There

BRAIN TYPES

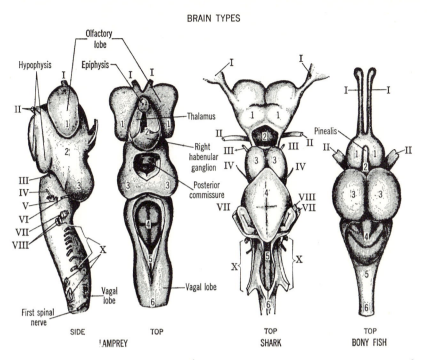

Fig. 3.20 The brain in major groups of living fishes: lamprey (*Petromyzon* based on Kingsley, 1926); shark (*Squalus* based on Haller v. Hallerstein in Bolk, Göppert, Kallius, and Lubosch, 1934) and bony fish (*Salmo* based on Healey in Brown, 1957). I-X, cranial nerves (see text); 1, forebrain; 2, 'tween brain; 3, optic lobes of midbrain; 4, cerebellum; 5, medulla oblongata; 6, spinal cord.

are enlarged plexi at the levels of paired fins. Segmentally along the spinal cord are the relay centers (ganglia) of the spinal nerves.

Olfactory Organ (Fig. 3.21)

The organ of smell in fish is a pouch that opens to the water through a naris or nares. The pouch is a blind sac in all fishes except a few of the living bony fishes (for example *Astroscopus*) and the lungfishes (Dipnoi: *Neoceratodus*, *Lepidosiren*, and *Protopterus*) in which the external nostril leads through the pouch to the oral cavity. The lining of the sac is typically a lamellar sensory epithelium, connected by the olfactory nerve with the forebrain.

Eye (Figs. 3.22 and 3.23)

In its gross anatomy, the vertebrate eye is a most constant feature. It is moderately soft to the touch and the bulk of is contents (humors), excepting the

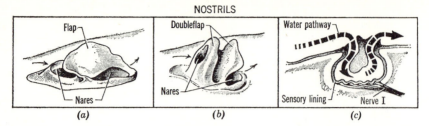

NOSTRILS

Flap
Nares
(a)

Doubleflap
Nares
(b)

Water pathway
Sensory lining   Nerve I
(c)

Fig. 3.21   Diagrams of nares and the olfactory organ in bony fishes: (a) round whitefish (*Prosopium cylindraceum*); (b) shallowwater cisco (*Coregonus artedii*); (c) water pathway and its relation to the sensory lining (olfactory epithelium) supplied by olfactory nerve. (Source for a and b, Hubbs and Lagler, 1958).

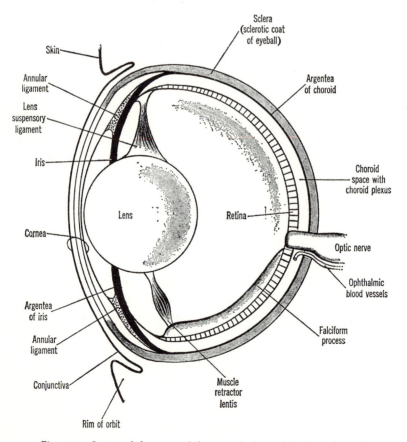

Skin
Annular ligament
Lens suspensory ligament
Iris
Cornea
Argentea of iris
Annular ligament
Conjunctiva
Rim of orbit

Sclera (sclerotic coat of eyeball)
Argentea of choroid
Choroid space with choroid plexus
Lens
Retina
Optic nerve
Ophthalmic blood vessels
Falciform process
Muscle retractor lentis

Fig. 3.22   Sectional diagram of the eye of a bony fish. (Based on Walls, 1942).

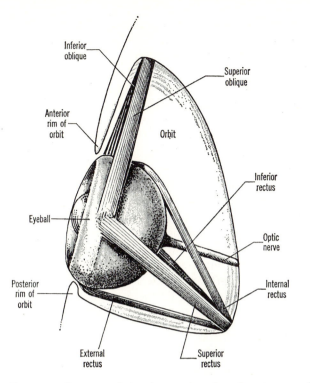

Fig. 3.23   Diagram of the three pairs of oculomotor muscles in a fish as viewed from above. (Source: Walls, 1942).

lens, is watery. Its basic structure is seen as well in a lamprey as in a bird, a cow, or a man. Nevertheless, there is great adaptive variation among fishes in the minute structure and visual capacity of the organ. These variations range from reductions that produce blindness (hagfishes, Myxinidae; cavefishes, Amblyopsidae; blind cave characin, *Anoptichthys*; blind deepsea ray, *Benthobatis*, etc.) through those that enable vision in air as well as water (climbing perch, *Anabas*; mudskipper, *Periophthalmus*). The eyes are lidless except in some sharks (as in the soupfin, *Galeorhinus*) which have lid-like nictitating membranes and in a few bony fishes which have fixed fatty (adipose) eyelids. The nictitating membranes clean the surface of the cornea and the adipose eyelid protects the eye; both help to streamline the surface. Among the most persistent similarities are the discreteness of the eyeball and the presence of a transparent cornea, an iris, a spheroid lens (often termed "crystalline lens"), a retina, a sclerotic capsule with vitreous filling in its cavity, and three pairs of oculomotor muscles (Fig. 3.23).

With all the gross similarity among fish eyes as indicated, many differences exist. In the cyclostomes, the skin and cornea have not fused and a corneal epithelium is thus developed only in higher fish groups. A peculiar innermost layer of tissue (autochthonous layer) is variously present in the teleost cornea. The iris lacks closing (sphincter) and opening (dilator) muscles in the cyclostomes, some shark-like fishes (Chondrichthyes) and primitive bony fishes, and has them ill-defined, when present, in higher bony fishes. The eyeball is composed of connective tissue in cyclostomes, but is reinforced by cartilage in all higher groups, with thin bony plates added among the teleosts. Peculiar to the teleost eye, with variations, are the annular ligament, falciform ligament, and choroid gland. The annular ligament is a circular connection between the cornea and outer rim of the iris. The falciform ligament (process) is a ridge, in the floor of the eyeball, that is variously connected by the lens retracting (retractor lentis) muscle to the lens; the process and its ligament are lacking in the lobefins and lungfishes, which have no accommodative structures. The choroid gland is a particular part of the choroid layer of the eye (between the retina and the sclera). In addition to all of the foregoing, there are many differences in the structure (and capabilities) of the visual layer of the eyeball, the retina. Furthermore, the two oblique eye muscles are not found in the South American lungfish (*Lepidosiren*); only the two pairs of rectus muscles are developed.

Organs of Hearing and Balance (Figs. 3.24 and 3.25)

As in all higher vertebrates, receptors of hearing and balance in fishes are concentrated in the ear. However, in fishes only an inner ear is present; lacking are the other two well-known regions of the ear in man, for example, the outer (that ends inwardly at the drum) and the middle (that is bridged by the chain of small bones—hammer, anvil, and stirrup). The cochlea of the inner ear of other vertebrates is also absent in fishes. Because there is no drum, vibrations to be received by the fish ear are conducted to it by the body tissues. Excepting the ear stones (otoliths) that it contains, the inner ear is a rather delicate membranous labyrinth filled with a fluid (endolymph) and bathed by perilymph. It is encapsulated in the posterior angles of the cranium (posterolaterally). Housed in the organ are receptors for vibrations and for stimuli that affect the maintenance of equilibrium.

From an evolutionary point of view, the inner ear is a specialized part of the lateral-line sensory system and together with it comprises the "acoustico-lateralis system." In many fishes, the lateralis system is housed in the lateral line canal, prominent along the sides of the body and head. In lampreys (Petromyzonidae), in some minnows (Cyprinidae), in herrings (Clupeidae), and in other fishes, however, the lateralis end organs are scattered over the body. The function is thought to be that of detecting water movements and

EAR TYPES

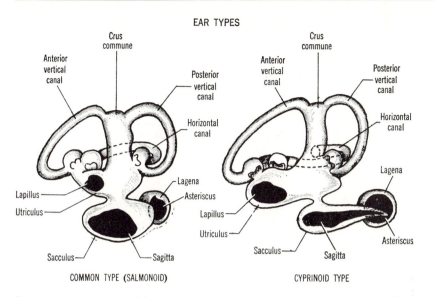

Fig. 3.24   Inner ear of fishes. Two principal types of membranous labyrinths as shown in view of organs of right side as seen from midline. (Source: von Frisch in Brown, 1957).

thus to complement, for orientation of a fish, the work of balance performed by the membranous labyrinth and otoliths of the inner ear. In the sharks and relatives (Chondrichthyes), the presence of an endolymphatic pore and duct, leading from the exterior of the fish to the interior of the membranous labyrinth of the inner ear, has been signaled as evidence of the obviously close relationship of the acoustical and lateralis organs.

The membranous labyrinth of fishes can be likened to a purse (Fig. 3.24). Leading from the main pocket (utriculus) of this purse are loops (the semi-circular canals) and a side pocket (sacculus) that has yet another pocket opening from it (lagena). Otoliths are present in the three main pockets; in bony fishes these structures are calcareous and often show growth zones (annual rings)—that of the utriculus is the lapillus, of the sacculus, the sagitta, and of the lagena, the asteriscus. In the sharks and relatives, the otoliths are small, numerous, and diffuse. In teleosts their discreteness and species individuality have made them of use in taxonomy. Each semicircular canal is tubular; both ends communicate with the cavity (vestibule) of the pouch system just described. Near one of its ends, just before its opening into the vestibule, there is an enlargement, an ampulla. Each ampulla contains sensory nerve endings (neuromast cells) in a gelatinous terminal cup as do the otolith organs and other regions of the compound vesicle that is the inner ear. Higher fishes have three semicircular canals on each side of the head—horizontal, anterior vertical, and posterior vertical. The system on the two sides of the

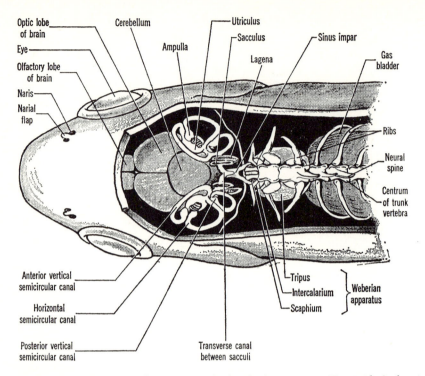

Fig. 3.25 Diagrammatic dissection of the head of a minnow (Cyprinidae) showing position of semicircular canals of inner ear, and relationships to Weberian apparatus and gas bladder in cypriniforms. (Based on von Frisch in Young, 1950).

head may be connected by a transverse canal connecting the two sacculi. On each side, in the lampreys (Petromyzonidae) there are only two semicircular canals and in the hagfish group (Myxinidae), only one.

In some fishes, notably the minnows, suckers, and catfishes (Cypriniformes = Ostariophysi), there is an elaborate connection of alternating ligaments and small bones (ossicles) between the gas bladder and the otic capsule (Fig. 3.25). Not only is this structure (Weberian apparatus) characteristic of the entire cypriniform order, but differences in individual ossicles are of value in species identification. The ossicles are derived from anterior vertebrae.

Autonomic Nervous System

As in the vertebrates generally, in fishes the autonomic nervous system is less well known than the central system in regard to both anatomy and function. In the cyclostomes, there are no strongly defined concentrations (ganglia) of autonomic-system nerve cells. Rather, the cells are widely distributed and obscurely connected to other nervous components. The sharks and their relatives exhibit a rather regular series of sympathetic ganglia in the body with fibers (preganglionic) from the spinal cord going to them and postganglionic fibers leaving them to innervate muscles of the digestive tube and of the arteries. The vagus (cranial nerve X) also sends involuntary nerve fibers to the gut as well as to the heart. A chain of sympathetic ganglia is shown best by the higher bony fishes (teleosts) in the trunk region and forward in the head as far as cranial nerve V, the Trigeminalis. Unlike the sharks and relatives, the bony fishes have a supply of postganglionic fibers to the skin (as well as to the gut and arteries). Sympathetic fibers to the iris are opposed in their action by parasympathetic fibers.

Most of the autonomic system is closely connected either to the brain or the spinal cord. The fact that it is listed as a division of the nervous system is a matter of convenience, not of true functional character.

SPECIAL REFERENCES ON ANATOMY

Bolk, L., E. Göppert, E. Kallius, and W. Lubosch. 1931–1939. Handbuch der Vergleichenden Anatomie der Wirbeltiere. Urban und Schwarzenberg, Berlin.

Bronn, H. G. 1940. Klassen und Ordnungen des Tierreichs. Band 6. Abteilung 1. Pisces. 2 Buchs. Akademie Verlagsgesellschaft MBH, Leipzig.

Daniel, J. F. 1934. The elasmobranch fishes. University of California Press, Berkeley, Calif.

Greene, C. W., and C. H. Greene. 1914. The skeletal musculature of the king salmon. *Bull. U.S. Bur. Fish.*, 33: 21–60.

Gregory, W. K. 1933. Fish skulls. *Trans. Amer. Philos.* Soc., 23 (2): 75–481. (Offset reprint, Eric Lundberg Publ., Laurel, Fla., 1959.)

Kingsley, J. S. 1926. Outlines of comparative anatomy of vertebrates. The Blakiston Company (McGraw-Hill Book Company, New York).

Mujib, K. A. 1967. The cranial osteology of the Gadidae. *Jour. Fish. Res. Bd. Canada*, 24 (6): 1315–1375.

Neal, H. V., and H. W. Rand. 1939. Comparative anatomy. The Blakiston Company (McGraw-Hill Book Company, New York).

Nursall, J. R. 1956. The lateral musculature and the swimming of fish. *Proc. Zool. Soc. London*, 126 (1): 127–143.

Stockard, A. H. 1949. A laboratory manual of comparative anatomy of the chordates. Edwards Brothers, Ann Arbor, Mich.

Tominaga, S. 1965–67. Anatomical sketches of 500 fishes. 3 Volumes (5 Books). Kadokawa-Shoten, Tokyo.

Walls, G. L. 1942. The vertebrate eye and its adaptive radiation. *Bull. Cranbrook Inst. Sci.*, 19: 785 p.

Winterbottom, R. 1974. The familial phylogeny of the Tetraodontiformes (Acanthopterygii: Pisces) as evidenced by their comparative myology. Smithsonian Contributions, Zoology, 155, 201 p.

Wunder, W. W. 1936. Physiologie der Süsswasserfische Mitteleuropas. Band II. B. Handbuch der Binnenfischerei Mitteleuropas. E. Schweizerbart'sche Verlagsbuchhandlung, Stuttgart.

4
Skin

As in other vertebrates, the skin of a fish is the envelope for the body and is the first line of defense against disease. It also affords protection from, and adjustment to environmental factors that influence life, for it contains sensory receptors tuned to the surroundings of a fish. Furthermore, the skin has respiratory, excretory, and osmoregulatory functions. Housed in the skin too are the devices of color and living light that either conceal the individual, advertise its presence, or afford sexual recognition. Also present, in the skin of the electric catfish (*Malapterurus*), are electric organs with prey-stunning discharge. The skin also contains venomous glands, as in the madtom catfishes (*Noturus*), and mucous glands, the secretions of which give the characteristic slimy touch and odor to a fish.

STRUCTURE

The skin of a fish consists of two layers (Fig. 4.1). The outer is the epidermis, and the inner, the dermis or corium. Fish epidermis is similar in many ways to the lining of the human mouth. Typically it is composed superficially of several layers of flattened, moist epithelial cells. The deepest layers are a zone of active cell growth and multiplication (stratum germinativum). Here cell multiplication goes on all the time to replace from within the outermost layer of cells as it is worn off, and to provide for growth. These epithelial cells from the epidermis are the first to close a surface wound.

The dermal layer of the skin contains blood vessels, nerves and cutaneous

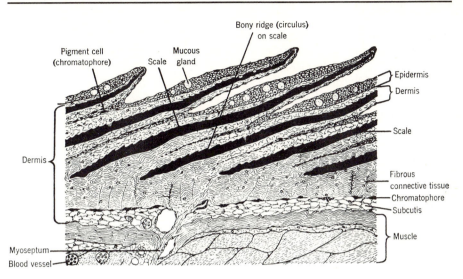

Fig. 4.1   Section of fish skin. (Source: General Biological Supply House).

sense organs, and connective tissue. While a fish is being skinned, the fibers of the connective tissue (membranous skeleton) that bind the skin to the underlying muscle and bone are very evident. The dermis plays the main role in the formation of scales and related integumentary structures.

Scattered among the flattened cells of the epidermis are numerous openings of the tubular and flask-shaped mucous gland cells that extend into the dermis. These cells secrete the slippery mucus that covers most fishes. In some, as in a foot-long hagfish (*Myxine*, sometimes called slime-eel) these cells work so well that a single individual is said to be capable of secreting a pail-full of slime. Presumably mucus lessens the drag on a fish when it swims through the water. As mucus is sloughed off, it may be envisioned as carrying away microorganisms and irritants that might be harmful if they accumulated. In some species mucus coagulates and precipitates mud or other suspended solids in water. Fish odors are present in mucus. Mucous cells are also probably a source of chemical communication among fishes (e.g., alarm substance or Schreck-stoffe).

SCALATION

Outstanding among the special features of the skin are its appendages, and outstanding among these, of course, are the very prominent scales that most fishes display. However, some kinds are "naked," in the technical sense of having no scales. Examples include the lampreys (Petromyzonidae) and North American freshwater catfishes (Ictaluridae). An intermediate structural category comprises fishes that are nearly naked and have scales only on a few places of the body, or have scales reduced to a few prickles. Some of these prickles are in very localized positions. Examples of partly scaled fishes include the paddlefish (*Polyodon*), a relative of the sturgeons, that inhabits streams in central North America and has also a near-relative, the freshwater swordbill, *Psephurus*, in China. *Polyodon* is essentially naked, except in the regions of the throat, pectoral girdle, and on the upturned base of its tail, where some scales are found. Some sticklebacks (Gasterosteidae), are naked including *Culaea* and *Pungitius*; others, notably the threespine stickleback, *Gasterosteus aculeatus*, have bony plates (rarely absent). Interestingly, the extent of bony plating is related to the degree of salinity of the water inhabited by this fish.

The sculpins (Cottidae) of the Northern Hemisphere, which inhabit streams, some of the large lakes, such as the Great Lakes, and Arctic seas, are essentially naked but may be variously adorned. Commonly among these cottids the axils of the pectoral fins and the head are scattered with little prickles that appear to be vestiges of scales. Other derivatives of scales include teeth, the spiny "sting" of the stingrays (Dasyatidae), and bony armoring plates such as in the seahorses and pipefishes (Syngnathidae) and the armored catfishes of South America (such as Loricariidae). One kind of the common carp (*Cyprinus carpio*), the mirror variety, has only a few large scales that are often separated by large areas of skin. Then there are a few fishes that have scales so small and/or so deeply imbedded that the species appear to be naked although they are not; freshwater eels (*Anguilla*) provide a legendary example. Other examples include the brook trout (*Salvelinus fontinalis*) and the burbot (*Lota*), a freshwater member of the cod family (Gadidae).

Scale patterns (Fig. 4.2) are fundamentally associated with body segmentation as manifested initially in embryonic development by the vertebrae and myomeres. In arrangement, scales are most often imbricated and thus overlap like the shingles on a roof with the free margin directed toward the tail in a manner that minimizes friction with the water. Rarely, aberrant individuals exhibit total or partial reversal of this arrangement. In some fishes, such as the burbot (*Lota*) and the freshwater eels (*Anguilla*) the pattern is mosaic; rather than overlapping one another, the scales are minutely separated, or meet their neighbors only at their margins.

Scale Shapes

Although they comprise only a few basic structural types, scales exhibit many modifications that are often characteristic of groups or species. First, let us classify fish scales on the basis of prevalent shapes, and then repeat on the basis of structure. On the basis of shape, one type is platelike (placoid), with

SCALE TYPES

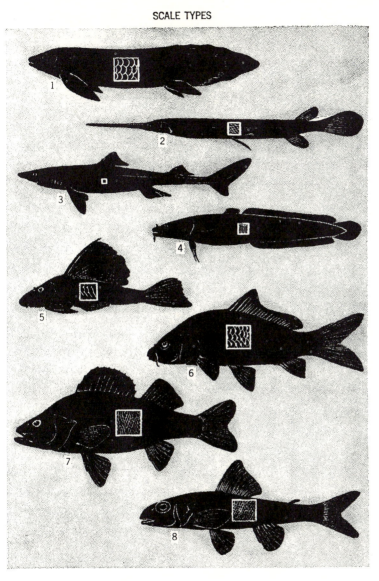

Fig. 4.2   Diversity in scalation among fishes.

each plate carrying a small cusp, as common among the sharks (Elasmo-branchii). A second type is diamond-shaped (rhombic) and characterizes the integument of the gars of North America (Lepisosteidae) and the bichirs (*Polypterus*) of the Nile. Such scales also occur on the tail of the sturgeons (Acipenseridae) of the Northern Hemisphere and of the American paddlefish

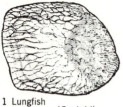

1 Lungfish
(Cycloid)

2 Gar   (Rhombic, Ganoid)

3 Shark   (Placoid)

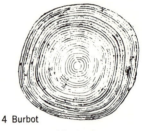

4 Burbot
(Cycloid)

5 Armored catfish   (Bony Plate)

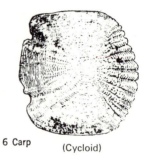

6 Carp
(Cycloid)

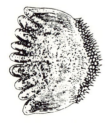

7 Perch
(Ctenoid)

8 Trout-perch
(Ctenoid)

Fig. 4.2   *Continued.*

(*Polyodon*). A third type of fish scale on the basis of shape is cycloid (Fig. 4.3). It gets its name because it is typically smoothly disc-like, and prevalently more or less circular in outline (the exposed, caudal surface or margin is entire, not toothed as it is in the next type of scale). In the fourth type, ctenoid (Fig. 4.4), the posterior surface or margin is toothed or comblike. Cycloid scales are found on most soft-rayed fishes (Malacopterygii); ctenoid scales almost universally characterize the spiny-rayed bony fishes (Acanthopterygii).

However, some of the soft-rayed fishes have ctenoid-like contact organs on their scales, for example, a few of the characins (Characidae), killifishes (Cyprinodontidae), and livebearers (Poeciliidae). Some spiny-rayed fishes have cycloid scales exclusively, for example, the brook silverside, *Labidesthes*. Many spiny-rayed species exhibit *both* cycloid and ctenoid scales. As an example, the common basses (*Micropterus*) of North American fresh waters have mostly ctenoid scales, but on the cheeks, in the axils of the paired fins, about the vent, and elsewhere, there may be patches of scales that altogether lack teeth. These fishes may still be considered as predominantly ctenoid in their scalation, but the degree and extent of development of the teeth on the scales vary from place to place on the body.

Structural Types

Structurally, there are two types of fish scales, placoid and nonplacoid. Nonplacoid scales are basically of three kinds—cosmoid, ganoid, and bony-ridge. The last term is proposed to fill a gap in the story (in need of refinement) which has been developed concerning scales over the past century. In addi-

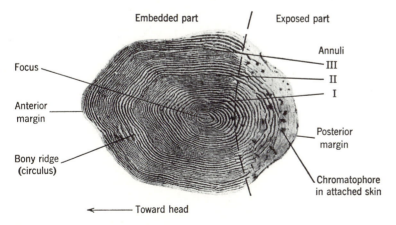

Fig. 4.3 Bony-ridge scale of the cycloid type. (Source: Michigan Institute for Fisheries Research).

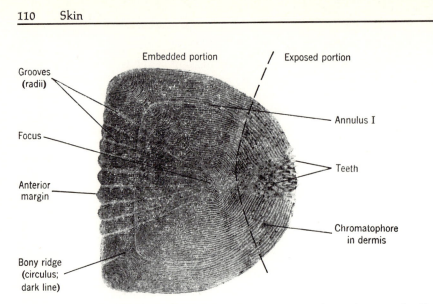

Fig. 4.4 Bony-ridge scale of the ctenoid type. (Source: Michigan Institute for Fisheries Research).

tion, bony fishes have many modifications of scales including reduction to tiny prickles or expansion and strengthening into strong armor, as previously described.

Placoid.   Placoid scales, also called dermal denticles, have an ectodermal cap that is usually of an enamel-like substance (as on human teeth) termed vitrodentine. The cap is underlain by a body of dentine (as in vertebrate teeth generally) with a pulp cavity and dentinal tubules emanating from it. Each scale has a disc-like, basal plate in the dermis with a cusp projecting outward from it through the epidermis. Placoid scales occur among sharks and their relatives (Chondrichthyes) (Fig. 4.2₃)

Cosmoid.   Cosmoid scales have a thinner, harder, outer layer than do placoid ones. Although the external material has a slightly different crystallographic makeup from that of enamel of placoid scales it also has been termed vitrodentine. The particular layer beneath this enamel is hard and noncellular and has been called cosmine. There follows inwardly a vascularized mid-layer that is a zone of perforate bony substance (termed isopedine). A further distinctive feature of this type of scale is reportedly that growth is at the edge of the scale from beneath—not from without, because no living cells cover the surface. Cosmoid scales are found both in the living (*Latimeria*) and extinct lobefins. In *Latimeria*, the scales have a denticulate outer surface, almost ctenoid in gross aspect. The lungfishes (Dipnoi) have scales which appear to be cycloid scales because the basic cosmoid structure is highly modified.

Ganoid. In a ganoid scale, the outer layer is a hard inorganic substance (ganoine), differing from vitrodentine. Beneath the ganoin cap there is a cosmine-like layer. The innermost lamellar bony layer is isopedine. Besides differing in structure from a cosmoid scale, a ganoid scale is alleged to grow not only at the edges and from underneath, but also on the surface. Among living fishes ganoid scales are best represented in bichirs (*Polypterus*) and the gars (Lepisosteidae) where they invest the entire body. In rhombic shape they are also present on the upturned lobe of the tail of such chondrosteans as the sturgeons (Acipenseridae) and the paddlefishes (Polyodontidae), as shown previously.

Bony-Ridge. Bony-ridge scales are typically thin and translucent, lacking both dense enameloid and dentinal layers of the three other kinds. Bony-ridge scales characterize the many living species of bony fishes (Osteichthyes) that have cycloid (Fig. 4.3) or ctenoid (Fig. 4.4) scales. The outer surface of such a scale is marked with bony-ridges that alternate with valley-like depressions. The inner part or plate of the scale is made up of layers of criss-crossing fibrous connective tissue. Growth of these scales is both on the outer surface and from beneath.

For describing the structure and development of the bony-ridge scale, there is need to standardize the names of parts because of inconsistencies of past usage. For the ridges on the scales, two terms seem permissible: ridges and circuli. Changes in the growth pattern of the individual may be reflected in the character and distribution of the ridges. Natal-, metamorphic-, breeding-, and year-marks (annuli) (Figs. 4.3 and 4.4) have been identified on this basis in many species. In most ordinary ctenoid or cycloid scales, a nuclear, central zone is recognizable. This zone may properly be called the focus of the scale. It is the first part to develop and it is often central in position, although later differential growth in the forepart or after-part of the scale may lead to the apparent (relative) displacement of the focus toward the posterior or the anterior margin. In many species, grooves (radii) radiate from at or near the focus toward one or more of the margins of the scale. On ctenoid scales, the most commonly used and acceptable term for the denticulations on the posterior scale margin is teeth.

In the use of scales for the classification and identification of fishes or for the study of life histories, it has sometimes been desirable to describe various regions of a scale. For most cycloid and ctenoid scales there is a readily recognizable anterior field that in many scales roughly corresponds to the imbedded portion of the scale. There is also a posterior field, often approximating the part of the scale that can be seen without removing it from the body of the fish. By what characteristics may one ordinarily recognize, given a single fish scale, the imbedded or anterior field from the exposed or posterior one? In ctenoid scales, as indicated, the teeth are typically part of the posterior field. In cycloid scales, the posterior field may often be identified by the fact that the ridges are least distinct there, often appearing as if they had been eroded or abraded. Frequently also, pigment cells (chromatophores) adhere in this zone.

In development, bony-ridge scales first appear in the dermis as tiny aggregations of cells, most often forming first on the caudal peduncle and spreading from there. Such an aggregation soon forms a scale platelet, the focus of the definitive scale. These platelets make their first appearance at different sizes of individuals in different fishes; an ordinary size at which scales first appear might be something under an inch for common species. Soon ridges are deposited on the surface of the growing scale, a fact that contradicts an antiquated idea that the ridges are the margins of superimposed laminae or layers of which scales were presumed to have been composed. The only ridges, then, which appear on the surface of ordinary fish scales, are deposited on the outward surface during development, and they are laid down somewhat in relation to growth. In Temperate-Zone waters, deposition undergoes yearly seasonal changes which makes circuli partial indicators of the annulus on a fish scale. The deepest part of the scale, the plate (also "basal plate," "lamellar layer"), is made up of successive layers of parallel fibers; those in one layer run at sharp angles to those in adjacent layer(s). Some calcification of this fibrillary plate may occur to strengthen the scale. Certain fishes, however, lack the fibrillary plate, as in the eel (*Anguilla*) where the scale is a composite of more or less concentrically arranged loculi of calcareous deposition.

Derivatives

Scales are held to be the starting point for several structures in fishes. Jaw teeth in the sharks and relatives as well as those of higher vertebrates are modifications of placoid scales, so too are spines such as those of the dorsal fins of spiny dogfish (*Squalus*) and the chimaeroids (*Chimaera, Hydrolagus,* etc.) and the "stinger" on the tail of the sting rays (Dasyatidae). The saw teeth of the sawshark (*Pristis*) are of like origin. The lancets of the surgeonfishes (*Acanthurus*) obviously originated from bony-ridge scales as did also the sawtooth-like belly scutes of herrings (Clupeidae). Without an ancestry in scales, soft-rays (lepidotrichia) may never have appeared to support fins in the Osteichthyes, nor superficial bones to invest the skull in these fishes and their vertebrate descendants. Both lepidotrichia and dermal-bones are derivatives of bony scales. Specially perforated and tubulated modifications of scales constitute a hardened channel for the lateral line sensory canal in many bony fishes. Armature too is various, ranging from spaced bony-plate bucklers in the sturgeons (Acipenseridae), through semirigid cases in pipefishes and seahorses (Syngnathidae), to encasements rivaling those of the most completely boxed turtles in the trunkfishes (Ostraciidae) for example, the cowfish (*Lactophrys quadicornis*). Extensions of dermal armature beyond the regular body surface are seen in the porcupinefish (*Diodon hystrix*).

Use in Classification and Natural History

Fish scales, which are useful to the fish for many purposes suggested above and for additional reasons, are also useful to ichthyologists for purposes of classification and natural history. How are scales useful in classification? In the major groups of living fishes the lampreys and hagfishes (Cyclostomata) are scaleless; the sharks and relatives (Chondrichthyes) are characterized by the placoid-scale type; the lobefins (Crossopterygii) and lungfishes (Dipnoi) have cosmoid scales; primitive bony fishes are associated with ganoid scales (Ganoidei of Agassiz); the higher bony fishes most often have bony-ridge scales. Besides being useful for broad aspects of fish classification, scales have characteristics that are useful for separating orders and families. If one is fortunate enough to be in an area where a good part of the fauna is made up by families represented by a single species (monotypic families), such as the bowfin (Amiidae) and the cod (Gadidae) families in the Great Lakes region, he or she can identify the species as easily by a single scale as by a whole individual. Scales have utility too, in classifying the remains of fishes in the waste heaps (kitchen middens) of prehistoric man, and in fossil deposits. Ability to classify on the basis of scales has proven worth also in the study of food habits of fish-eating animals.

The diagnostic value of scale morphology for the separation of species varies among families or genera. Where there are many closely related species, perhaps of recent origin, scale structure is likely to be as much alike as entire fishes. For example, in the whitefishes (*Coregonus*), the many closely similar forms in the cisco subgenus *Leucichthys* cannot be satisfactorily distinguished on the basis of scales. The same is true within the large genera of minnows (Cyprinidae), both in Asia and in North America. Thus, scale morphology is often relatively useless for effecting fine separations, although it may be indicative of affinities or of memberships in major groups. Highly useful, however, in taxonomic work have been scale counts such as numbers in the lateral line, and rows along or around the body (Chapter 13).

In addition to values in identification and classification, scales have some utility in the interpretation of life history. Recall, for example, that from annuli on the scales (as well as on other bony structures, Fig. 5.11) one may determine age in years. From the spacing of annual rings and a knowledge of the length of the individual at capture, fish length at each previous year mark may be calculated, after proportionality of scale growth to body growth has been determined. Sometimes scales also disclose how many times a fish has spawned. In the Atlantic salmon (*Salmo salar*), for example, there are spawning marks on the scales. Thus, one can tell how old such a fish was when it first went to sea, its age at capture and at first spawning, and also how many times it has spawned. The time when the individual first went to sea can be discerned by an abrupt change in the distance between the circuli, for when growth is accelerated in the sea the circuli are more widely spaced.

BARBELS AND FLAPS (FIGS. 4.6 f and g)

Besides smooth nakedness and some relatively minor divergences from basic scalation, fishes have evolved a wide variety in integumentation. Included are weird extensions of the skin into barbels and flaps. Barbels have appeared independently in many major groups as accessory feeding structures that carry sensory organs. Barbels of different structure and location are present on sturgeons (*Acipenser*), marine and freshwater catfishes (nematognath or siluroid Cypriniformes), goatfishes (Mullidae), and many others. The sargassum fish (*Pterophryne*) and the alga-resembling seadragon (*Phyllopteryx*) are the most frequently cited examples of the extension of the skin into flaps. The function ascribed in both these fishes is protective resemblance; at least, it cannot be denied that fishes with flaps are camouflaged by their adornment.

COLORATION

From the large variety of coloration that fishes exhibit, certain generalizations may be drawn concerning basic pattern in relation to habit. Free-swimming, open-water fishes such as the many marine herrings and relatives (Clupeidae) as well as the early planktonic stages of many bottom fishes such as the flounders (Pleuronectidae) are mostly of simple coloration grading from a whitish belly, through silvery lower sides, to upper sides and back that are irridescent blue or green. Bottom dwellers and weed-bed occupants are often very strongly and intricately marked above and pale beneath. The most brilliantly, elegantly, and bodaciously colored fishes are among the frequenters of the tropical coral reefs. Included are cardinalfishes (Apogonidae), butterflyfishes and angelfishes (Chaetodontidae), wrasses (Labridae), damselfishes (Pomacentridae), surgeonfishes (Acanthuridae), parrotfishes (Scaridae), and triggerfishes (Balistidae). Indigenous to streams of eastern North America are the darters (Percidae; Etheostomatinae) the males of many of which are beautifully colored, as are many of the small freshwater tropical fishes sold in the aquarium trade. Three types predominate in the color of oceanic fishes:

silver in the upper zones; red in the middle range; and violet or black in the greater depths.

Coloration in fishes is primarily due to skin pigments. Background color or complexion is due, of course, importantly to underlying tissues, to body fluids, and even to gut content. Background color is the essential hue of the blind cavefishes (Amblyopsidae). In other fishes, hues ranging from bright to dull that mask background color are dermal in origin. The common ground of coloration in fishes is the prevalent lightness on the ventral body surface, darkness on the back, and gradual shading on the sides from light below to dark on the back as described above. This plan illustrates the primary principle of camouflage by obliterative countershading; students of animal coloration call this Thayer's principle. Beyond this general plan, there are many extraordinary features of color dress in fishes. One of these is color uniformity due either to lack or excess of pigment in albino or in melanistic fishes.

Lack of pigment and resultant transparency characterizes the pelagic, free-swimming young of many kinds of fishes. Another extraordinary feature is almost uniform coloration by some one hue or another. An example of this would be the revealing uniform gold color that characterizes certain genetic strains of goldfish (*Carassius*) or the magnificent orange shown by the garibaldi (*Hypsypops*) of the southern California coast. Still other examples approaching uniform coloration would include the over-all blackness or melanism that characterizes many of the deep-sea fishes, although some are reversely countershaded. Other fish species, however, successfully rival the most colorful of the butterflies or birds in their combinations of multiple hues on single individuals.

Sources of Color

Coloration in fishes is due to schemachromes (colors due to physical configuration) and biochromes or true pigments (Fox, 1953). White schemachromes are seen in the skeleton, gas bladder, scales, and testes; tyndall blues and violets are in the iris; and, iridescent colors are in the scales, eyes and intestinal membranes. The foregoing schemachromes are also to be seen in the integument. Biochromes include carotenoids (yellow, red, and other hues), chromolipoids (yellow to brown), indigoids (blue, red, and green), melanins (mostly black or brown), porphyrins and bile pigments (red, yellow, green, blue, and brown), flavines (yellow, often with greenish fluorescence), purines (white or silvery), and pterins (white, yellow, red, and orange). Carotenoids, melanins, flavines and purines appear in fish skin. The liver, eggs, and eyes have carotenoids. Melanins occur in the endoderm and skin. Muscle and blood have porphyrins whereas the skeleton and bile have bile pigments. Flavines are widespread in blood, muscle, spleen, gills, heart, kidneys, eggs, liver, and eyes. Purines are in the scales and eyes. Pterins are in the eyes, blood, liver, kidneys, and stomach.

The special cells that give color to fishes are of two kinds, chromatophores (Fig. 4.5) and iridocytes. Chromatophores are assorted in hue and impart true color. They are located in the dermis of the skin, either outside or beneath the scales. Also, they are often found in the peritoneum and even deeply around the brain and spinal cord. Cytoplasmic inclusions in chromatophores, called pigment granules, are the actual sources of the color. The pigment migrations within the chromatophores—the granules can disperse through the cell or concentrate in the center—account for many of the color changes that fishes exhibit. The pigment in the granules reflects some wave lengths of light and absorbs others—the ones that are reflected are the ones that are seen; those that are absorbed cannot be seen. The basic chromatophores according to the colors of their pigment granules are red and orange (erythrophores), yellow (xanthophores), black (melanophores), and white (leucophores). Red, yellow, and orange pigments are most often carotenoids that are acquired through food and are related to vitamin A. The black or brown pigments, melanins, are highly polymerized compounds derived from the amino acid tyrosine and other phenols. In addition to these colors, fishes exhibit others—

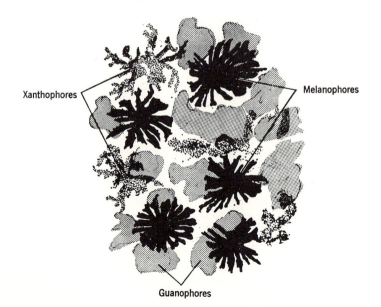

Fig. 4.5 Color cells in the skin of a flounder, *Paralichthys*. (Based on Kuntz, 1917).

such as green and brown. These and other hues come from various associations of the three basic kinds of chromatophores. Black and yellow chromatophores interspersed among one another, for example, give ostensibly green hues. Yellow and black make brown, but orange and blue may also associate to give brown.

Iridocytes could be called mirror cells, because they contain reflecting materials that mirror colors outside the fish. Both leucophores and iridophores contain purines, primarily guanine, but the former contain small crystals that can move back and forth in the cytoplasm, whereas in the latter there are large crystals incapable of movement and usually stacked in layers. Guanine itself is a breakdown product of nucleic acids (genetic material).

Significance

The functions of coloration and other visual signals (bioluminescence) are for communication with other members of the same species (intraspecific signals) and communication with other kinds of animals (interspecific signals). Intraspecific signals serve social (recognition, threat, warning) and sexual purposes, and may also offer cues by the host fish to other small fishes that browse on the body surface and clean it. Young cichlid fish travel in swarms near their parents, to which they flee when danger approaches. Recognition of their parents is based partly on color. For example, in *Hemichromis*, which are bright red during the period of caring for the young, the small fish have a pronounced preference for red and orange over all colors. Interspecific signals may be for warning or intimidating potential predators and other assailants or for decoying or masking purposes directed toward either prey or predators. Cott (1940) grouped principal functions of coloration under three headings—*concealment, disguise,* and *advertisement* (Fig. 4.6). The various kinds of concealing color are suggested as being: (1) general color resemblance; (2) variable color resemblance; (3) obliterative shading; (4) disruptive coloration; and (5) coincident disruptive coloration.

Color Resemblance.   General color resemblance between fish and the background is the basic characteristic of fishes to resemble the shades and hues of the habitats which they frequent. As previously stated, many coral-reef fishes are as very highly colored as the coral heads that they frequent, but some at least may be revealingly colored to advertise their presence to other fishes as removers of parasites from them or to afford recognition to congeners or species mates. Weed-bed inhabitants, living where there are zones of brightness and shadow, are commonly mottled. Littoral dwellers, living over light-shaded bottoms are light-colored, but over dark bottoms (or at greater depths or at night) the same species are dark.

Variable color resemblance is the ability of a fish (or other animal) to change color, gradually or rapidly in order ostensibly to match its background more perfectly. Some such variations in color resemblance occur with life history stage. The stream resident phase of a migratory rainbow trout (*Salmo gairdneri*) is multicolored, including dark spots and parrmarks (young), and rosy sides (adult). In the sea, the same individual grades from steel-blue above, through silver, to white below. There is also variation in color among fishes with the seasons, with day and night and even with momentary changes in habitat. Trout (*Salmo*) occupying bright, or partly shaded riffles in the summer have a mottled aspect. The same individuals in pools or in ice-covered waters in the winter would be rather evenly dark.

One form of variable color resemblance alters with change in body form or structure. It is to be seen in the kind of a shift that takes place when the transparent leptocephalus larva of the eel (*Anguilla*) comes from the ocean into a stream. This eel hatches far at sea in the tropical Atlantic. The transparent, pelagic larvae migrate to coastal streams of North America and Europe. With arrival at the streams, glassy band-like individuals (leptocephali) change body shape to a pencil-like form and become opaque (elvers), pigmented following Thayer's principle (Fig. 4.6a).

Rapid, adjustable color change may result in quickly matching a fish to changes in its surroundings. This type of color adaptation is usually brought about by rearrangement of pigment granules within chromatophores with the chromatophores themselves becoming differentially concealed or prominent. Speedy alteration in appearance may be of special value for concealment when the fish moves over various backgrounds. This ability to change color by fishes has been termed the most wonderful automatic cryptic device in existence. The internal regulation of these rapid changes is very complex. It involves reflex activities induced through the visual stimulation of the eyes and/or the pineal body, through hormones, and through direct action of light on skin and/or chromatophores (Odiorne, 1957; Waring, 1963).

Adjustment of coloration in fishes may be semi-permanent as well as rapid. A semi-permanent adjustment is brought about by an increase in the number of chromatophores—as an artist increases the number of stipples for shading. So, a fish, living in a darkened area, or over a dark substratum for a long time, may gradually add chromatophores. Such an adjustment takes time to mature, and though it is partly reversible by contraction of the chromatophores, the chromatophores themselves may not be destroyed.

In contrast, rapid adjustment is brought about through rearrangement of the pigment granules within the chromatophores by expansion and contraction. The changes that are achieved by fishes in this way exceed in rapidity and extent that in the famed chameleons. Some of the characins (Characidae) of aquarists flash light and then darken instantly when their nests are threat-

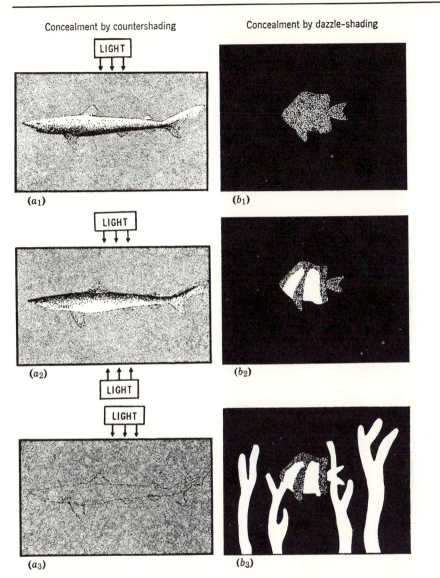

Fig. 4.6   Camouflage in fishes. (a_{1-3}) Thayer's principle of obliterative shading: (1) light and shade produced by a uniformly colored fish illuminated from above; (2) countershaded fish illuminated uniformly; (3) combined obliterative effect of top-lighting and countershading.

(b_{1-3}) Principle of maximum disruptive contrast (dazzleshading) showing distractive effect of a color pattern that contrasts violently with tone of background—the observer tends to see the pattern, not the fish.

Concealment through disruptive coloration

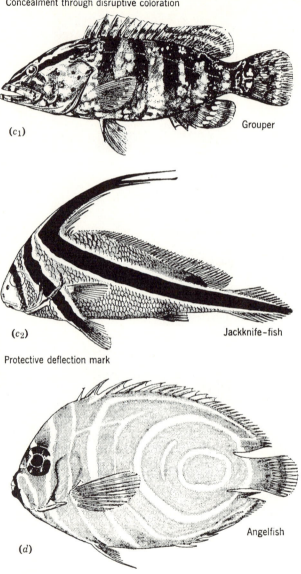

(c₁)   Grouper

(c₂)   Jackknife-fish

Protective deflection mark

(d)   Angelfish

Fig. 4.6 *Continued.*

(c₁₋₂) Concealment through disruptive coloration: (1) in a Nassau grouper (*Epinephelus striatus*) and (2) in a young jackknife fish (*Equetus lanceolatus*).

(d) Disruptive coloration and mark that may deflect predatory attacks in a young angelfish (*Pomacanthus*).

Advertisement by light organs

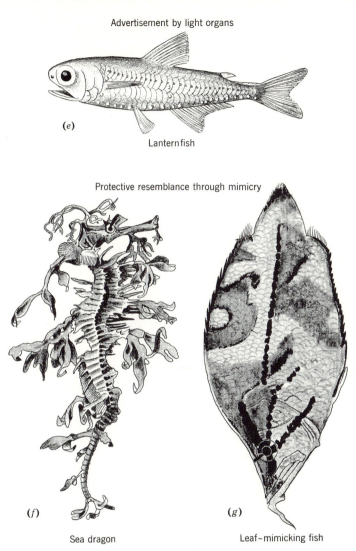

(e)

Lanternfish

Protective resemblance through mimicry

(f)

Sea dragon

(g)

Leaf-mimicking fish

Fig. 4.6   *Continued.*

(e) Advertisement through possible protection or recognition characters in light organs (light-colored spots) on a lanternfish (*Myctophum*).

(f) Protective resemblance in both color and form to sea algae in a sea dragon (*Phyllopteryx*).

(g) Protective resemblance in color (note "venation") and form to a leaf in the leaf-mimicking fish (*Monocirrhus*).

Part (e) based on Jordan and Evermann, 1900; all others on Cott, 1940.

ened. The reaction is similarly instantaneous in certain reef fishes when they are moved from a brightly lighted and colored area to subdued light with dull hues. A classical example of physiological color change in fishes is among the flatfishes (Pleuronectiformes). These fishes are adapted to life on the bottom by the great morphological adaptation that results in perpetual living on one side of the body and in having the side turned toward the bottom unpigmented and that turned toward the light pigmented (rare anomalies are the reverse of this). Equally as striking as the physiological adaptation of flatfishes to rapid color change is their matching ability. A flatfish placed on an illuminated checkerboard will, after some time, afford a background almost suited for a game! The mystery remains, however, as to how the retina in flatfishes (Pleuronectidae) can initiate this remarkable matching of pattern.

Thus the coloration of most fishes is not constant throughout life since there are both short-term and long-term changes. Such pigmentary changes or responses are both morphological and physiological. For melanophores, morphological change is slow and exhibits buildup of pigment and cells. Physiological change can be rapid (often within a few minutes) and involves a redistribution of pigment granules within the melanophores. Both of these changes are responses either to visual or nonvisual stimuli, with the latter being either coordinated by nerves and hormones or uncoordinated where the melanophore is its own receptor and effector (not known for fishes). In eyeless sharks (*Mustelus*) and blinded catfishes (*Ictalurus*), darkening takes place in light and paling in darkness, whereas intact individuals darken by melanin dispersal when placed on a dark background. Similarly, by pituitary hormone responses, removal of the pituitary results in paling in another shark (*Scyllium*) and pituitary implants result in darkening in a skate (*Raja*). The role of nerve impulses in chromatic regulation has been demonstrated by cutting peripheral nerves and having the denervated area darken, and by electrically stimulating the peripheral nerves. In *Scyllium*, at least, darkening as a black background response involves MDH (melanin dispersing hormone) and paling on a white background involves MDH or secretion of MAH (melanin aggregating hormone), both from the pituitary. There is also good evidence for MDH in some teleosts (*Anguilla*, *Ictalurus*, and *Fundulus*). In the relatively few species of teleosts studied, there is uneven information on the color change mechanism but there is general agreement that coordination of color change is by interaction of nervous and hormonal control.

Obliterative Shading (Fig. 4.6a). Yet another kind of coloration for concealment is obliterative shading. The optical principle upon which this type of concealing coloration depends is that of countershading. Light and shade give an observer the third dimension of objects seen. Shadow, then, is a serious matter when trying to reduce the visibility of any object, as one may try to do, say, in time of war, to obscure a vehicle, a ship, or an aircraft. Practical countershading is evidenced by most fishes—also, by most birds, mammals, reptiles and amphibians—since they possess dark dorsal surfaces and light bellies, tending to make them appear optically flat, like a shadow (Thayer's principle). The surfaces normally directed toward a source of light are counter-shaded by darkening, whereas those which would normally be in shadow are counterlighted, and properly graded tones between render the object flat and thus reduced in visibility. Countershading has disappeared in cave fishes (*Amblyopsis* and *Anoptichthys*) and reversed in the depths of the sea, as previously stated.

Disruptive Coloration (Fig. 4.6b,c,d). Yet another means of concealment that has evolved in fishes is disruptive coloration, a further means of camouflage. Since it is the continuity of surface, bounded by a specific contour or outline, that enables us chiefly to recognize an object with the shape of which we are familiar, disruption of this contour tends to conceal. Thus, for effective concealment, it is essential that the telltale appearance of form should be destroyed. The function of camouflage for a fish, then, may be thought to be to prevent, or to delay as long as possible, recognition on sight. When the surface of the fish is covered by irregular patches of contrasted color and tones, these patches tend to catch the eye of the observer and to draw attention away from the shape that bears them. In contradicting the form, the patterns concentrate the attention upon themselves, and thus the patterns may cause the object that bears them to pass for part of the general environment.

Coincident disruptive coloration is a special kind of camouflage. It may appear to join together separate, unrelated parts of the body in order further to reduce chances of recognition. Appendages are concealed by this device and eyes are too. The eye, particularly its staring black pupil, is made to appear another shape by various forms of coincident disruptive eye masks and thus, joined to some other part of the head, ceases to resemble an eye. However, it has also been proposed that the eye-line—horizontal or tear-drop vertical—is a sight-line for aiming attack (Fig. 4.6, c_1, c_2).

Related Forms of Concealment and Disguise (Fig. 4.6d,f,g)

Associated with concealment by use of color are other means of masquerade or disguise, often relating structure with habit as shown well by Breder (1946). Habits such as cryptic attitudes are used to conceal position and may complete the role of color disguise.

Gars (*Lepisosteus*), both young and adult, have the habit of basking motionless near the surface. In posture, form, and coloration they strongly resemble floating twigs or logs and are often mistaken for them. *Strongylura*, a genus of needlefishes (Belonidae), looks much like the stems of plants in its habitat. Young of some halfbeaks (*Hemiramphus*) and pipefishes (Syngnathidae) in profile look like leaves of the eelgrass (*Zostera*). Other fishes that resemble plant leaves as young or adults include the leatherjacket (*Oligoplites saurus*), the tripletail (*Lobotes surinamensis*), a filefish (*Monacanthus polycanthus*), certain spadefishes (Ephippidae; *Platax*), and the orange filefish (*Alutera schoepfi*). The young of some flyingfishes (*Cypselurus*) look very much like the blossoms of a plant (*Barringtonia*) in their habitat. Resembling algal fronds (i.e., thalli) or fragments of them, are several fishes including seadragons (Syngnathidae; *Phyllopteryx*), the dwarf wrasse (*Doratonotus megalepis*), and the giant kelpfish (*Heterostichus rostratus*). In addition, the Atlantic spade-fish (*Chaetodipterus faber*) resembles a seed pod (*Rhizophora*), and the lump-fish (*Cyclopterus lumpus*) looks like an algal flotation capsule.

Disguise is also accomplished by various conspicuous localized characters, which simply tend to reduce the resemblance of the fish to itself. Deflective and directive marks are important here. Deflective marks are those which have been thought of as deflecting the attack of an enemy from a more or less vital part of the body to some other part. A simple illustration is the dark spot on the tail of the bowfin (*Amia*); another is in the young of certain angelfishes (*Pomacanthus*, Fig. 4.6d). During ritualized territorial fights of one angelfish (*Pterophyllum scalare*), there are changes in the darkness of the vertical bar pattern and of the eye spot on each of the gill covers. A directive mark, contrariwise, has been thought of as diverting the attention of prey from the most dangerous part of the predator's body. Stargazers (Uranoscopidae) have fringed mouths that may well obscure the organ when the fish is mostly buried in the sand. Prey fishes move into the region of the obscured mouth, perhaps directed or attracted there by fleshy flaps that resemble food, and are engulfed (even though the flaps may serve primarily to keep sand out of the mouth). The luring of prey to the mouth region is another function of directive marks, usually a combination of structure and color. Goosefishes and anglers (Lophiiformes) have one or two slender, stalked appendages, each a virtual "bait-rod" (illicium), extending forward over the snout, sometimes with a little luminous bulb or variously tasseled lure at the tip of the structure. One stargazer (*Uranoscopus*) of the Mediterranean is said to have a little red "worm" in its mouth that lures prey to their demise.

Advertisement (Fig. 4.6e)

Some forms of coloration of fishes appear to advertise or to reveal rather than conceal their presence. Outstanding examples of this are among the darters (Etheostomatinae) of American streams. Among these small members of the perch family (Percidae) are the most brightly hued freshwater fishes of this continent. Coloration of this type may be of significance for sexual recognition; hence, it is a form of advertisement. Experiments with sticklebacks (*Gasterosteus*) suggest value of color in sexual recognition for some fishes, but trials with other species show that color may not be discriminated or may be valueless for this purpose.

Use in Classification

Color pattern in fishes is used often as a character to distinguish smaller taxonomic units such as species and subspecies (Chapter 13). In spite of variation among individuals of one size and kind, and variability with sex, age, habitat, and so on, fundamental color pattern, often the exact placement of individual chromatophores, is under genetic control (Chapter 12). Not uncommonly, therefore, color characters appear in the descriptions of the smaller taxonomic categories.

LIGHT ORGANS

Although absent in freshwater fishes, living light (bioluminescence), like that in the common fireflies or glowworms (Lampyridae), is not uncommon in the oceans and appears in many groups of animals, including many marine fishes. It is most extensively developed in many of the midwater and bottom dwelling deepsea species. The suspicion is that light organs may serve in species and mate recognition by the fishes themselves. Certain light organs on the head have been implied to act as lures to attract prey. Most of the luminescent lights are blue or green. The organs are useful as taxonomic characters, for example, in the lanternfishes and relatives (Myctophiformes).

Luminescence in fishes is of two types (Harvey, 1957): (a) that which results from the presence of luminous bacteria living on the fish in a symbiotic relationship; (b) that which arises from self-luminous cells on the fish, the photophores. Self-luminescence may be either light generated by the animal within its own tissues (intracellular luminescence) or by discharge of luminous secretion (extracellular luminescence). An example of the latter is a searsiid, *Searsia schnakenbecki*.

Fish families in which luminescence is of bacterial origin are represented in both shoal and deep waters and include the grenadiers (Melanonidae), cods (Gadidae), pine-cone fishes (Monocentridae), anomalopids (Anomalopidae), luminous cardinalfishes (Acropomatidae), slipmouths (Leiognathidae), sea basses (Serranidae), cardinalfishes (Apogonidae), swallowers (Saccopharyngidae), and deepsea anglerfishes and some of their relatives in the order Lophiiformes. Elasmobranchs with self-luminescence include some sharks

(*Spinax, Centroscyllium, Etmopterus*) and an electric ray (*Benthobatis moresbyi*). The greatest number of self-luminous species among the bony fishes belong to the deepsea scaly dragonfishes (Stomiatidae) and to the lanternfishes (Myctophiformes). Photophores are also present in the toadfishes and midshipmen (Batrachoididae).

Many hypotheses have been advanced to explain the value of bioluminescence among fishes. Evidence supports the importance of light organs in illuminating the dark waters of fish such as the lantern fish *Gonichthys* and stomiatoids. Bioluminescence is a vital part of the courting behavior of the midshipman. Other hypotheses awaiting confirmation are the use of light as lures for prey, to deter or confuse predators, for advertising, or to obliterate the fish's silhouette. Further possibilities also include the spacing out of the territories of midwater fishes. The ventral prevalence of luminous organs and tissues in such fishes suggests protection from the larger number of predators above.

POISON GLANDS

Poisonous glands are present in many fishes as skin derivatives, ostensibly as adaptations of mucous glands. They secrete a venom that when injected by puncture into man may be painful or, rarely, lethal. The study of these glands and their secretions is a part of the field of ichthyotoxism so well advanced by the work of Halstead (e.g., 1959 and 1970). The field of ichthyotoxism includes the various forms of intoxication resulting from eating poisonous fishes (ichthyosarcotoxism) or being stung by venomous fishes (ichthyoacanthotoxism).

As an integumentary derivative, venom glands have evolved independently several times in widely separated families of fishes. Ichthyoacanthotoxism is known to result from the stings of many fishes including the following (selected to show spread through several orders), for which the stinging apparatus is briefly described and venomous examples given.

a. Sharks, fin spine with venomous glandular epithelium in groove—horn shark (*Heterodontus francisci*) and spiny dogfish (*Squalus acanthias*).
b. Rays (Dasyatidae and Myliobatidae), caudal stinger with glandular skin of its sheath—stingrays (*Dasyatis*), spotted eagle ray (*Aetobatis narinari*), bat stingray (*Myliobatis californicus*).
c. Chimaeras (Chimaeridae), dorsal fin spine with venom from glandular epithelium composing its sheath and lining its groove—*Chimaera* and *Hydrolagus* (the ratfish).
d. Scorpionfishes, lionfishes, and rockfishes (Scorpaenidae), dorsal, anal, and pelvic fin spines with venom glands in their grooves—including species

of scorpionfishes proper (*Scorpaena*), bullrouts (*Notesthes*), lionfishes (*Pterois*), and stonefishes (*Synanceja*).

e. Weeverfishes (Trachinidae), opercular stinger and spines of dorsal fin, all with venom glands in their grooves—*Trachinus*.

f. Toadfishes (Batrachoididae), opercular stinger and dorsal fin spines with glands at their bases—*Batrachoides*, *Thalassophyryne*, and *Opsanus*.

g. Catfishes, spines of dorsal and pectoral fins with glands beneath skin, opening through pores at bases of spines—stonecats and madtoms (*Noturus*), bullheads (*Ictalurus*), seacatfishes (*Galeichthys felis* and *Bagre marinus*), and Indo-Pacific catfishes (*Heteropneustes*, *Clarias*, and *Plotosus*).

h. Surgeonfishes (Acanthuridae), stinger on each side of caudal peduncle presumably with venom glands in sheath of spine—*Acanthurus*, *Naso*.

i. Dragonets (Callionymidae), fin spines with venom glands—*Callionymus*.

j. Rabbitfishes (Siganidae), dorsal, anal, and pelvic fin spines with venom glands on them—*Siganus*.

k. Stargazers (Uranoscopidae), shoulder stingers with venom glands at bases.

Venom from stings of the following fishes may be fatal as well as having painful symptoms to humans: rays, chimaeras, scorpionfishes, weeverfishes, and stargazers. Nonfatal but nevertheless painful to humans (sometimes very much so) are stings of the venomous sharks, toadfishes, catfishes, surgeonfishes, rabbitfishes, and dragonets. All sting puncture wounds convey to a recipient human the chance for secondary infection including gangrene and tetanus. Nothing is known of the biological significance of the venom apparatus in fishes although it is easy to surmise roles in food getting, offense, or defense.

SPECIAL REFERENCES ON SKIN

Breder, C. M. 1946. An analysis of the deceptive resemblances of fishes to plant parts, with critical remarks on protective coloration, mimicry, and adaptation. *Bull. Bingham Oceanogr. Coll.*, 10 (2): 1–49.

Cott, H. B. 1940. Adaptive coloration in animals. Oxford University Press, New York.

Crozier, G. F. 1974. Pigments of fishes. *In* Chemical Zoology, 8: 509–521. Academic Press, New York.

Fox, D. L. 1953. Animal biochromes and structural colors. Cambridge University Press, Cambridge.

Fox, D. L. 1957. The pigments of fishes. *In* The Physiology of fishes, 2; 367–385. Academic Press, New York.

Halstead, B. W., L. S. Kuninobu, and H. G. Hebard. 1953. Catfish stings and the venom apparatus of the Mexican catfish, *Galeichthys felis* (Linnaeus). *Trans. Amer. Microsc. Soc.*, 72 (4): 297–314.

Halstead, B. W. 1959. Dangerous marine animals. Cornell Maritime Press, Cambridge, Md.

————. 1970. Poisonous and venomous marine animals of the world. U.S. Gov't. Printing Office, Washington, D.C. 3 volumes.

Harvey, E. N. 1957. The luminous organs of fishes. *In* The physiology of fishes, 2: 345–366. Academic Press, New York.

Nicol, J. A. C. 1969. Bioluminescence. *In* Fish Physiology, 3: 355–400. Academic Press, New York.

Odiorne, J. M. 1957. Color changes. *In* The physiology of fishes, 2: 387–401. Academic Press, New York.

Van Oosten, John. 1957. The skin and scales. *In* The physiology of fishes, 1: 207–244. Academic Press, New York.

Wallin, O. 1957. On the growth structure and developmental physiology of the scale of fishes. *Rept. Inst. Freshwater Res. Drottningholm*, 38: 385–447.

Waring, H. 1963. Color change mechanisms of cold-blooded vertebrates. Academic Press, New York.

5

Foods, Digestion, Nutrition, and Growth

Fishes, as all animals, require adequate nutrition in order to grow and survive. Through observation in the field and examination of the contents of digestive tracts, and through physiological studies in the laboratory, researchers have learned much concerning feeding behavior and the kinds of organisms that are eaten as well as the mechanisms that have developed for digestion. Experiments have also disclosed dietary values of various classes of foods and have analyzed factors that affect growth. Particularly significant has been the information on nutrition gained as a corollary of man's attempts to propagate fishes in the most efficient manner possible.

The gross anatomy of the digestive tract is described in Chapter 3.

NATURAL FOOD OF FISHES

Varieties of Foods

Nature offers a great diversity of foods to fishes including nutrients in solution and hosts of different plants and animals. Little is known of the direct uptake of soluble nutrients by fishes, although there is some evidence of direct uptake of glucose from ambient water. Many essential and inessential compounds and ions in the water are absorbed directly (as through the gills) or are swallowed with food and then absorbed in the digestive tract. For example, extra-enteral absorption exists for calcium, vital to scale and bone formation.

It is to be expected that a group as diversified as the fishes has become adapted to a wide variety of foods. Some fishes feed exclusively on plants, others feed only on animals, whereas a third and larger group derives its proteins, carbohydrates, and fats, as well as vitamins and most minerals necessary for growth and upkeep, from both plant and animal sources. Specialized parasitic fishes such as the sea lamprey (*Petromyzon marinus*) live on the blood and the tissue fluids of other fishes. Many young fishes, with tiny mouths and often still with yolk in their yolk sacs, begin to feed on plankton, the microscopic plant and animal life of water, including bacteria. Some fishes, adapted with numerous, elongate, and close-set gill rakers such as the gizzard shads (*Dorosoma*) feed for their entire lives principally on phytoplankton. Other fishes that have cutting teeth, such as parrotfishes (Scaridae) and surgeonfishes (*Acanthurus*), as well as the freshwater tilapia (*Tilapia melanopleura*) and the grass carp (*Ctenopharyngodon idella*) take parts of plants as food, nibbling off leaves, fronds and small stems. The parrotfishes are particularly well fitted for this habit with incisor-like teeth that are more or less fused into a cutting-edged beak. The beak serves to bite off pieces of algae whereas a set of pharyngeal grinding teeth mills the pieces to break open the cells and expose their contents to the action of digestive juices (Fig. 5.6e). The grinding teeth also mill pieces of coral.

A wide range of kinds and sizes of animals is important in the food chains of fishes (Fig. 5.1). Among the earliest animal foods to be consumed are animal plankton organisms, or zooplankton. Zooplankton includes many different kinds of protozoans, microcrustaceans and other microscopic invertebrates, and the eggs and larvae of many animals including those of fishes themselves.

Of outstanding importance as fish food among the larger invertebrate animals are: annelid worms (Annelida); snails, mussels, and clams (Mollusca); and crustaceans and insects (Arthropoda). All of the classes of vertebrates are prey for fishes—birds, mammals, reptiles, amphibians, and, of course, fishes themselves. It is not uncommon to find a small rodent, snake, or turtle in the stomach of a bass (*Micropterus*), a pike (*Esox*), a gar (*Lepisosteus*), or a bowfin (*Amia*). Sometimes a duckling or a hapless songbird will be found in one of the foregoing fishes as well, or in a goosefish (*Lophius*). As every angler knows, frogs are a natural food of predatory freshwater fishes; hence, they make good bait for many game species. Not even humans escape, for they are attacked by several fishes, including barracudas (*Sphyraena*), certain sharks, and piranhas (*Serrasalmus*) of South America.

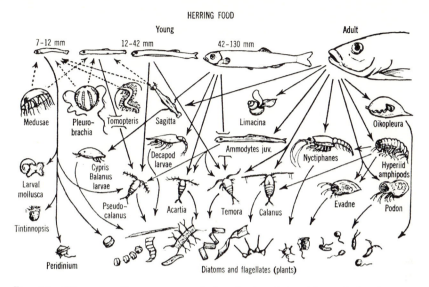

Fig. 5.1 Diagrammatic summary of the food relationships of the herring (*Clupea harengus*) at different sizes. (Source: Hardy, 1959).

Abundance of Food

Because of natural fluctuations in abundance, any one food organism is not of constant numerical availability to fishes. Such fluctuations of forage organisms are often cyclic and due to factors of their life histories or to climatic or other environmental conditions. Fish migrations often reflect a search for a particularly abundant source of food. The herring (*Clupea harengus*) in the North Sea, for example, follows plankton concentrations, and predators of the herring, such as certain mackerels (Scombridae), follow the fish about the North Sea. Both growth and survival in fishes have been shown repeatedly to reflect changes in abundance in food organisms. Availability of plankton at the time fish hatch is important to survival of the year class. Trouts (*Salmo*; *Salvelinus*) in American streams accomplish most of a year's growth in a few weeks of the late spring or early summer, when there is a peak of insect emergence. This emergence takes place at the time when the aquatic stages of many different kinds of insects metamorphose and leave the water preparatory to reproduction. During this interval the prey is seasonally concentrated and particularly vulnerable to predation.

Abundance of a potential food species often determines whether or not it will be eaten by fishes, for indeed availability is a key factor in determining what a fish will eat. Most fishes are highly adaptable in their feeding habits and utilize the most readily available foods. Relatively few kinds approach being strict herbivores or carnivores, and perhaps none at all feed solely on any one organism. Some, such as the Bermuda angelfish (*Holacanthus bermudensis*), may even change their diet with the season and may be quite herbivorous in winter and spring and become predominantly carnivorous in summer and early fall.

Food Chains, Food Webs, and Food Pyramids (Fig. 5.1)

Fishes are tied to other forms of life in their environment by food webs. If one considers the relative positions of the eaters and the eaten, herbivores and carnivores are at different vertical positions (trophic levels) of a food pyramid. Usually the largest carnivore or top predator can be placed at the apex of the pyramid. Such a pyramid represents the decreasing numbers of these animals in a community as their individual size increases; it may also represent decreasing biomass in the different levels of the pyramid with the least at the top (the largest predator).

The bottom level is, of course, occupied by green plants that bind the sun's energy for further transfer through the living world. Then comes an intermediate level composed of the herbivores—mollusks, certain insect larvae, and many crustaceans and fishes. Finally there is the highest trophic level, occupied by carnivores, in which there may be several tiers of fishes that prey on fishes that prey on fishes, and so on.

To complete the round of the wheel of life or to close the food cycle, excretions of living fishes and other organisms and their remains when dead are acted upon by bacteria. The materials composing them are returned into solution to give again to plants the raw materials to harness radiant energy into new protoplasm during the process of photosynthesis.

Food relationships may appear relatively simple when we look only at the adult stage of a fish, say a black bass (*Micropterus*) or a bluegill (*Lepomis macrochirus*), but when the entire life cycle of a fish is considered, from fry to juvenile to adult, food webs are likely to be very complex. They appear as a web of interrelations of eating and being eaten involving many phyla of aquatic organisms as the food changes with the life history stage.

A food cycle is fairly easy to visualize in the upper sunlit waters of lakes, rivers, and the oceans, as having a producing or synthetic phase, possessing consumer links and having return of raw materials to the synthetic phase through excreted wastes, death, and decay to be used again. But what of the dark, deepwater abyssal zones of the oceans, the deeps of great inland lakes such as Superior and Baikal? Here the organisms, including the fishes, depend on items that move into the deeps from above, either by migration or in the form of a rain of carcasses. Furthermore, in the deepwater zone of standing waters, the transfer may involve a great series of layers through which nutrients are relayed by organisms from the manufacturing zone of the upper waters to the purely consuming zones of the deepwaters. In this connection, one may visualize an upper zone in which synthesis is going on, followed downward by a deeper zone in which there are some crustaceans and other invertebrates and fishes that browse up into the lower part of the synthetic level. Beneath the level of the browsing, pelagic crustaceans may be another one where there is so little light that no synthesis is possible but in which perhaps there are some small fishes that feed upon the crustaceans, to bridge the light and dark areas, and so on through successive fishes and other animals to the deeps (Chapter 14). Here all organic matter that sinks accumulates into concentrations of ooze to become food for deep sea bacteria and food webs built upon them. At each successive downward level (the sea bottom excluded) the total biomass becomes smaller and smaller.

FEEDING HABITS

In any discussion of the feeding of fishes, both the manner and the stimuli for feeding need to be included. The feeding habits or the feeding behavior of fishes is the search for and ingesting of food. These should be distinguished from food habits and diet, which are the materials habitually or fortuitously eaten.

Major Feeding Types

As for the manner of feeding, only one broad common characteristic prevails —the food is taken into the mouth. Other than this the feeding habits and adaptations of fishes are very diverse. Nevertheless, certain very broad types of food getting are recognizable either by species or by life history stages. On this basis, fishes can be classified, although somewhat arbitrarily, according to their feeding habits as predators, grazers, food strainers, food suckers, and parasites.

Predators.   Fishes that feed on macroscopic animals have certain adaptations in common. They usually have well-developed grasping and holding teeth, as in many sharks (Elasmobranchii), the barracudas (*Sphyraena*), the pikes, pickerels, and muskellunge (*Esox*), or the gars (*Lepisosteus*). In predatory fishes there is well-defined stomach with strong acid secretions, and the intestine is shorter than that of herbivores of comparable size. Many predators such as the voracious bluefish (*Pomatomus saltatrix*) and many deepsea fishes actively hunt their prey, whereas others, like the groupers (*Epinephelus*) often lie in wait till an animal passes and then dart out to grasp it. The anglerfishes (Lophiidae and Antennariidae) have developed an anterior ray of the first dorsal fin into a lure to attract their prey. The archerfish (*Toxotes jaculator*) of Southeast Asia even shoots down insects from plants at the water's side by spitting at them. The accurate aim of this fish testifies to its well-developed aerial vision. Some predatory fishes hunt by sight whereas others, including many sharks (Squaliformes), nocturnal fishes (such as bullheads, *Ictalurus*) and morays (Muraenidae) rely largely on smell, taste and touch and probably also on their lateral-line sense organs to locate and catch their prey.

Grazers.   In grazing, the actual taking of the food is by bites, often by individual small ones. Sometimes organisms are taken singly or at other times in small groups in a rather continual type of browsing much like cows or sheep in a pasture. Grazing characterizes many fishes that feed on plankton or on bottom organisms. For example, a bluegill (*Lepomis macrochirus*) feeds along the bottom of a lake, taking this and that, often individual dipteran midge larvae. Many young fishes that grow to be predators of other fishes feed on plankters which they hunt individually.

Parrotfishes (Scaridae) or butterflyfishes (Chaetodontidae) browsing on a coral reef, may scrape off coral particles to ingest polyps or bite off pieces of algal fronds as previously indicated. A very specialized kind of grazing is that in which fish browse on parts of one another. Scales that are plucked from others and ingested (lepidophagy) constitute the bulk of food of an Indian catfish (Schilbeidae) and an African cichlid. Furthermore, it is not uncommon in the crowded waters of fish hatcheries to see trouts (*Salmo*; *Salvelinus*) nip parts of fins from one another.

Strainers.   The straining of organisms from the water is a generalized type of feeding, inasmuch as food objects are selected by size and not by kind. Plankton filtration and swallowing of the pea-soup-like concentrate illustrates this feeding act as in many herring-like (clupeoid) fishes including the gizzard shads (*Dorosoma*), as previously indicated. The menhaden (*Brevoortia tyrannus*), which occurs in large schools along the American Atlantic coast, swims mouth agape through the rich plankton beds. An adult menhaden is capable of straining as much as 1–2 gallons of water a minute through its gill rakers. The fish can thus swallow during the same short time, several cubic centimeters of plankton concentrate, mainly diatoms and crustaceans. Other fishes, some attaining large sizes, such as the paddlefish (*Polyodon*), the basking shark (*Cetorhinus*) and the whale shark (*Rhincodon*) also have efficient food straining or filtering adaptations. A principal adaptation of the strainers is the development of numerous, close-set, and elongated gill rakers (Fig. 5.2).

Suckers.   The sucking into the mouth of food or food-containing material is often practiced by bottom feeding fishes such as the sturgeons (Acipenseridae) and suckers (Catostomidae) (Fig. 5.3). Old World minnows (such as fringelips, *Labeo* and *Osteochilus*), with inferior mouths and sucking lips, have similar habits. In a few carp (Cyprinidae) species in Southeastern Asia the sucking response depends so strongly on the stimulus of touch on the frilled, fleshy lips that fishermen bite the lips off these fishes before crowding

GILL RAKERS

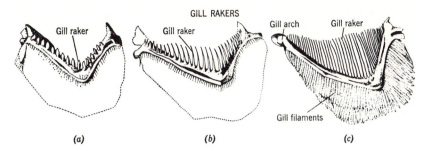

(a)   (b)   (c)

Fig. 5.2   Relationship of gill rakers to feeding habits, showing increasingly efficient sieve potential from left to right: (a) round whitefish, *Prosopium cylindraceum*; (b) lake whitefish, *Coregonus clupeaformis*; (c) blackfin cisco, *Coregonus nigripinnis*. (Source: Hubbs and Lagler, 1958).

SUCKER MOUTHS

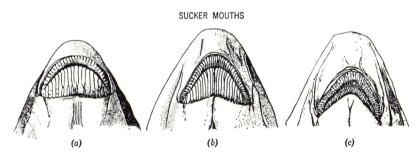

(a)   (b)   (c)

Fig. 5.3   Variations in the development of the suctorial lips and mouth in three redhorse suckers (*Moxostoma*); (a) shorthead redhorse, *M. macrolepidotum*; (b) golden redhorse *M. erythrurum*; (c) silver redhorse, *M. anisurum*. (Source: Hubbs and Lagler, 1958).

them into holding pens with other species. The concern is that the other fishes will be damaged by being mistaken for the algae- or moss-covered stones on which the minnows usually feed by scraping and sucking.

Fishes that suck in mud to extract the organisms in it may or may not get a good mouthful of food with each ingestion. In some, the food items are separated from the sediments before being swallowed but in others, such as some oriental catfishes (Schilbeidae, Siluridae), remains of flocculent bottom deposits can be found in the digestive tract together with high concentrations of bottom organisms. This feeding habit indirectly plays havoc with machinery for oil extraction from fish wastes because metal parts become covered with a sticky emulsion of fish fats and clay.

Parasites.   Parasitism* is perhaps the most unusual and highly evolved feeding habit among animals. In the fish world, an outstanding example of this

* The semantics of neither the words parasite nor predator precisely connotes the sarcophagic feeding habits of the "parasitic" lampreys or the hagfishes. Thus we use the term parasitic loosely here.

practice is represented by parasitic lampreys (some Petromyzonidae) and hag-fishes (Myxinidae) that suck body fluids from the host fish after rasping a hole in the side of the body. A species example is the sea lamprey (*Petromyzon marinus*; Fig. 5.4) of the western North Atlantic and adjacent continental waters. Another is the Pacific lamprey (*Lampetra tridentata*) that includes whales among its hosts. A deepsea eel (*Simenchelys parasiticus*) is also para-sitic. The males of some of the deepsea anglerfishes (*Ceratias*) are obligatory parasites on the females of the same species (Fig. 10.4). Shortly after hatch-ing, the male finds a female and attaches by his mouth to her body. The female obligingly responds by developing a fleshy papilla from which the male fish can absorb nutrients since subsequent to attachment he takes no free-living food at all. The male remains relatively small, little more than an animated gonad. The intra-uterine absorption of nourishment (and respiratory exchange) by the embryos of certain livebearing fishes represents an even greater nutritional specialization than the foregoing extreme of parasitism.

Feeding Adaptations

The diversity in feeding habits that fishes exhibit is the result of evolution leading to structural adaptations for getting food from the equally great diversity of situations that have evolved in the environment.

Lips.   A significant advance in vertebrate evolution was the appearance of true jaws to border the mouth opening. Most generally the jaw-equipped mouth has a biting function and fishes that swallow large morsels of food usually have unmodified relatively thin lips. Suctorial feeders have an inferior mouth and fleshy modification of lips (Figs. 5.3 and 5.5d). Notable among these are the sturgeons (Acipenseridae) of the North Temperate Zone and the suckers (Catostomidae; Fig. 5.3) of North America and Asia. The lips of the sturgeons and suckers are mobile and described as plicate (having folds) or papillose (having small tufts of skin or papillae). Many suctorial feeders also have well developed barbels more or less bordering the mouth, as in the sturgeons (*Acipenser*) and Asiatic hillstream loaches (Homalopteridae). The barbels have many sensory end organs (Chapter 11) and help to locate food grubbed from soft bottom materials.

Suctorial lips of free-living fishes may also serve as holdfast organs in fast flowing mountain streams, as in one of the Southeastern Asiatic sisorid cat-fishes, *Glyptosternum*, some of the armored catfishes of South America (Loricariidae), and in the loach-like gyrinocheilid (*Gyrinocheilus*) of South-eastern Asia. The last fish has an extremely specialized suctorial adaptation in having developed in addition to suctorial lips, an opercular structure with a separate water inhalent and exhalent device; for these purposes the fish

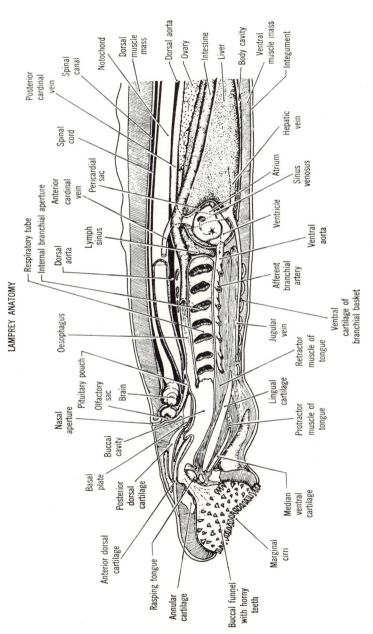

LAMPREY ANATOMY

Fig. 5.4 Jawless mouth, foregut, and related structures of an adult sea lamprey (*Petromyzon marinus*) as seen in sagittal section. (Source: Lagler, 1954).

137

MOUTHS

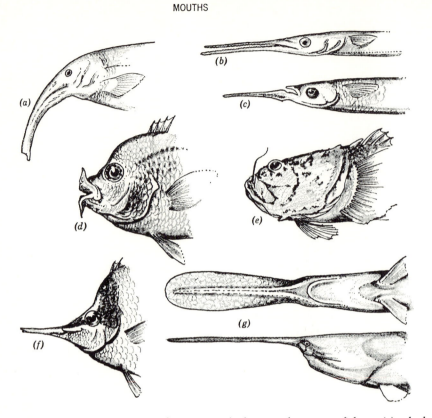

Fig. 5.5 Some structural adaptations of the mouth among fishes: (a) elephantfish, *Gnathonemas elephas*; (b) garfish, *Strongylura longirostris*; (c) halfbeak, *Hyporhamphus unifasciatus*; (d) thick-lipped mojarra, *Cichlasoma lobochilus*; (e) stargazer, *Zalescopus tosae*; (f) longnose butterflyfish, *Forcipiger longirostris*; (g) paddlefish, *Polyodon spathula*. (Source: Norman, 1931).

has two branchial openings on each side. The mouth is relieved thereby of taking in respiratory water when engaged in suction for holdfast or feeding. The mouth is small and the inner surface of the lips has rasp-like folds to facilitate the scraping of algae from the stones to which the fish adheres.

In the parasitic members of the lamprey family (Petromyzonidae) and in the hagfishes (Myxinidae) the jawless suctorial mouth serves both as a hold-fast for attachment to the host and as a food-remover from the host. The sucking disc of lampreys is also used to dislodge and transport stones from the nest pit in streams. Later the disc serves to anchor spawning individuals into a side-by-side position with head directed upstream and genital pores located over the nest excavation.

Modifications in the Shape of the Mouth (Fig. 5.5).   Among the grazers and suctorial feeders there exist not only specially developed lips but also adaptations of other mouth parts. The trumpetfishes (Aulostomidae), the cornet-fishes (Fistulariidae), and the pipefishes (Syngnathidae), as well as many butterflyfishes (Chaetodontidae) of coral reefs, have mouths that resemble elongated beaks. This adaptation is achieved by a protraction of the hyomandibular bone rather than by a lengthening of the lower jaw bones (dentaries) themselves.

Among the fishes, the method of feeding may be by suction such as a syringe would exert in the case of the trumpetfishes, cornetfishes, and pipefishes, or it may be a selective grazing action with sharp teeth where the long snout enables the butterflyfish to reach into small crevices of the coral.

Some predators, such as the dories (Zeidae), certain wrasses (Labridae), and the European bream (*Abramis brama*, Cyprinidae), can form temporary tubes in which to engulf their prey from close range by forward extension of the jaws enabled by special articulation of the premaxillaries and other skull bones. Other adaptations, also of skull bones and their articulations, to increase the gape of the mouth exist among predatory deepsea fishes such as the deepsea viperfish (*Chauliodus*).

A peculiar structure among mouth modifications has arisen in the halfbeaks (Hemiramphidae) where the lower jaw projects into a beak, often a third of the length of the fish itself, with the mouth opening above it. Halfbeaks are usually surface-feeding fishes, and it has been surmised that the "beak," as well as being an aid in food getting, serves for steering and maintenance of equilibrium in conjunction with a posteriorly inserted dorsal fin.

Teeth (Fig. 5.6).   Outstanding among the obvious adaptations for feeding in fishes are the teeth. They are thought to have arisen from scales covering the lips, as represented in living sharks (Squaliformes), where the placoid scales of the skin visibly grade into teeth on the jaws. In the bony fishes (Osteichthyes), teeth are of three kinds, based on where they are found: jaw, mouth, and pharyngeal. Jaw teeth are variously those on the maxillary and premaxillary bones above, and on the dentaries below. In the roof of the oral cavity, teeth are variously borne by the median vomer and by the palatine and ectopterygoid bones on each side. In the floor of the mouth, the tongue often has teeth on it. Pharyngeal teeth occur as pads on various gill arch elements in many species. In the carps (Cyprinidae) and suckers (Catostomidae), the only teeth are those deep in the pharynx that develop from modifications of lower elements of the last gill arch. Tooth-like modifications of gill rakers and dermal bones (perhaps scales) are not uncommon supplemental adornments found on the inner surfaces of the pharyngeal arches in many predacious fishes such as the northern pike (*Esox lucius*).

Based on their form, some major kinds of jaw-teeth are the following: cardiform, villiform, canine, incisor, and molariform. Cardiform teeth are numerous, short, fine, and pointed. A pad of them on a bone resembles the

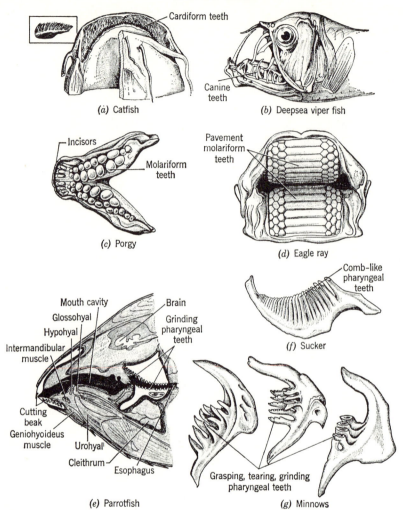

Fig. 5.6 Some variations of teeth in fishes by structure and location: (a) upper jaw of flathead catfish, *Pylodictis olivaris* (source: Hubbs and Lagler, 1958); (b) deepsea viperfish, *Chauliodus sloani* (based on Goode and Bean, 1895); (c) lower jaw of a porgy, Sparidae; (d) pavement teeth on roof and floor of mouth of an eagle ray, *Myliobatis*; (e) sagittal section of a parrotfish (Scaridae) (based on Monod in Grassé, 1957) to show "beak" from fusion of jaw teeth and to show pharyngeal teeth; (f) pharyngeal teeth of a golden redhorse sucker, *Moxostoma erythrurum* (based on Hubbs and Lagler, 1958); (g) types of minnow pharyngeal teeth, from left to right, showing pointed, grasping teeth of a barb (*Barbus*), crenulated, tearing teeth of a rudd (*Scardinius*), and molariform grinding teeth of the tench (*Tinca*) (based on Grassé, 1958).

140

multiple-toothed wool card of spinning-wheel days, hence the term cardiform. Such dentition with variations is found in very many fishes that have multiple-rowed teeth; included would be North American catfishes (Ictaluridae), perches (Percidae), and many seabasses (Serranidae). Villiform teeth are more or less elongated cardiform ones, in which the length-to-diameter relationship resembles that of intestinal villi (as in the needlefishes, *Belone*, and lion-fishes, *Pterois*). Canines are dogtooth-like, often even quite fang-like. They are elongated and subconical, straight or curved and are adapted for piercing and holding; they are possessed by the walleyes (*Stizostedion*), among many others. In certain fishes, such as the morays (Muraenidae) and the pelagic marine predator, the so-called handsawfish (*Alepisaurus*), the canines are hinged, yield to backward pressure, but snap into a locked position when pushed toward the mouth openings—obviously an adaptation to retain living, moving prey. Sharply edged cutting teeth are called incisors. In some fishes such as the sea bream (*Archosargus*) they look almost human; in others they are crenulated (sometimes saw-edged), and in still others they have become variously fused into cutting "beaks" as in the parrotfishes (Scaridae). Molari-form teeth are for crushing and grinding, and hence have flattened, often broadly occlusal surfaces; they are generally characteristic of the bottom-dwelling skates (including *Raja*), the chimaeras (Holocephali), and several drums (Sciaenidae).

There is strong correlation among kind of dentition, feeding habits, and food eaten. Predacious fishes, such as the pike (*Esox*), gars (*Lepisoteus*), and deepsea gulpers and swallowers (Saccopharyngoidei) have sharply pointed teeth of apparently great use in grasping, puncturing, and holding prey. Feeders on plankton and scrapers of encrusting periphyton characteristically have tooth-less jaws, although by special adaptation the jaw itself may have a cutting edge as in the stoneroller minnow (*Campostoma anomalum*) where the den-taries have a hardened, gristly leading edge that is used in scraping surfaces for food. In skates (Rajidae) and in drums (Sciaenidae) there are grinding (molariform) teeth in oral or pharyngeal cavities in association with a diet that includes snails, clams, and hard-bodied crustaceans. Razor-like cutting teeth have developed in predacious fishes such as the formidable piranha (*Serrasalmus*) of the Amazon and the barracuda (*Sphyraena*) of warm seas. Some vegetable feeders, such as many parrotfishes (Scaridae), chop their food into a veritable salad with cutters resulting from the fusion of individual teeth into sharp-edged ridges on the jaws. Initial cutting is then followed-up with fine trituration by means of pharyngeal grinding plates.

Within a single group the diversity of dentition on identical bones may be almost as great in principle as the foregoing gamut. In the carps and minnows (Cyprinidae), the pharyngeal teeth range from sharp in the carnivores such as the creek chub (*Semotilus*) to molariform in the common carp (*Cyprinus carpio*) and have almost disappeared in some species. In these and many other fishes, teeth are of value in classification. Many species of fossil sharks have been described solely on the basis of teeth.

In general, teeth are absent in plankton feeders and in some of the more generalized omnivores. They are present on increasing numbers of bones in the more and more predacious fishes. The premaxillary bones are toothed in most fishes that have any jaw teeth at all. This is true of many such soft-rayed species as the bowfin (*Amia*), the gars (*Lepisosteus*), the salmons and trouts (Salmoninae) and of tooth-bearing spiny-rayed fishes in general (for example, the large order Perciformes).

The maxillae are typically toothed in those soft-rayed fishes that carry premaxillary teeth. However, the maxillae are characteristically toothless in otherwise tooth-bearing, spiny-rayed fishes. In the process of evolution from soft-rayed fishes to spiny-rayed ones, the maxillae are tooth-bearing bones only in some soft-rayed kinds. In those with spiny rays, the maxilla is edentulous, and is excluded from the gape, no longer forming part of the margin of the mouth.

Gill Rakers. Besides protecting the tender gill filaments from abrasion by ingested materials that are coarse in texture, gill rakers are also specialized in relation to food and feeding habits, as previously described. They are very stubby and unadorned in omnivores such as the green sunfish (*Lepomis cyanellus*) and the pumpkinseed (*L. gibbosus*). In many plankton feeders, the gill rakers are elongated, numerous, and variously lamellated or ornamented, presumably to augment efficiency in straining. Simple, but very numerous rakers are possessed by gizzard shads (*Dorosoma*) and the paddlefish (*Polyodon*). Ornamented structures, however, are found in flatfishes (Pleuronectidae) and are of taxonomic use. Here each raker resembles a feather in having a main axis with lateral processes (and even branches on these lateral elements). When these processes on adjacent rakers overlap, a sieve is formed that can strain very finely.

The Digestive Tube. Another adaptation that fishes have for feeding is the great distensibility of the esophagus. Seldom does an individual choke to death because it cannot swallow something that it got into its mouth. Only occasionally is a predacious fish found in mortal distress because of a prey fish lodged in its throat. Catfishes (Ictaluridae) and sticklebacks (Gasterosteidae) are frequently offenders to their predators in this regard. Their spinous fin rays, when erected, cause them to become stuck. Yet most often if these spines had been depressed, the engulfed and lodged fish would have slipped down the esophagus very easily. In general the esophagus is so distensible that it can accommodate anything that the fish can get into its mouth and can sometimes even accommodate the item if it happens to double on itself two or three times on its way to the stomach.

The stomach, too, shows various adaptations, one of which is shape (Fig. 5.7). In fish-eating fishes, the stomach is typically quite elongate as in the gars (*Lepisosteus*), bowfin (*Amia*), pikes (*Esox*), barracudas (*Sphyraena*), and the striped bass (*Morone saxatilis*). In omnivorous species, the stomach is most often sac-shaped, similar to that in humans. A very special adaptation is the modification of the stomach into a grinding organ, as in the sturgeons (*Acipenser*), gizzard shads (*Dorosoma*), and mullets (*Mugil*) (Fig. 5.7). Here the stomach is reduced in overall size but its wall, greatly thickened and muscularized. The lining too is heavily strengthened with connective tissue, and the lumen (the space within), made very small. The organ is not a bin for mixing and primary digestion but rather a food grinder, not unlike the gizzard of chickens and other fowl. Great distensibility is the adaptation of the stomach in the predatory deepsea swallowers (Saccopharyngidae) and the gulpers (Eurypharyngidae) enabling these fishes to take relatively huge prey.

A remarkable modification of the stomach exists in the puffers (Tetraodontidae) and porcupinefishes (Diodontidae) which can inflate themselves with water or air to assume often an almost globular shape. Either the stomach itself or an evagination from its anterior portion are filled by action of cardiac and pyloric sphincters and by another sphincter in the evagination itself. When inflated, this evagination or gastric sac is so distended that its walls are paper thin. Emptying of the sac proceeds, again by sphincter action, assisted by the abdominal body musculature. The presence of a distensible stomach in these families goes hand in hand with the absence of pelvic fins; the pelvic bones may be absent also. The adaptive value of this modification of the digestive tract is probably mainly one of defense, for many puffers and porcupinefishes have spines all over the body which can thus be erected. It ought to be mentioned that the gastric sac of puffers, porcupinefishes and some of their relatives is neither related to, nor derived from the gas bladder.

Not all fishes have a stomach, that is a portion of the digestive tube with a typically acid secretion and a distinctive epithelial lining different from that of the intestine. In the plant-feeding roach (*Rutilus rutilus*), a cyprinid of the Old World, for instance, epithelial tissue of the esophagus grades directly into that of the intestine. In other grazers such as the parrotfishes (Scaridae) analogous conditions are found. Some carnivores also have lost their stomachs, as the sauries (*Scomberesox*), as have certain plankton-feeding species in the pipefish-seahorse family, Syngnathidae. The primary criterion for being able to do without the stomach does not seem to be whether a fish is an herbivore or a carnivore but whether accessory adaptations for trituration and very fine grinding of food exist either in the form of teeth or a grinding

STOMACHS

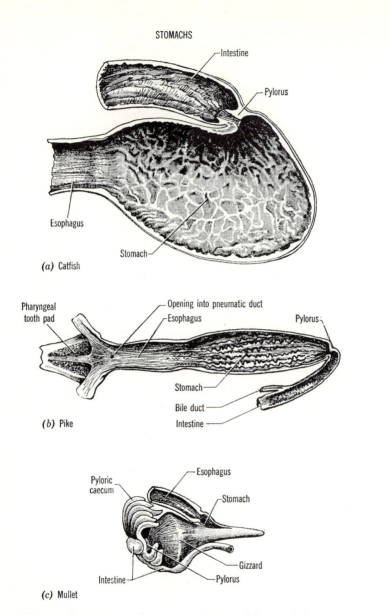

(a) Catfish

(b) Pike

(c) Mullet

Fig. 5.7   Variations in shape, appendages, and lining of the anterior portion of the digestive tract in three fishes: (a) an omnivore, a catfish, the European Wels (*Silurus glanis*);
(b) a carnivore, the northern pike (*Esox lucius*); (c) a bottom-grubber, the striped mullet
(*Mugil cephalus*). (Based on Pernkopf in Bolk, Göppert, Kallius, and Lubosch, 1937).

apparatus such as a gizzard. Where stomachs exist, most pronouncedly in carnivores, they are characterized by a low pH and the prominent presence of pepsin among other digestive juices.

The intestine too, has many variations (Fig. 5.8). It is shortened in essential carnivores such as the pike (*Esox lucius*; Fig. 5.8c) perhaps because meaty foods can be digested more readily than vegetable ones. In opposite fashion, it is often elongated and arranged in many folds in predominantly herbivorous species, as in certain loricariid catfishes (Fig. 5.8a). The sharks and relatives (Chondrichthyes), and a few other fishes have substituted a spiral valve (Fig. 5.8b) or a large fold of absorptive tissue loosely wound in a roll for folds of the intestine and have ostensibly improved efficiency of digestion and/or absorption and visceral compactness thereby. The intestine itself seems to undergo digestion (autolysis) in fishes that cease feeding as sexual maturity and breeding arrive. The once-functional digestive tract of an adult sea lamprey (*Petromyzon marinus*) becomes a mere thread with practically no lumen by the time spawning is over and death approaches. Similar reduction has taken place in migrant Pacific salmons (*Oncorhynchus*) when they have reached their freshwater spawning grounds.

Stimuli for Feeding

Of interest both to ichthyologists and anglers are considerations of the stimuli to feeding in fishes. However, the mechanism of feeding behavior is a very complicated one, as we shall see. The stimuli to feed are of two kinds: (*a*) factors affecting the internal motivation or drive for feeding, including season, time of day, light intensity, time and nature of last feeding, temperature, and any internal rhythm that may exist; (*b*) food stimuli perceived by the senses like smell, taste, sight, and the lateral-line system, that release and control the momentary feeding act. The interaction of these two groups of factors determines when and how a fish will feed and what it will feed upon.

One important factor in feeding is time of day. Some fishes, such as the bullheads (*Ictalurus*) that find food by smell and taste, are predominantly night feeders. Still others, such as the pikes (*Esox*) and other predators that feed largely by sight, are most active during daylight hours. Season, strongly influencing water temperature in the nontropical areas and water levels in the tropical region, seems to have something to do with feeding also. Some fishes cease feeding altogether during their spawning seasons, including salmons (*Salmo, Oncorhynchus*) and lampreys (Petromyzonidae). Others such as the swamp eels of southeastern Asia (Synbranchidae), while estivating in moist burrows in the mud, live for weeks on accumulated fat. Most fishes of the Temperate Zones feed very actively as environmental conditions change in the spring, near the beginning of a new growing season. Fishing lore

INTESTINES

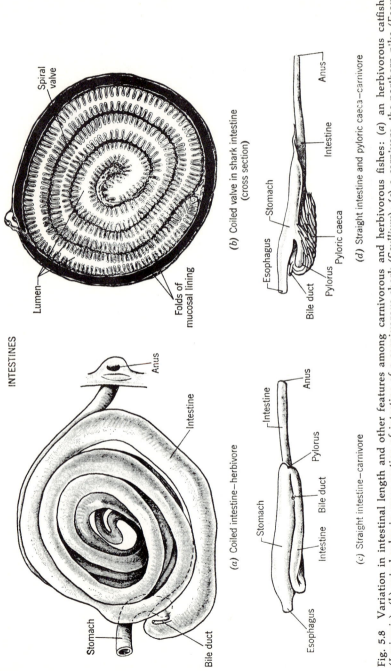

Spiral valve

Lumen

Folds of mucosal lining

(b) Coiled valve in shark intestine (cross section)

Anus

Intestine

Stomach

Esophagus

(a) Coiled intestine—herbivore

Stomach

Bile duct

Anus

Intestine

Pylorus

Bile duct

Intestine

Esophagus

(c) Straight intestine—carnivore

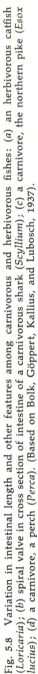

Anus

Intestine

Stomach

Esophagus

Pyloric caeca

Pylorus

Bile duct

(d) Straight intestine and pyloric caeca—carnivore

Fig. 5.8 Variation in intestinal length and other features among carnivorous and herbivorous fishes: (a) an herbivorous catfish (*Loricaria*); (b) spiral valve in cross section of intestine of a carnivorous shark (*Scyllium*); (c) a carnivore, the northern pike (*Esox lucius*); (d) a carnivore, a perch (*Perca*). (Based on Bolk, Göppert, Kallius, and Lubosch, 1937).

146

is replete with contradictory statements regarding climatic conditions or stimuli that cause fish to bite. However, all that can be said truthfully is that very much remains to be discovered regarding the regulatory role of climatic factors in feeding activity.

Since fishes have senses of smell and taste, many workers have tested chemical factors as determinants of feeding. For many species, the chemical senses do indeed first cue the feeding act. The dogfish sharks (*Squalus*) as well as some of the man-eating sharks (*Carcharodon*) are led to feed importantly by chemical attraction, and moray eels (*Gymnothorax*) have been shown to find and select food primarily by smell. Catfishes (Ictaluridae) and goatfishes (Mullidae) feed mainly by taste and also by touch.

Visual stimuli trigger the feeding act in fishes with movement, color, and the shape of objects perceived each having its role as releasers of the act. Among schooling fishes of one species, as well as in aggregations of several species in a common environment, nonfeeding individuals may be triggered to feed by the discovery of food by one individual. In fact, it has been suggested that one adaptive value of the schooling habit is to facilitate feeding. Aquarium fishes can be trained secondarily to associate feeding with some auditory, olfactory, or visual stimuli. Similarly, fishes learn to avoid certain foods because of their texture, taste, or other quality. A pike (*Esox*), for example, soon learns to avoid a sharp-spined stickleback (*Gasterosteus*) as prey although the innate response of the pike may be to feed on an object with the dimensions, movement, and color of a stickleback.

Rapid changes in light intensity are also important in feeding. The yellow perch (*Perca flavescens*) and many other fishes exhibit circadian rhythms in feeding activity and movement with peaks at dawn and dusk. There are many subtle changes in the environment tied to day and night or tidal conditions; included are not only light, but also temperature, salinity, pH, and current. The variations of any one of these factors by itself or in combination with others are discernible by fishes and may influence activity and feeding pattern. Innate timing mechanisms, termed physiological clocks, independent of, or unrelated to environmental stimuli may exist in fishes. In nature, however, timing landmarks such as diurnal light, pH, temperature, and other changes are rarely absent and usually trigger levels of activity.

Promptings for feeding may include those of actual contact and textural feel. In aquaria, when both the bluegill (*Lepomis macrochirus*) and the pumpkinseed (*L. gibbosus*) are offered small snails, only the pumpkinseed will crush the shells, eject them through their gill openings, and swallow the soft parts. However, if soft parts of small snails alone are offered the bluegill, they will be swallowed. This may be evidence of the significance of texture in feeding and a sense of touch in the mouth. It should be pointed out in this connection that although the jaw teeth are much the same in these two sunfishes, the pharyngeal teeth of the pumpkinseed alone are truncated and thus adapted for a crushing function.

In addition to factors such as the foregoing, there are others that doubtless lead fishes to feed. These include hunger, gluttony, curiosity, and other less defined factors. After a long interval of enforced deprivation between feeding bouts in sticklebacks, the feeding that follows is in some respects more rapid, suggesting increased motivation. Hunger and satiation may be controlled by the hypothalamus. Regarding gluttony, the inclination to eat long after the normal capacity of the animal has been reached has been described for the bowfin (*Amia*) in aquaria. During some food utilization experiments, small serranids (*Epinephelus guttatus*) gorged themselves far beyond a sustained need for optimal growth. Gluttonous feeding results in the passage of apparently less well-digested material in the feces than is usual. It may, of course, be prompted by the shortage of some nutrient in the food.

Even under constant environmental conditions experimental brown trout (*Salmo trutta*) show cyclic fluctuations in annual growth rates, reflecting variations in their feeding and metabolism. Growth has a spring maximum, is rapid in summer, and is slow or negligible in autumn and winter. Another aspect of changes in metabolism, apparently determined by internal factors, concerns the occurrence of periods of rapid growth in length alternating with periods of rapid growth in weight of this trout. The periods in question are of 4 to 6 weeks in duration, but the reason for the fluctuations remains unexplained.

To temper the action of all stimuli to feed, there is the relation between internal motivation or drive for feeding and the degree of selectivity controlling the proper feeding act, such as the capture of prey. For example, the less active the fish is, the more the stimuli emanated by the prospective prey regarding such things as dimensions, form, color, and movement must correspond to the ideal prey experience, perhaps also innately, or both.

DIGESTION, ABSORPTION, AND UTILIZATION OF FOODS

The basic function of the digestive system is to dissolve foods by rendering them soluble so that they can be absorbed and utilized in the metabolic processes of the fish. The system may also function to remove dangerous toxic properties of certain food substances.

Movement of Food in the Tract

The movement of foodstuffs and their undigested remains in the digestive tract of a fish is accompanied, as it is in man and other vertebrates, by peristaltic waves of muscular contraction. In the anterior part of the tract, at least, the movement is voluntary due to the presence of skeletal muscles in the wall of the gut. Elsewhere it is involuntary and involves smooth muscles. The tongue, in contrast to that in higher vertebrates, is not mobile by itself; it has no skeletal muscle components and moves only when the fish moves its underlying visceral skeleton. The typical route of a bolus of food in a fish is essentially that of higher vertebrates, as already described.

Many predatory fishes appear to regurgitate large food items from the stomach with great facility. It has been suggested that this is made possible by the pronounced development of striated muscles in the walls of the esophagus extending to the stomach.

Intestinal Surfaces

The mouth cavity, esophagus and stomach are lined with a soft mucous membrane as is the rest of the tract. There are no salivary glands (except in some specialized fishes such as the parasitic lampreys of the family Petromyzonidae). However, the tract wall is liberally supplied with glands that secrete mucus which lubricates passage of food materials, and protects the gut lining. The gut is highly elastic and permits the passage of outsized masses of food. The linings of the intestines, small and large, are highly absorptive and pick up digestive products in solution. The absorptive capacity of these areas is increased by throwing the walls into lengthwise folds (such as the typhlosole in lampreys, Petromyzonidae), transverse folds (rugae), or multitudinous fingerlike projections (villi) in such fishes as mullets (*Mugil*). Along the course of the tract, there are many gland cells that contribute digestive enzymes.

Glands and Digestive Enzymes

The general, widespread distribution of the mucous glands both outside and inside the fish has already been described. Additionally, gastric glands occur in the stomach at least in predacious fishes. These secrete hydrochloric acid and pepsinogen, effective in combination to split large protein molecules. In typically carnivorous fishes such as the northern pike (*Esox lucius*) gastric acidities of pH 2.4 to 3.6 have been measured. Consequently there exists in the stomach a suitable substrate for protein-splitting enzymes or peptidases that have been demonstrated in many fishes. Evidences for stomach enzymes other than peptidase are not clear-cut. Some minnows (Cyprinidae) lack gastric glands and, on this basis, may not possess a true stomach. Similarly the fish gizzard (Fig. 5.7c) does not have digestive glands.

Pyloric caeca (Fig. 5.8), when present, may have a digestive function, an absorptive one, or both. The caeca (Fig. 5.8) are finger-like pouches that extend outward from the intestine near the pylorus; the enzyme lactase has been found in them in trout (Salmoninae). A higher level of saccharase (invertase) activity has been found in the pyloric caeca and intestine of carp and bluegill compared to pickerel. The former ingest a considerable amount of vegetable matter whereas the pickerel ingest practically none. The pyloric caeca and the intestinal mucosa are sources of lipase which breaks down fats into fatty acids and glycerine. Absorption of fats may occur in the anterior part of the intestine. The copepods upon which anchovies, sardines, and herring prey contain up to 50 percent wax. The pyloric caeca of these fish produce a bile-activated wax lipase, an enzyme that catalyzes the metabolism of wax. The morphology of the pyloric caeca as single or branched (bifid) as well as their numbers are characters of value in the classification of some fishes.

Several discrete or diffuse glands are associated with the digestive tract. The thyroid, the thymus, and the ultimobranchial are of endocrine function but arise embryologically from the developing digestive tract (Chapters 10 and 11) and have no direct bearing on digestion.

The liver, as in other vertebrates, arises in the embryo as a ventral evagination of the developing intestine. The anterior portion develops into the liver proper and the posterior portion into the gall bladder and its ducts. Some holostean and some teleostean fishes have more than two liver lobes but there are always only two hepatic ducts from the liver to the cystic duct which terminates in the gall bladder. The gall bladder is a temporary storage organ for secretions of the liver. From the junction of the hepatic and cystic ducts there arises the common bile duct which enters into the beginning of the intestine near the pyloric region.

Some deepsea fishes have a vestigial gall bladder only and in others, such as the burbot (*Lota lota*), the gall bladder is completely absent. The liver adjusts its form to that of the body cavity and consists of lobules of tubular glands where the blood bathes tiers of cells on one side while bile is secreted into the bile canals on the other side. The bile contains the fat-emulsifying bile salts along with the bile pigments, biliverdin and bilirubin, that originate from the breakdown of red blood cells and hemoglobin. The bile salts may not only help to hydrolize fats but also to adjust the digestive juices of the intestine to the proper alkalinity for the action of digestive enzymes.

Besides its role in digestion, the liver also acts as a storage organ for fats and carbohydrates (glycogen). It further has important functions in blood cell destruction and blood chemistry, as well as other metabolic functions such as the production of urea and compounds concerned with nitrogen excretion (Chapter 9). Fishes vary greatly in the amount of fat that is stored in the liver but two general types can be distinguished. In flatfishes (Pleuronectiformes) and cods (Gadidae), for example, fat is stored predominantly in the liver. Fat is to a great extent stored in the muscle of fishes such as tunas (Scombridae) and herrings (Clupeidae). Sharks and bony fishes differ in the relative amounts of unsaturated fatty acids in their liver oils. In some sharks the liver may take up as much as 20 percent of the body weight. Besides storing fat, the livers of fishes also store vitamins A and D. The content of these vitamins in the tunas for instance, is so high that persistent eating of their liver may lead to severe disturbances in human metabolism, which presumably result from hyper-vitaminosis.

The pancreas is really two organs rather than one; it has both exocrine and endocrine functions (Chapter 11). It varies greatly in type of development and position in the various groups of fishes. In sharks and rays (Elasmobranchii) the pancreas is relatively compact, often two-lobed, but in bony fishes (Osteichthyes) the pancreas is often diffuse (Table 2.3). There are smaller or larger nodules of pancreatic tissue in the omentum, the suspensory tissues of the intestine, in addition to dispersion of the pancreas into the liver to form a hepatopancreas among many spiny-rayed fishes (Acanthopterygii). The arrangement of endocrine cells that secrete insulin, the pancreatic islets or the islands of Langerhans, also varies in different groups of fishes which possess a discrete pancreas.

In most fishes the surface folds inside the intestine show a rectangular pattern, apparently instrumental in slowing down the passage of the food. The enzymes secreted by the small intestine, as well as bile and pancreatic secretions that pour into this part of the gut, work best at a pH ranging from neutral to alkaline. Different proteases, affecting terminal and inner bonds uniting the amino acids of proteins, are secreted either from the intestinal mucosa, pancreas, or pyloric caeca. Alkaline protease has been found in the intestine of rainbow trout and acid protease in the stomach of this species. Intestinal enzymes are secreted in an inactive form as zymogens, a general term for inactive enzymes, before chemical changes in the lumen of the intestine make them active in digestion. This change is brought about by other enzymes such as enterokinase. This adaptation prevents self-digestion (autolysis) of the intestinal mucosa.

Various enzymes that digest specific carbohydrates have been found in the intestine as well as in the pancreatic juice of fishes. In the predominantly herbivorous *Tilapia*, amylase activity was distributed throughout the gastrointestinal tract, but in carnivorous perch, the pancreas was the only source of amylase activity. There is evidence from several fishes, for example menhaden (*Brevoortia*), silverside (*Menidia*), and silverperch (*Bairdiella*) that fishes have endocommensal bacteria possessing cellulase which can break down the cellulose plant materials both for their own nutritive value, small as that may be, and to make the contents of the plant cells available for utilization by the fish. Still herbivorous fishes largely rely on mechanical breakdown of plant cell walls, hence the extensive development of cutting, tearing, and grinding teeth among them, and the habit of feeding almost incessantly during daylight. Examination of the feces of an herbivore such as the Asiatic grass carp (*Ctenopharyngodon idella*) reveals that cells not broken down by the teeth appear intact and even retain their green pigments. Whole algal cells are also passed intact through the gut of some plankton feeders such as the gizzard shad (*Dorosoma cepedianum*).

The innervation of digestive organs is both sympathetic from paired ganglia lateral to the spinal cord and parasympathetic through branches of the vagus nerve (Chapter 11). Experiments with sharks and rays have shown that sight and smell of food do not cause gastric secretion. This is in contrast to observations on mammals that have so-called psychic or appetite flow of digestive juices mediated through the vagus nerve. The pancreatic, and probably also intestinal secretion and bile flow are under both hormonal and nervous control. Several hormones have been isolated from fish intestinal tissue and shown to act on the flow of zymogens as well as other fluids such as bile and pH-buffering pancreatic secretion.

Absorption of Digested Materials

Absorptive Process.   In order for digested foods to be absorbed, they must be in aqueous solution; hence, they themselves must be soluble. The component molecules must further be of a size that will enable them to cross the membranes of the cells lining the tract, pass into the circulatory system, and ultimately be carried to and enter cells that need them or store them. It is interesting to note in this context that fat absorption is intensified in the pyloric stomach of some fishes (as in the basking shark, *Cetorhinus*) and the pyloric caeca of others (as in the genus *Salmo*). Fats have been shown to enter into the lymph ducts in these regions without being split into their component fatty acid and glycerol molecules upon which intestinal absorption depends.

Although most digested proteins are absorbed in the intestines, some absorption of protein derivatives has been shown to occur in the shark stomach. It has also been shown that the smallest units into which proteins are broken down in intestines are dipeptides (compounds consisting of two amino acids). This apparently leaves to intracellular digestion the final breakdown of proteins into single building stones (amino acids) from which the fish proteins are resynthesized. Thus evidence accumulates to render less distinct one of the earlier suggested differences between lower invertebrates and vertebrates, that based on the presence of intracellular digestion in one and not the other group of animals.

Toxification of Flesh. Apparently because of the absorption of materials from foods, the flesh of a great many fishes is toxic to man (ichthyosarcotoxism). Some fish species are usually so, whereas others may be consumed safely hundreds of times and suddenly cause grave illness or even death when eaten again. The strongest evidence as to the origin of ichthyosarcotoxism now points to the feeding habits of fishes. The poison is thought to originate in marine plants and to be conveyed to man when the fish is eaten, perhaps after accumulation, concentration, or even alteration in the fish. Herbivorous fishes may transmit the poisons to the flesh of carnivorous fishes with no effect on the predator. Yet, when the predator is consumed by man, toxicity may result. Although widely distributed throughout the world, the greatest numbers of poisonous fishes are known from tropical waters. Temperate waters, however, have many puffers (Tetraodontidae), among others, which may be very poisonous with substances that they themselves produce. The Greenland shark (*Somniosus microcephalus*), sometimes poisonous, inhabits Arctic seas.

Poisonous fish flesh is known to exist in several sharks, including some requiem sharks (Carcharhinidae), cow sharks (Hexanchidae), dogfish sharks (Squalidae), and mackerel sharks (Lamnidae). In the bony fishes it occurs frequently among reef-dwellers. Families with known toxic representatives include the morays (Muraenidae), the mackerels and tunas (Scombridae), trunkfishes (Ostraciidae), puffers (Tetraodontidae), and porcupinefishes (Diodontidae). More than three hundred kinds of fishes have been incriminated in the ciguatera-type of fish poisoning, which is fatal to about 7 percent of the humans who eat the flesh of these fishes. There is no specific treatment, but a ciguatera antibody is being developed. The reactions are mostly nervous, with a reversal of cold and heat sensations being common in severe cases. Fish families periodically involved include the bonefishes (Albulidae), herrings (Clupeidae), anchovies (Engraulidae), goatfishes (Mullidae), basses (Serranidae), snappers (Lutjanidae), jacks (Carangidae), porgies (Sparidae), wrasses (Labridae), parrotfishes (Scaridae), barracudas (Sphyraenidae), surgeonfishes (Acanthuridae) and trunkfishes (Ostraciidae). Unfortunately, the occurrence of ciguatera is unpredictable and may arise very suddenly.

NUTRITION OF FISHES

Once food has been ingested and digested it may enter into various functions in the body of a fish. It may provide energy for life processes or provide material for the restoration or replacement of worn out cell components, or for growth and reproduction. A few fishes (tunas) can regulate their body temperature; in all others it fluctuates with that of the environment (poikilothermous condition). Thus their rate of metabolism and hence their nutritional requirements are, above all, dependent on the temperature of the surrounding water. Since waters in the Temperate Zones may have seasonal fluctuations of more than 25°C (77°F), a wide range of metabolic rates and food consumption can occur in one and the same individual throughout the year.

Most of our knowledge of food requirements of fishes stems from man's endeavors to raise fishes for food and for stocking in lakes and rivers. Detailed information is therefore restricted to relatively few species, mainly trouts and salmons of the family Salmonidae. Only here and there are the basic nutritional data on these salmonids supplemented by studies from natural freshwater and marine environments.

Food Conversion and Efficiencies

Fishes, like other animals, can be compared to complex machines with regard to the efficiency with which they utilize their food. One may simply measure the weight of food that brings about the increment of a weight unit of fish (be it pounds, grams, or kilograms) and arrive at a "conversion factor." For instance, hatchery-raised rainbow trout (*Salmo gairdneri*) may, under certain conditions and with certain foods, require 3.5 pounds of food for each pound of weight increase; this particular kind of food would, with these fishes, have a conversion factor of 3.5.

Values for conversion factors differ with the nature of the diet, the species and size of the fish, temperature, and other variables. Conversion factors range from 1.5 when certain dry artificial diets are fed to trout (Salmoninae) in hatcheries to 8 or more when diets with a high vegetable component such as corn and lupines are fed to carp (*Cyprinus carpio*) or other omnivorous fish species. The conversion factor is 2.5 to 3.0 for most hatchery diets. Weight-conversion factors in nature often tend to be higher than those measured in hatcheries because of the presence of much indigestible roughage, such as insect or crustacean shells, combined, perhaps, with greater energy-expending activity of fish in the wild than in confinement. The relation between food and weight gain can also be expressed as a weight conversion coefficient. For the example given in the above paragraph, this would be

$$\frac{1 \text{ weight unit gained}}{3.5 \text{ weight units fed}} = 0.2857,$$

or a weight-conversion efficiency of 28.57 percent, the above ratio times 100.

Knowledge of efficiency of food conversion becomes more useful to the nutritionist, the physiologist, or the ecologist when it is expressed in calorie equivalents of the weights of food and fish flesh respectively. Carbohydrates, fats, and proteins can be burned to furnish heat energy as they combine with oxygen in the burning process. Metabolic processes can be thought of as similar to slow combustion or other chemical transformation where energy is derived from carbohydrates, fats, protein, and other compounds. Heat energy is expressed and measured as calories. One calorie is the amount of heat required to warm one gram of water one degree on the Centigrade scale. After suitable transformations other forms of energy such as chemical, mechanical, or electrical can also be expressed in terms of calories.

A fish transforms one compound into another when it produces flesh, scales, bones, and conducts its movements and other vital functions. The energy used for these purposes is derived from the breakdown of various compounds that all have different energy equivalents depending on their chemical composition and structure. For example, 1 pound of average fish flesh has about 600 large Calories (a large Calorie, written with a capital C, is a thousand times larger than the small calorie referred to above), sugar has 1695 and butter, a fat with a high energy equivalent, has 3260 Calories.

It is well known that some of the energy of gasoline exploding in the cylinder of the engine of an automobile is dissipated in overcoming the friction of the different parts of the engine and becomes lost as an effective forward propellant. Thus, the efficiency of an engine is never 100 percent. By very loose analogy we may compare such losses to those we find in animals when a portion of the energy they liberate in their metabolism goes into maintenance and such processes as respiration, digestion, excretion, and other vital functions and is not to be found in weight gain.

We can measure or weigh the amount of new material a fish adds to its own body during growth as well as we can measure or weigh the amount of food it eats. We can also express both of these quantities in calories instead of in pounds or grams, and compare them, namely

$$\frac{\text{calories contained in flesh added per unit of time}}{\text{calories contained in various foods eaten per unit of time}} \times 100$$

This fraction, expressed as a percent, is called the over-all or gross efficiency coefficient or the first-order efficiency coefficient. For pike, between 14 percent to 33 percent of the calories consumed were found to be contained in the flesh. We can often go further and break down the denominator by determining the calories used in cell maintenance, respiration, excretion, and other metabolic processes that go on while new flesh is laid down and therefore represent a part of the intake of energy not returned in growth (growth being the transformation of ingested material into new tissue beyond that required for repair or replacement). Then,

$$\frac{\text{calories in flesh added per unit of time}}{\substack{\text{(calories contained in food eaten per unit of time)} - \\ \text{(calories used in metabolism, maintenance, etc.)}}} \times 100$$

becomes the net or second-order efficiency coefficient that is always larger than the first-order one and provides a more exact measurement of the nutritive properties of various foods than the first-order efficiency coefficient.

Thus the first-order, or over-all coefficients, should be distinguished from the second-order coefficients or efficiencies. Second-order efficiencies are those in which the gain in weight, potential energy, or protein accretion is compared to the quantities of material or energy from which this gain is realized. In the denominator, the term *assimilated* may be substituted for *eaten* to reflect only food that passes through the gut wall and excludes egestion of materials not digested. Comparisons on the basis of calorie equivalents permit appraisal of the effects on food utilization of the age of fishes, temperature, the amounts and kinds of food, and other variables.

Overall energy efficiencies decline with age and size of fish. They range from 58 percent for fish embryos (as in the mummichog, *Fundulus heteroclitus*) to 10 percent and less in large, old fishes (such as a flounder, *Pleuronectes*). Most energy coefficients lie between 15 percent for old fishes and 45 percent for young ones. Negative values are possible for short periods such as during the winter when many fishes of the Temperate Zones lose weight.

The building of body proteins is influenced by the kinds of food a fish eats. This was shown in trouts (Salmoninae) by a comparison between lots fed artificial food or natural food. Despite the fact that the artificial diet was superior to natural food in proteins and vitamins, the trout fed natural foods (insects, etc.) were twice as efficient as those on the artificial diet in the hatchery in converting food proteins into fish flesh, and, in addition, these natural-food fed trout had meat of a lower-water and higher-protein content.

Values for assimilation, expressed as percentage utilization of foods retained in the body, have a mode between 80 and 90 percent; low water temperature and environmental stress may reduce these efficiencies. Provided diets are of the same composition, feeding experiments with the brown trout (*Salmo trutta*) and the northern pike (*Esox lucius*) indicated that fish utilize food more efficiently after fasting than when they are satiated and also that food is absorbed less completely when fish are overfed.

Quantity of Food Consumed

Predominantly herbivorous reef browers, such as parrotfishes (Scaridae) and surgeonfishes (Acanthuridae), feed more or less uninterruptedly throughout the day. In this way they, and many plankton feeders as well, ingest quantities of food much greater than do many carnivores. However, most attached algae have only half the caloric value and less than a quarter of the protein content of animal flesh. Furthermore in spite of the fact that unicellular planktonic algae and diatoms, and even parts of rooted aquatic plants may have adequate nutritive values, they contain a higher percentage of water. For these reasons, herbivores usually consume greater weights of food than carnivores.

Carnivores of different feeding habits also vary in the amounts they eat because their prey are often of very different compositions and sizes. Macro-crustacean feeders ingest relatively more indigestible roughage than piscivores. Some fishes may, at times, get a square meal of such a size that it lasts them for a week or longer. A northern pike (Esox lucius) after having swallowed a fish of nearly its own size may have the tail of the prey hanging out of its mouth for many hours while the anterior portion is being digested in the stomach. The aptly named gulpers (Saccopharyngidae) and swallowers (Eurypharyngidae) of the deep seas often take prey fish that are larger than themselves, having provided ample storage in their highly distensible foregut.

Fishes have a remarkable capacity for surviving long periods of starvation since they first mobilize stored glycogen and fats before body protein. Lungfishes (Dipnoi) and swampeels (Synbranchidae) survive a prolonged season of estivation by such resorption of bodily substances. Freshwater eels (Anguilla) can survive for more than a year without taking food. The intestines of such starved fishes become nonfunctional and partly degenerate.

Quantities of food eaten by fishes over a certain period are often expressed as average rations per day or per a longer interval. Such an average can be a good illustration of the primary influence of temperature on feeding activity. The bluegill (Lepomis macrochirus) may consume average weekly rations up to 35 percent of the body weight (about 5 percent per day) in the summer at a mean water temperature of 20°C (75°F) and less than 1 percent (about 0.14 percent per day) during the winter when the water averages between 2 and 3°C (36°F, Fig. 5.9). Not only the amount of food eaten but also the assimilation efficiencies are influenced by temperature. The metabolism of fishes increases with temperature up to an optimum temperature, above which the metabolism decreases. Marine fishes also compensate in their feeding rates for the higher energy and repair requirements of higher temperatures; the red hind (Epinephelus guttatus) more than doubles its feeding rates per day between 19° and 28°C.

Daily rations also vary with the size of fishes. Small fishes, like other small animals, have a higher metabolic rate than large ones of the same species. They, therefore, require relatively more food to maintain a unit of weight of their bodies than do large individuals. The daily maintenance ration of the red hind (Epinephelus guttatus) varies from 1.7 to 5.8 percent of the body weight for specimens that average 250 grams and between 1.3 and 3 percent for fish that average 600 grams as the water temperature rises from 19° to 28°C. Individuals of a species that average smaller in size than the red hind may have still higher daily ration requirements than the foregoing.

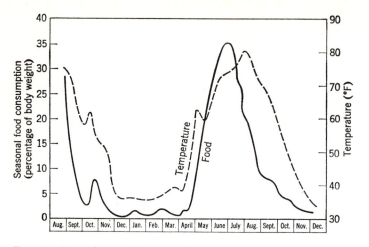

Fig. 5.9 Dependence of food intake on temperature as shown in the bluegill (*Lepomis macrochirus*). (Based on Anderson, 1959).

Availability of food to a fish also influences the amounts consumed. The red hind gorges itself with daily intakes of over 10 percent of the body weight at the onset of satiation feeding experiments but soon tapers off to more conservative rates of feeding. Yearlings of the carp (*Cyprinus carpio*) in ponds may have a daily ration as high as 16 percent of their body weight when starting to feed at the onset of a growing season. Unlike most terrestrial animals, the majority of fishes experiences severe depletion for a part of every year of their lives.

Food Quality

True growth is the accretion of flesh (protein) and bone; it does not include the less permanent fat and water deposits in fish tissues. Nevertheless, the diet of fishes must be balanced and contain the primary or basic food components—proteins, carbohydrates, and lipids (fats)—in requisite, though differing amounts for different species of fishes. Vitamins and minerals are also required for growth, sustenance and replacement of tissues as well as for normal metabolism.

Basic Foods. Evolutionary adaptations among fishes have led to differing capacities in utilizing foods with varying amounts of proteins, carbohydrates, and fats. Trout (Salmoninae) for instance, develop fat-infiltrated livers when the fat content of their diet exceeds 8 percent (dry weight) whereas the Atlantic menhaden (*Brevoortia tyrannus*) normally feeds on algae that contain at least 25 percent fat. The common carp (*Cyprinus carpio*) and tilapias of the family Cichlidae can be raised on a diet of even higher lipid content. Not only do relative amounts of fat in the food affect food utilization, but the composition of the fat also has an effect; unsaturated fatty acids with a low melting point (high iodine number) are digestible more readily than saturated fatty acids.

Permissible carbohydrate content in artificial diets, which, after all, are manufactured to be similar if not superior to a natural food of fishes, vary similarly for different species. When trouts are fed a diet with more than 12 percent digestible carbohydrate (wet basis), the results are abnormal deposition of liver glycogen and excessive mortalities. By contrast, the various minnows (Cyprinidae) raised for food throughout the world often grow well on food mixtures with 50 percent or more of starchy matter. Different carbohydrates vary in their digestibility by trout, for example, 99 percent of glucose was digested, whereas 73 percent of sucrose was digested. The most rigorous requirements of the three basic foodstuffs probably pertain to proteins. A certain protein content is mandatory in the diet of any animal. The protein requirement changes with changes in the fish's life cycle, being greatest for small, fast growing fish and for prespawning fish that are forming eggs and sperm. Invertebrate fish food averages approximately 11.5 percent protein. It appears that food with less than 6 percent protein (wet weight) cannot sustain any fish, let alone produce growth.

Not only are amounts of protein ingested important but so also are their relative amino acid content as intimated by the highest conversion efficiencies obtained when a fish eats others of its own kind. Like all animals, fishes can synthesize certain amino acids readily, but those of others only at rates too slow for satisfactory replacement. Therefore, these have to be supplemented in the diet. Yet a third group of amino acids is absolutely essential and has to be supplied in the food because fishes cannot synthesize them. Detailed studies of this aspect of fish nutrition have been made with the chinook salmon (*Oncorhynchus tshawytscha*). Arginine, histidine, isoleucine, leucine, lysine, methionine, phenylalanine, threonine, tryptophane and valine were the amino acids that this salmon either did not synthesize at all or manufactured in such small amounts that no growth was accomplished when they were absent from the diet. In this respect the chinook salmon resembles the rat (*Rattus*) rather than man who can dispense with two of the above compounds in his diet (histidine and arginine) because they can be synthesized in the human body.

Water, though not a basic food proper, is necessary in fishes, as in other animals, for metabolic processes including extracellular digestion. An animal may lose practically all of its fat and half of its protein and live, but a loss of only 10 percent of its water causes death. The amount of water taken in by fishes incidental to feeding is not known precisely. The presence of a set of sphincter muscles at the entrance to the stomach in some fishes and the straining and concentrating adaptations in the mouth and pharynx of others such as the plankton feeders suggest that accidental water intake may be insignificant. Freshwater fishes absorb water through exposed semipermeable membranes of the gills and other exposed surfaces. Marine fishes counteract osmotic water losses (Chapter 9; Fig. 9.1) in part by swallowing sea water and in part by deriving water from food. It is probable that much of the water required for digestive processes by freshwater fishes is also extracted from food. The range of water content of foods taken in by fishes lies between 70 and 90 percent of wet weight of the foods, hard parts such as shells and bones not included.

Vitamins. Knowledge of the role of vitamins in fish nutrition is derived primarily from feeding experiments with several species of trouts and salmons (Salmoninae) and also channel catfish and carp. In the natural diet of animals, vitamins occur in sufficient amounts to supply the very limited quantities required for normal metabolic functions. However, when fishes are raised on artificial diets, the selection of food constituents, the property of rapid deterioration of some vitamins, and other factors may lead to specific deficiencies with their attendant impairments of body functions. Trout and salmon require ten members of the vitamin B complex (Table 5.1). For carp the requirement has been established for pyridoxine, riboflavin, and pantothenic acid and for channel catfish some of the fat- and water-soluble vitamins. Trout need water-soluble vitamin C. Requirement of fat-soluble vitamins A, K, and E, but not D has been established for some fishes.

Minerals. Minerals, like vitamins, occur in most natural diets in sufficient quantities to satisfy the metabolic requirements of fishes; even in fish hatcheries mineral deficiencies are rare and easily alleviated when discovered.

Mineral and other elements can be classified according to their three major functions in the body: (a) *structural*—calcium, phosphorus, fluorine, and magnesium (tooth and bone formation); (b) *respiratory*—iron, copper, and cobalt (function and formation of hemoglobin); (c) *general metabolism*—many elements (body and cell functions). In the last category, sodium, potassium, calcium, and chlorine regulate osmotic balance and cell turgor. Boron, aluminum, zinc, arsenic and other trace minerals may also be required here; chlorine ions are needed for the manufacture of gastric juice; and magnesium, phosphorus, and chlorine ions are needed to activate digestive enzymes. Copper and zinc are required cofactors for some enzymes and sulfur and phosphorus occur in phospholipids and other nuclear constituents. Traces of phosphorus are important in adenosine triphosphate (ATP) and other high energy compounds. Bromine figures in the maturation of the gonads whereas iodine is a constituent of thyroid hormones (Chapter 11). A deficiency of iodine can lead

Table 5.1. **Effects of Vitamin Deficiencies on Fishes**[a]

Substance	Deficiency Symptoms
Vitamin B₁ (Thiamine)	Poor appetite; muscle atrophy; convulsions; nervous malfunctions; death
Vitamin B₂ (Riboflavin)	Eyes become opaque; growth ceases
Vitamin B₆ (Pyridoxine)	Nervous disorders; ataxia; anemia; growth reduction; death (in 4 to 6 weeks in trouts)
Vitamin B₁₂ (Cobalamine)	Erratic hemoglobin and erythrocyte counts; growth reduction
(Unnumbered vitamins in B complex)	
Biotin	Mucous production impaired; convulsions; "blue slime disease"
Choline	Growth reduction
Folic Acid	Hemorrhagic kidney and intestine; growth reduction; anemia
Inositol	Growth reduction and probably anemia
Niacin	Increased sensitivity to ultraviolet light; "sunburn"
Pantothenic Acid	Gill disease; mucous production impaired; "blue slime disease"; sluggishness

[a] Certain effects have been ascribed to other trace factors but these have been resolved to be synergistic actions of several B-complex vitamins and to lack of some essential amino acids.

to goiter-like conditions in trouts at hatcheries and is alleviated by its addition to the artificial foods. Radioactive isotopes of calcium are absorbed directly from the water in increasing amounts as the respective mineral is lacking in experimental diets. Lithium, sodium, chlorine, and bromine also enter a fish through the gills, but most elements are adequately represented in the natural food of fishes and appear to be taken in through the food.

The Chemical Composition of Fishes.   Most data on the elementary composition of fishes refer to the composition and nutritive values of the flesh among commercially important species, some 350 to 400 in number, mostly marine bony fishes. The data are in terms of water, fat, proteins, carbohydrates, and minerals per unit of weight.

Water content of fish flesh averages 80 percent, slightly more than for birds and mammals. For fishes, extreme values of 53 and 89 percent have been recorded for certain species, seasons, and localities.

Water content of fish flesh varies with life history stage and species. Sex differences in water content are small as reported for a few species, as for example the herring, *Clupea harengus.* Males of this herring often contain

1 to 2 percent more water than the females. This difference may be correlated with the higher water content of testes than of ovaries. Sac fry (larvae with yolk sacs) embryos have a lower percentage of water than adults at maturity. The leptocephalus larvae of freshwater eels (*Anguilla*), however, contain more water than after transformation into elvers of adult body form and structure. Among adults, however, progressive dehydration follows with age. Anadromous species such as the salmons (*Oncorhynchus* and *Salmo salar*) and marine-run races of the sea lamprey (*Petromyzon marinus*) increase their water content perceptibly at spawning as a result of having replaced with water large amounts of fat and protein used for energy to reach the spawning sites in rivers. Cods (Gadidae), flounders (Pleuronectidae) and goosefishes (Lophiidae) may contain over 80 percent water, whereas freshwater eels (*Anguilla*) in the sea, the marine herrings and sardines (Clupeidae), and the tunas and mackerels (Scombridae) generally have a water content below 75 percent. The first group, represented by the cods, flounders, and goosefishes, is demersal, and its members have specific gravities slightly above that of sea water, to be raised, if necessary, with the help of their gas bladders. The second group represented by the herring and mackerel families, is largely pelagic, has flesh of a higher fat content than the first, and has some representatives with only small or vestigial gas bladders. The specific gravity of the latter is approximated to that of sea water by enlarged fat deposits in the muscle.

Fat content of fish flesh varies not only according to families but also with the season. Plankton feeders of Temperate seas such as the herrings and sardines (Clupeidae) have more fat and less water in their tissues after the vernal plankton pulses than after spawning, the time of their leanest condition. Water and fat content of the tissues of individual fish vary inversely to one another and fat content even varies from place to place on one fish. Muscle of the herring (*Clupea harengus*) has unusually high fat content in the abdominal region counterbalanced by another place of concentration near the dorsal fin.

Freshwater fishes also show group differences in relative fat content and generally have fats of different chemical characteristics than marine species. Tropical freshwater catfishes such as the labyrinthine catfishes (Clariidae) and the schilbeids (Schilbeidae) contain more fat than other groups in the same region, reflecting specific feeding habits and adaptations to utilize certain foods. Most freshwater fishes of the Temperate Zones store fat for the winter and/or in preparation for spawning. Important storage sites are the connective tissue and mesenteries of the intestines.

Only even-numbered fatty acids occur in the fats and oils of fishes. These fatty acids are less saturated than those occurring in land animals and in vegetable oils. Their oleic acid content, a precursor of cholesterol, is relatively low. For this reason they have been held to be beneficial in the diet of humans in relation to pathological hardening of the arteries.

Proteins make up between 14 and 23 percent of the wet weight of fishes. Here again, variations occur in species, seasons, and life history stage. The total nitrogen in the fish body is usually between 2 and 3 percent of the living weight. Shark-like fishes (Elasmobranchii) contain more nitrogen than bony fishes (Osteichthyes) because elasmobranch tissues have much osmo-regulatory urea in them which bony fish tissues lack (Chapter 9). On a dry weight basis, nitrogen is most highly concentrated in the liver, followed by the skin and then by the flesh.

Carbohydrates present in muscle for immediate energy release make up less than 1 percent of the wet weight of fishes; they are more concentrated than this in the liver where they are stored as glycogen.

Some of the many chemical elements found in fishes are more concentrated in marine than in freshwater species. The elevated content of these elements is thought to reflect their richness in sea water and their subsequent passive incorporation by marine fishes through concentration in all links of their food chains. Ash content in soft tissues of fishes is about 1 percent of their living matter. When bones and scales are included in the analysis, ash amounts to 3 percent or more because calcium phosphate, $Ca_3(PO_4)_2$, and calcium carbonate, $CaCO_3$, make up the bulk of these hard parts of fishes. Scales are about 30 to 35 percent ash on a dry weight basis.

GROWTH OF FISHES

The growth pattern of fishes is remarkable inasmuch as most of them have the capacity of sustained though diminishing growth (indeterminate growth) throughout their entire lives if sufficient food is available. Thus members of one species may be of variable sizes at the same ages in contrast to more definite sizes at any one age (determinate growth) as among a few fishes, such as males of the guppy (Lebistes), and generally among the individuals of terrestrial vertebrate groups. In other words, most fishes, in contrast to birds and mammals, do not cease growth after they have reached sexual maturity. Whereas the growth of fishes is thus much more variable and flexible than that of birds or mammals, there still exists an ultimate genetic limit to the growth of a species that is most evident in the smallest kinds. It has been proposed in explanation of the foregoing contrast that fishes living in a fluid medium that supports them mechanically can continue growth throughout their lives because there are more biotic than mechanical limits imposed on their maximum sizes.

Growth in animals has been defined above as the addition of structural or fleshy, hence protein, elements; fish growth in particular is of great importance to man as the exploiter and potential manager of fish populations for sport or sustenance. In fact it has been shown in controlled calorimetry experiments with trout and domestic livestock that fish are far superior in protein-building than the latter. This is because birds and mammals have to expend part of their caloric intake for the maintenance of body temperature and in supporting themselves; fish in contrast are poikilotherms and get such support from the surrounding water. Certain special aspects of growth and factors that influence it are discussed below.

Age and Growth Determinations

When the accretion of fish length (or weight) is plotted against time throughout the life span, the growth rate is represented by a curve (Fig. 5.10). Seasonal fluctuations in growth are often disregarded and usually empirical data on growth are available only for the postembryonic or even postjuvenile periods of the life history.

Since many commercial fishes are available for measurements only once or a few times in their life cycle, man has sought to reconstruct their rates and patterns of growth. Most useful for this endeavor—at least for fishes of the Temperate Zones—are the year marks on bony parts (Chapter 4), such as scales, otoliths, spines, and opercular bones (Fig. 5.11). Yearly zones of growth on these parts are due to a slowing down of temperature-dependent

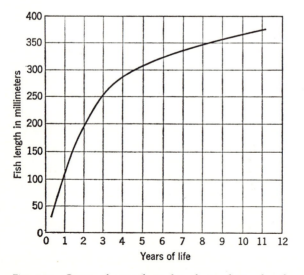

Fig. 5.10 Curve of growth in length as shown by the shallowwater cisco (*Coregonus artedii*). (Based on Hile, 1936).

YEAR MARKS ON BONES

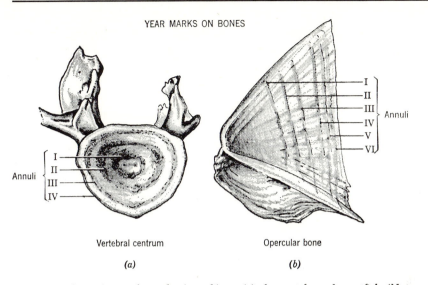

Vertebral centrum   Opercular bone

(a)   *(b)*

Fig. 5.11   Annual growth marks (annuli) on *(a)* the vertebra of a catfish (*Noturus*) and *(b)* the operculum of a perch (*Perca*).

processes during the winter and a resumption of a more rapid metabolism at a time in the spring. Certain freshwater fishes of the monsoon tropics also show seasonal growth marks that correspond to the onset of the dry seasons. The growth zones can be measured and the correspondence between body growth and growth of bony parts can be established. For example, in certain positions on the body, scales grow at the same rate as the body (isometric growth) at least in the postjuvenile period. The growth history of an individual can be reconstructed by calculations of proportionality. Although sometimes large numbers of individuals of different sizes are available for construction of life-table-like computations (length-frequency diagrams) and although sometimes the growth of fishes of known ages can be observed, the construction of growth patterns based on reading of year marks on hard body parts has come to be the prevalent method of growth determination among fishes of the Temperate Zones. In the uniformly warm waters of equatorial oceans, however, fish growth shows little or no seasonal fluctuation and the age of fishes there is extremely difficult to assess.

The length of a fish is often easier obtained than its weight or age. The following relation exists between the length and the weight of a fish: $W = a \cdot L^n$, where a is constant and the exponent n fluctuates between 2.5 and 4, most frequently near 3, since growth represents an increase in three dimensions whereas length measurements are taken in one dimension. This relationship also lends itself to comparison of individuals within and between different fish populations. The value sought then is the constant a (above), expressed as "ponderal index" or condition factor K in the formula $K = W/L^3$ which varies to provide the index when the exponent n is fixed at 3.

The Longevity of Fishes

Unconfirmed accounts mention that carp in ponds have reached an age of 200 or even 400 years, yet authenticated records of ages of captive fishes suggest that the most venerated old carp do not exceed 50 years.

By and large few fishes in the wild will live longer than 12 to 20 years and even in the shelter of an aquarium 30 years are rarely exceeded. Generally species with large individuals may be expected to exhibit older ages than species composed of little fish. Most fishes that grow longer than a foot (30 centimeters) have a life span of at least 4 or 5 years. Many minnows (Cyprinidae) of the Temperate Zones and other small fishes of the Temperate and Tropical Zones frequently have a life span of less than 2 years including some that are annual.

Genetic Factors (Chapter 12)

In nature the presence of genetic variations in growth potential in different populations of the same species is usually masked by environmental factors such as temperature or food supply. Nevertheless, some selected strains of the brook trout (*Salvelinus fontinalis*), rainbow trout (*Salmo gairdneri*), and the carp (*Cyprinus carpio*) have growth that is decidedly superior to that of other strains.

Fish growth usually slows down after the onset of sexual maturity when large amounts of nutrient material periodically go into egg or sperm formation. The gonads of the yellow perch (*Perca flavescens*) before spawning may make up more than 20 percent of the body weight of the female and more than 8 percent of the body weight of the male; after spawning the ovaries or testes may shrink to about 1 percent or less of the body weight.

The onset of sexual maturity and with it the change in growth rate in the life cycle of a fish may also be genetically determined as it is in the brown trout (*Salmo trutta*). Egg size also varies with genetic strains and with it varies the initial size of the fish at hatching. Large eggs produce large larvae that may also have the greatest growth potential of the lot. Advantages in growth pattern due to larger egg size may be noticeable for a short time until other factors such as food supply intervene; the advantages, however, may be lifelong.

Frequently there are size differences between mature male and female fish within a species. Size differences between the sexes that affect growth rates and growth patterns may be due to genetic factors, often correlated with inherited behavior patterns. Among the minnows (Cyprinidae) of North America, for example, the majority of species have larger females than males but in some species such as the hornyhead chub *(Nocomis biguttatus)* and the fathead minnows (Pimephalinae), the males are the largest. It is very likely that reproductive habits such as nest building effected these size differences between the sexes. The greatest size appears most often to be associated with the most "important" parent.

Seasonal and Temperature Factors

Fishes with a wide latitudinal distribution such as the largemouth bass *(Micropterus salmoides)* commonly have differences in rate of growth and in age at which they reach sexual maturity in northern versus southern waters. In Louisiana and Arkansas, for example, the largemouth bass may spawn after one year, but in southern Michigan or Wisconsin only after an average of three years. In Ontario and other comparable latitudes and climate, sexual maturity of this bass may be delayed until the fourth or fifth year. The ultimate size and age of the fish may not differ between northern and southern representatives of the species, but the shape of the curve that represents the plotting of weight against age will reflect a temperature-dependent difference in the growth pattern. Temperature also induces seasonal differences in growth since temperature affects metabolism and food consumption.

The seasonal rise in the temperature of fresh waters coincides with an increase of natural food for fishes but there are also seasonal and temperature dependent variations in activity that influence the amount of food diverted to growth and/or maintenance requirements. In one experiment (Brown, 1957), the brown trout *(Salmo trutta)* grew optimally between 7° and 9°C, and also between 16° and 19°C. Appetite was high once the water warmed to 10°C. Maintenance requirements also increased most rapidly between 9° and 11°C, then tapered off and became almost constant when 20°C was reached. Thus, optimum temperatures for rapid growth are those at which the appetite is high and maintenance requirements low whereas minimum growth occurs at intermediate temperatures when maintenance requirements are high because fish are most active. Below 7°C, maintenance requirements are low, but so is appetite; above 19°, appetite again falls off while maintenance requirements remain at the same level. Thus, the brown trout achieves optimum growth outside of the temperature range where the fish show maximum activity.

Light, mostly through day length, may affect the rate and pattern of fish growth. Since changes in day length coincide with seasonal rise or fall in temperature, it is difficult to appraise the respective effects of these environmental factors. Illumination is known to act indirectly on thyroid metabolism and peaks in thyroid activity coincide with maximum day length but these peaks also cause increased swimming activity, resulting in a reduction in growth rate in experiments with the brown trout (*Salmo trutta*) in midsummer. If food is plentiful, shorter hours of day length may be most advantageous for rapid growth because of dark-induced rest patterns, but where strong competition occurs and feeding proceeds by sight, long days can be of advantage.

Biotic Factors

These factors have been called density-dependent variables but it would also be valid to term them "social." Among them is included competition or the influence of numbers on available food, nest sites, feeding behavior, and so forth. Since the expression of the biotic factors is most often studied in nature, it is very frequently confounded with the effects of environmental (physical and chemical) density-independent factors. Relatively few controlled experiments exist where density-dependent factors have been successfully segregated for scrutiny, either one by one or in combination.

The Indirect Effect of Fish on Fish.   Experiments with the goldfish (*Carassius auratus*) grown in water previously inhabited by other goldfish have shown that such experimental fish grew better than their control running mates in "unconditioned" water. The improved growth may have been due to the uptake of regurgitated food from the conditioned water. More likely, however, it was due to the presence of a growth-promoting substance that could be isolated from the slime in the tanks with conditioned water. The substance was found to be effective in stimulating growth at dilutions as low as 1.2 parts per million (ppm).

Direct Effects of Fish on Fish (Group Effects).   Order of dominance or peck order has been observed as a factor in growth among captive fishes including sunfishes (Centrarchidae), trout-like fishes (Salmonidae), and cichlids (Cichlidae). One individual is at the top of the order (often the largest individual) and will regularly secure the best and/or most morsels when food is supplied to the group. If there are but a few fish in the holding tank, these hierarchal effects are often broken down, but there are indications that they persist in nature. The lake trout (*Salvelinus namaycush*) of northern lakes and rivers, among other fishes, has shown definite dominance of one individual over others. Whether growth differences due to peck orders result directly from increased food intake of the dominant individual or whether it comes about indirectly through the psychological stress placed on the lower members of the order is not known.

It has also been suggested that there may be a group effect in growth, over and above that of the hierarchy among individual fish. The suggestion is based on the following experiment. When different numbers of trout fry were placed in tanks of the same size and reared in groups of 25, 50, 80, 100, and 150 individuals with each group given an oversupply of food, the most successful group was the one which contained 80 fish. It appeared that the growth rate of individuals was influenced by group size as well as by the numbers of larger and smaller fish present.

The most pronounced group effect on growth is to be found in crowded fish populations. Fish grown in crowded conditions may release a chemical "crowding substance," which depresses growth. Many instances are known where fish in a body of fresh water become so numerous that their specific growth rate is reduced as compared to that of the average for the area or species. The condition has been grouped under the term "stunting." When some of the stunted fish or fish of a competing species are removed, the remainder responds to the new conditions and the specific growth rate increases. Among the fishes for which such stunting has been described are the yellow perch (*Perca flavescens*), European perch (*Perca fluviatilis*), bluegill (*Lepomis macrochirus*) and other members of the sunfish family (Centrarchidae), and a tilapia (Cichlidae—*Tilapia mossambica*).

SPECIAL REFERENCES ON FOODS, DIGESTION, NUTRITION, AND GROWTH

Al-Hussaini, A. H. 1949. On the functional morphology of the alimentary tract of some fishes in relation to differences in their feeding habits. Anatomy and histology. *Quart. Jour. Micr. Sci.*, 90: 109–139.

——. 1949. On the functional morphology of the alimentary tract of some fishes in relation to differences in their feeding habits. Cytology and physiology. *Quart. Jour. Micr. Sci.*, 90: 323–354.

Bagenal, T. B., ed. 1974. Proc. Intern. Symp. *Aging of Fish.* Unwin Bros., England.

Barrington, E. J. W. 1957. The alimentary canal and digestion. The physiology of fishes, 1: 109–161. Academic Press, New York.

Brown, M. E. 1957. Experimental studies on growth. The physiology of fishes, 1: 361–400. Academic Press, New York.

Gerking, S. D. 1954. Food turnover of a bluegill population. *Ecology*, 35 (4): 490–498.

Gregory, W. K. 1933. Fish skulls. *Trans. Amer. Phil. Soc.*, 23 (2): 75–481. (Offset reprint, Eric Lundberg Publ., Laurel, Fla., 1959.)

Halver, J. E. 1972. Fish nutrition. Academic Press, New York.

Hile, R. 1941. Age and growth of the rock bass, *Ambloplites rupestris* (Rafinesque), in Nebish Lake, Wisconsin. *Trans. Wisconsin Acad. Sci., Arts and Lett.*, 33: 189–337.

Ivlev, V. S. 1961. Experimental ecology of the feeding of fishes. Yale University Press, New Haven, Conn.

Love, R. M. 1970. Chemical biology of fishes. Academic Press, New York.

Phillips, A. M., Jr., 1969. Nutrition, digestion, and energy utilization. Fish physiology, 1: 391–432. Academic Press, New York.

Ricker, W. F. 1946. Production and utilization of fish populations. *Ecol. Monogr.*, 16: 373–391.

Vonk, H. J. 1927. Die Verdauung bei Fischen. *Zeitschr. Vergl. Physiol.*, 5: 445–546.

6

Skeleton, Build, and Movement

Body form and locomotion in fishes are the result of interaction of the skeleton and musculature. In adapting to the vast array of habitats that have arisen in the aquatic world, a variety of body forms have evolved, with corresponding skeletal and muscular modifications. Accompanying adaptation in body form, the basic serpentine pattern of locomotion has differentiated into many other patterns, all of less, though specialized, significance.

SKELETON

Gross internal and external skeletal features of fishes have already been described and we may now consider aspects of the skeletal basis of form and movement in fishes, along with the adaptations and evolutionary trends that are involved.

Functions

One of the functions of the skeleton is to impart support to the parts, hence, indirectly, to determine the form of fishes, as already indicated. Other functions are to provide protection, to give leverage, and questionably to take a role in red blood cell formation (hematopoiesis). In addition, in some fishes, skeletal modifications of the fins expedite the placement of sperm into the reproductive tract of the female.

Symmetry

Fundamental bilateral symmetry is the structural ground plan of skeletal organization in fishes; that is, the right and left sides of each individual are mirror images of one another. Although principal exceptions to such a body plan in fishes (as in other vertebrates) are in certain digestive organs, mention of a few skeletal departures will serve again to illustrate the great variation to be found in anatomical detail among fishes. A classical example of vertebrate departure from bilateral symmetry is the adult of flatfishes (Pleuronectiformes). Both of the bony eye sockets and the eyes are on one side, to which the eye of the blind side has migrated during growth stages from a bilaterally symmetrical larva to the adult.

Jaws and Face

The skeletal system of fishes exemplifies many evolutionary trends and adaptations. One group of these trends and adaptations deals with the formation of jaws and their supports and thus largely determines the facial appearance. The agnathous fishes possessed no true jaws, but first evolved some skeletal structures for supporting the gills. Later in evolution jaws of fishes were modified from anterior gill-arch components of jawless ancestors. In similar broad vein, it has also been proposed that the visceral or pharyngeal arch following the one that gave rise through specialization to jaws became the hyoid apparatus of modern fishes. The hyoid arch supports jaws, tongue, and gill arches in almost all jawed fishes.

The jaws of fishes have evolved into many different types, both as to position and details of structure—all apparently in relation to habits. One extreme variation of the generalized terminally jawed condition is the ventral suctorial device or holdfast organ formed by the mouth of such fishes as some of the armored catfishes (*Plecostomus* and other genera), in which thickened, protrusible lips on true jaws make efficient suckers. The jaws are frequently inferior (ventral), especially among bottom-feeding fishes, such as the skates (Rajidae; Fig. 2.8), and the sturgeons (Acipenseridae; Fig. 2.13). Other fishes have superior (dorsal) jaws, adapted to surface feeding, or to feeding in an upward direction. The mouths are superior in most of the topminnows (Cyprinodontiformes; Fig. 2.33), including the gambusias (*Gambusia*) that feed effectively at the surface of the water on the wiggling larvae of mosquitoes. The anglerfishes (Lophiiformes; Fig. 2.50) illustrate dorsally situated jaws that permit the fishes to lie on the bottom and feed on prey approaching from above.

In some fishes, as in the halfbeaks (Hemiramphidae; Fig. 5.5c), the lower jaws are greatly elongated. This modification, too, is related to a surface feeding habit, as previously described. In other fishes, the upper jaw is elongated to form a beak or a long, flattened rostrum. The swordfishes (Xiphiidae) and the marlins and sailfishes (Istiophoridae) of the oceans, as well as the paddlefish (*Polyodon;* Fig. 5.5g) of inland American waters, all show this type of elongation, but there is difficulty in ascribing function related to facility in feeding for such a structure. For example, the swordfish (*Xiphias*) and other fishes with a similar spear apparently do not often stab their prey, but have been known to drive the structure into the hulls of wooden boats. The sawfishes (*Pristis*), however, mutilate prey with sidewise swipes of the laterally strong-toothed rostrum. In the paddlefish (*Polyodon*), the rostrum seems to have evolved as an organ for housing a large number of sensory devices that presumably serve to detect plankton swarms, which are strained from the water by the numerous gill rakers as the fish swims slowly back and forth through them with mouth agape. It is possible that the "paddle" serves as a stabilizer and/or a scoop when the fish swims, mouth agape, while feeding on plankton.

Still other fishes have both upper and lower jaws greatly elongated. Among these are the predacious gars (Lepisosteidae; Fig. 2.15) of North American fresh waters. Included here also are the oceanic garfishes (Belonidae) and the longnose butterflyfish (*Forcipiger;* Fig. 5.5f), which has an elongated beak with "forceps" jaws at the end that are used for picking little polyps and invertebrates out from among coral heads. The shape of head and jaws, we may conclude, determine many of the things a fish can do or cannot do in ways of feeding and combat.

Fins

Fins of fishes are of two major kinds, median and paired. Both the shape and the position of these fins are related to body shape and location of center of buoyancy of the fish (in turn related to location and shape of the gas bladder, habits, and character and speed of swimming).

Origins. It has been theorized that fins of fishes all evolved from ancestors that had no fins at all. Presumably, it was advantageous for fish to respond to turbulence in water that would require stabilizing, together with stability problems in swimming; as a result a median ridge or keel developed on the body, from near the head around one end of the tail to the anus. This ridge, perhaps initially without supporting rays, gave rise to the median fins. Similar ridges appearing on the sides and then coalescing at the proper sites have been presumed further to be the ancestral antecedents of the paired fins. The finfolds of amphioxus (*Branchiostoma*), one of the possible vertebrate ancestors, and the finfolds that appear in many fish embryos tend to give credence to the foregoing finfold theory. As one might expect, however, there are finless fishes of specialized habit (for example, the burrowing, worm-like, synbranchid eel, *Typhlosynbranchus*).

Differentiation.   Variations in the skeleton have resulted in differences in both the median and paired fins of fishes, many of which appear to be of adaptive significance for locomotion and maneuvering. The internal skeleton of the caudal fin and some of its modifications have already been described. More superficially, the lower lobe of the caudal fin is reduced in many bottom dwellers, such as the skates and rays (Rajiformes; Figs. 2.8 and 6.2), but the lower lobe is enlarged in the flying fishes (Exocoetidae) that may use the tail as a propelling organ at the surface of the water just before take-off. In still others, the original tail fin is replaced by a backward growth of rays from the dorsal and anal fins (as in the cod, *Gadus*; Fig. 2.29). Only infrequently is it absent as in eel-relatives (Anguilliformes) such as *Ophisurus* and *Sphagebranchus* and in the seahorse (*Hippocampus*; Fig. 10.6). Even more rarely are both the caudal and anal wanting as in the mormyrid, *Gymnarchus*.

Dorsal fins vary greatly on the basis of support and thus in size, extent, and location. The fin may be absent (*Xenomystus*), only rudimentary (*Notopterus*), or extend over all or most of the length of the back as in the bowfin (*Amia*; Figs. 2.14 and 6.2), the dolphin (*Coryphaena*), the ribbonfishes (Trachipteridae) and the oarfish (*Regalecus*). Or, the fin may be continuous with the caudal and anal ones as in the lampreys (Petromyzonidae; Figs. 2.1 and 2.5), freshwater eels (Anguillidae; Fig. 2.25) and the swampeels (Synbranchidae). It is usual for a spiny-rayed fish to have two dorsal fins. The Temperate basses (Percichthyidae; Fig. 2.41), and the percids (Percidae) contain many common examples of this. However, some species have three or even more dorsal fins. The cods (*Gadus*; Fig. 2.29), for example, have three. The multiple condition reduces itself finally to a series of more or less single, isolated finlets (separate spines each with its own membrane) in the sticklebacks (Gasterosteidae; Fig. 2.30) and in the bichirs (*Polypterus*; Fig. 2.12). In some fishes, such as the trouts and their relatives (Salmonidae; Fig. 6.6), and the catfishes (Ictaluridae), one of the dorsal fins has no rays at all and is a fleshy structure termed an adipose fin.

Among the unusual modifications of the dorsal are the elongation of its last rays in leaping fishes such as the tarpon (*Megalops*) and the gizzard shad (*Dorosoma*). It is strongly modified in the remora or sharksucker (Echeneidae; Fig. 2.46), where it comprises the suction disk with which this world-famous hitchhiker attaches to other fishes, sea turtles, and ships. In the anglerfishes (Lophiiformes; Fig. 2.50) the dorsal fin provides the elements for the prey-enticing "rod and lure" that adorns the top of the head. Only rarely is the dorsal fin greatly reduced or absent, as in the South American gymnotid eels (Gymnotidae; Fig. 6.2b), the stingrays (Trygonidae), and the eagle rays (Myliobatidae).

Among the median fins, the anal is perhaps the most stable one, showing the least variation. In all except one of the families of North American freshwater fishes, for example, the anal fin is fairly uniform. In the male livebearers (Poeciliidae), however, it serves as an intromittent organ, the gonopodium (Fig. 10.5b). This structure is used to guide spermatozoa (either as milt or as sperm balls, technically termed spermatophores) for internal fertilization from the vent of the male into the vent of the female during breeding. Sexual difference in size of anal fin is shown by several fishes, including the white sucker (*Catostomus commersoni*). The anal is lacking in some fishes such as the modern rays and relatives (Rajiformes).

The pectoral fins of fishes are little more variable than the anal fin just discussed. They have as their principal functions locomotor maneuvering. Examples of primary locomotor function for the pectoral fins are found in percid darters (Etheosotomatinae), sticklebacks (Gasterosteidae), sculpins (Cottidae), and pipefishes and seahorses (Syngnathidae), as well as in parrotfishes (Scaridae), porgies (Sparidae), and other fishes which have the pectoral girdle strengthened and modified accordingly. In at least one topminnow (*Gambusia*) the pectoral fin is modified into a clasper, which the male fish uses to guide his gonopodium into the vent of the female. Other variations of pectoral fins include those of size; commonly they are enlarged in bottom dwellers such as the darters, sculpins, and skates (Rajidae). They are very greatly enlarged in flying fishes (Exocoetidae), by which they are used as support in soaring. They are absent in living lampreys (Petromyzonidae), as an extreme variation.

In addition to its use in locomotion, the pectoral fin is adapted as an offense-defense structure in some fishes. In the North American freshwater catfishes (Ictaluridae), the sharp, hardened ray at the leading edge of this fin has a locking structure that enables the catfish to erect it and hold it erect, presumably as an instrument of combat. In the madtoms (*Noturus*), this spinous ray has a special gland at its base. The secretion of the gland, injected by the spine, has a stinging, paralyzing effect on man.

The pelvic fins are mostly to be regarded as accessory maneuvering organs, if we ascribe to the pectorals the role of major maneuver. The pelvics are absent, of course, from some fishes (lampreys, Petromyzonidae; eels, Anguillidae; certain killifishes, Cyprinodontidae—*Orestias*, *Tellia*; snakeheads, *Channa*), reduced to a single spine in others (sticklebacks, Gasterosteidae; Fig. 2.30) or to a bundle of just a few rays in many other fishes (sailfish, Istiophoridae). Occasionally the pelvics fail to appear as a developmental abnormality of an individual. In the least darter (*Etheostoma microperca*), however, the pelvic fins are enlarged in the males and constitute a noticeable secondary sexual character, although no exact function has been ascribed to this enlargement. In male sharks the pelvics are modified into clasping structures, the myxopterygia. Each myxopterygium has three principal parts (Fig. 10.5a) in addition to the rayed portion. Like gonopodia, the claspers function to guide spermatozoa into the reproductive tract of the female. Some extreme variations of the pelvic fins are perhaps best exemplified by the suction disc that they form in some of the hillstream fishes of India (*Glyptosternum*) and some of the gobies (Gobiidae) that inhabit the surge zones of the continental coastal waters of the world.

A common characteristic of pelvic fins in spiny-rayed fishes is that each possesses a single spine. The ray numbers of the paired fins of fishes are more constant than those of the vertical fins. Hence the taxonomist seldom finds here characteristics for separating species or smaller groups of fishes but rather larger categories. For example, the typical pelvic fin formula for spiny-rayed fishes is one spine and five soft-rays (I, 5) and from this, there are in general only minor variations.

Fin Rays

In bony fishes (Osteichthyes), the fin supporting structures are of two principal kinds (Fig. 6.1): spines (single rays) and soft-rays (segmented rays). Spines in the fins of fishes are unsegmented, uniserial structures, as in the first dorsal fin of the sea basses (Serranidae) and the perch and relatives (Percidae) among others. Soft-rays are typically segmented, often branched, and always biserial (two lateral components paired on the midline) and may be present in any fin but always compose the rays of all caudal fins of bony fishes (Osteichthyes). The character of the fin rays, whether soft or spiny, early gave rise to the concept of two corresponding groups of bony fishes, respectively the Malacopterygii and the Acanthopterygii. Representative of exclusively soft-rayed families in the North American freshwater fauna are the sturgeon (Acipenseridae), paddlefish (Polyodontidae), bowfin (Amiidae), gar (Lepisosteidae), herring (Clupeidae), smelt (Osmeridae), trout (Salmonidae), mooneye (Hiodontidae), mudminnow (Umbridae), pike (Esocidae), sucker (Catostomidae), carp (Cyprinidae), North American freshwater catfish (Ictaluridae), eel (Anguillidae), killifish (Cyprinodontidae), livebearer (Poeciliidae), and cod (Gadidae).

The only deception in all of the foregoing is in the carp and catfish families. In both these groups some of the soft-ray elements have fused embryonically to give rise to a spinous ray (morphologically not a spiny-ray as described above). The carp (*Cyprinus*) and the goldfish (*Carassius*) have such a hardened bundle of soft-ray elements in the dorsal and anal fin.

FIN RAYS

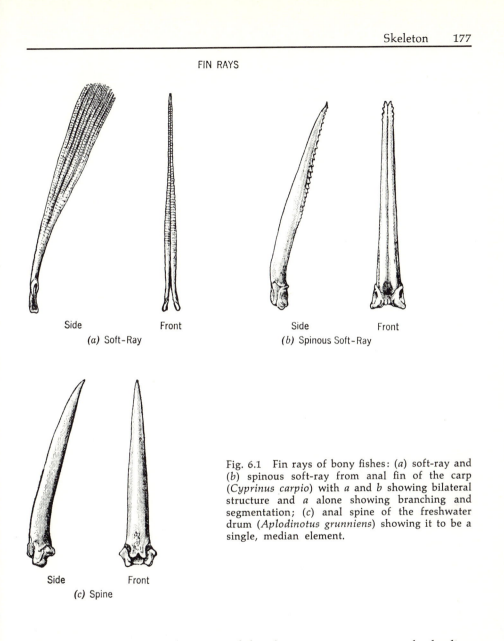

Fig. 6.1 Fin rays of bony fishes: (*a*) soft-ray and (*b*) spinous soft-ray from anal fin of the carp (*Cyprinus carpio*) with *a* and *b* showing bilateral structure and *a* alone showing branching and segmentation; (*c*) anal spine of the freshwater drum (*Aplodinotus grunniens*) showing it to be a single, median element.

The North American freshwater catfishes have a spinous ray in the leading edge of each pectoral fin.

To consider briefly the fin supporting elements in the major groups of fishes, we will take up the Agnatha, as one group, and then the Chondrichthyes and Osteichthyes as others.

In the lampreys and hagfishes (Cyclostomata), the fin supports are cartilaginous, unsegmented rods. These rods connect through membranous skeletal tissues to the imperfect vertebral column and feebly support the median fins.

In the sharks and their relatives and in the bony fishes, the median and paired fins have internal skeletal supports (Chapter 3) and dermal fin rays. The dermal fin rays are collectively called dermatotrichia. The dermatotrichia are mostly of three kinds: (a) ceratotrichia, which are horny and often un-branched and unjointed and found in the sharks and their relatives; (b) actinotrichia (Fig. 6.1), horny rays in the form of spines, which develop, embryonically at least, in all bony fish fins, and may persist as the spines in the spiny-rayed fishes (Acanthopterygii); or (c) lepidotrichia (Fig. 6.1) which are replacements of actinotrichia in the soft-rayed fishes (Malacopterygii), or in the soft-rayed fins or parts of fins of spiny-rayed fishes. Lepidotrichia are soft-rays, which, as noted before, are typically branched and jointed and are more superficial in their development than the spiny-rays. They may be thought of as receiving their superficial components from scale elements. Almost to be envisioned here is the embryonic migration of scales, up over embryonic fin spines, to cover them and ultimately to replace them. The term camptotrichia has been proposed for the rays of fins in the lungfishes (Dipnoi) that have been thought to differ microscopically from those of all other fish groups.

Number of rays, particularly in the vertical fins, are of considerable use in the classification of fishes. Rays of the dorsal and anal are related in their number to the embryonic segments of the developing animal and are termed meristic (along with numbers of scale rows along the side of the body, numbers of vertebrae, etc.) (Chapters 10 and 13). Enumeration of fin rays (and other meristic elements) has been most useful in characterizing species and in the study of subspecific and racial problems.

Oddities

Numerous skeletal oddities have arisen. Some have occurred in single species, others characterize entire groups. For example, in the bowfin (*Amia*) there is a bony gular plate, under the chin between the two sides (rami) of the lower jaw; in the bichirs (*Polypterus*) there are a pair of such gular plates. In salmonoid fishes (Salmoninae, Coregoninae, Osmeridae) there are spur-like processes, in the angles (axillae) of the pelvic fins. A similar process (gular process) is on each side of the throat in *Amia*. To such divergent structures no functional significance has been attached.

Included among abnormal skeletal occurrences are developmental and pathological anomalies. These structural aberrations are less frequently seen in nature than in fish of protected hatchery waters, because so many of them are monstrosities that do not favor survival. Often they impede the mobility required to escape predators. Included are such conditions as two-headedness, reversal of scale pattern, failure of one or more fins to appear, and sigmoid flexure of the vertebral column.

LOCOMOTION

Our interest here centers in locomotion by which travel or progression is achieved. Such locomotion may be either passive or active.

Passive Travel

The pelagic eggs and young of many fishes are often passively transported by currents. This happens in oceans, lakes, and streams, where such eggs and young are indeed a part of the plankton and often appear in tow-net samples. Livebearing fishes afford passive conveyance to their young inside the mother (Fig. 10.6). After birth or hatching there is also some protective transport of the young among fishes by temporary protective engulfment in the mouth cavity of a parent, as in some tilapias (such as *Tilapia mossambica*), catfishes (cypriniform suborder, Siluroidea; for example, the brown bullhead, *Ictalurus nebulosus*), sticklebacks (Gasterosteidae), and other species which actively guard the young. A brood pouch (Fig. 10.6) is also a sort of tonneau in which the young ride about, as in the pipefishes (*Syngnathus*) and seahorses (*Hippocampus*).

Passive travel may also be by hitchhiking on another organism or object. The most renowned hitchhikers among fishes are the marine suckerfishes, the remoras (Echeneidae), which attach harmlessly to sharks and other fishes, sea turtles, and even to whales. The organ of attachment is on top of the head and is a modification of the dorsal fin. The most bizarre hitchhikers are the parasitic males of certain deepsea anglerfishes (Ceratiidae) which attach by the mouth to the body surface of the females (Fig. 10.4). The parasitic adults of several kinds of lampreys (Petromyzonidae) are also carried about by their hosts. Most parasitize only fishes, but the Pacific lamprey (*Lampetra tridentata*) also attacks whales. The sea lamprey (*Petromyzon marinus*) is known to attach itself to the hulls of boats. Transport by the host during parasitic feeding may be suspected for the slime eel (*Simenchelys parasiticus*) and for some of the snake eels (*Pisodonophis*, Ophichthidae).

A most unusual kind of passive transport of fishes occurs when water masses containing fish are picked up into the air by strong wind currents resulting in water spouts. Later the fishes may come down again in a rain. There are authenticated accounts of such rains of fishes although the method seems so unlikely. There are no known records of fish egg transport by water birds; however, some fish eating birds, such as the herons (Ardeidae), have been known to drop live fish from their beaks while flying.

Man accounts for considerable passive transport of fishes by land, sea, and air. On all continents, fish cultural activities inevitably involve the moving about of indigenous species. Furthermore, most continents have introductions of exotic species, someties involving movement of the species halfway around the world from its native range.

Swimming

The most characteristic movements of fishes are for swimming. These are the movements for feeding and breeding, for offense and defense, and for active travel. They result in the commercial catches made in the world's great variety of stationary devices that capture fish by impounding them. Fishes also swim to make voyages of 1000 kilometers and more.

BODY FORM AND LOCOMOTION

In fishes, body form is the result of interaction of the skeleton, of the muscle mass, and of the evolutionary adaptation of both these systems to the lives of individual kinds of fishes. Primarily the adaptations are for various kinds of swimming although some fishes have also become adapted to limited locomotion on land.

As previously shown, the generalized body shape is fusiform for most aquatic animals of a free-swimming kind, be they fish, mammals, or invertebrates. This shape is a streamlining adaptation for movement through a fluid medium. The rounding of the leading (anterior) edge of such a shape and the tapering of the posterior part minimizes drag. In addition, in most fishes the body is covered with mucus, which further reduces drag by smoothing surface irregularities and reducing viscosity.

If the assumption be made that the primitive body shape of fishes is more or less fusiform, departures from it become adaptations to specialized modes of life. Close approaches to the ideal fusiform shape are retained by many pelagic fishes (Fig. 3.1), often by the swift swimmers such as the tunas and mackerels (Scombridae) and several of the requiem sharks (Carcharhinidae). Fish forms deviate from the ideal in three directions: compression, depression, and elongation. Each of these deviations in turn runs to extreme adaptations.

The laterally compressed body form (Fig. 3.1) prevails in quiet-water habitats of relatively dense cover. Many typical inhabitants of weed beds in lakes, ponds, and riverine backwaters are of this nature. Included are the North American sunfishes (*Lepomis*), the African tilapias (*Tilapia*), some of the cyprinids of both the Old and New Worlds such as the European bream (*Abramis*) and golden shiner (*Notemigonus*) respectively, and several cichlids such as the Amazonian angelfishes (*Pterophyllum*). It is also common among dwellers of the coral "forests" on tropical marine reefs, as for example in the butterfly fishes (Chaetodontidae) and many others. However, strong lateral compression in body shape with its utility in making short quick turns, is by no means confined to fishes that frequent dense cover. In sharp contrast, it occurs abundantly among open-water, often schooling species of herrings (Clupeidae). The weed dwellers are, however, relatively shorter and deeper-bodied than the herrings. The open-water ocean sunfishes or molas (Molidae) are also flattened from side to side but swim using pairs of median fins. The flatfishes (Pleuronectiformes), bottom dwellers *par excellence*, are the most striking laterally compressed fishes of all.

However, the depressed body prevails generally among bottom dwellers. In the highly secretive skates and relatives (Rajiformes), the fishes not only often rest on the bottom, but with their strongly depressed forms they advantageously burrow shallowly into it. Among the bony fishes this type of flattened form is typified by two of the most specialized families in the group —the goosefishes (Lophiidae) and the batfishes (Ogcocephalidae). Dorso-ventral flattening is characteristic in varying degrees of many stream fishes, ostensibly as a means for place holding in stream currents, as well as of those that take shelter beneath stones or other objects on the bottom. Common examples of such flattening, especially in the head region are among the sculpins (Cottidae) of the Northern Hemisphere and the widely distributed catfishes (cypriniform suborder Siluroidea) of all the major continents.

The losses in swimming efficiency by depression are perhaps offset by gains in mimetic resemblance to the bottom, in efficiency of certain secretive habits, or in other adaptations to life on the bottom. Extreme modifications of body form that are accompanied by loss of mobility are many among fishes. Globiform fishes such as the puffers (Tetraodontidae) and the porcupinefishes (Diodontidae) are examples as are also the leaf-mimicker (*Monocirrhus*; Fig. 4.6) and seahorses including the alga-like, leafy seadragon (*Phyllopteryx*; Fig. 4.6), the seahorses themselves and pipefishes (Syngnathidae; Fig. 10.6), and the trunkfishes (Ostraciidae).

FINS AND LOCOMOTION

Although fins are very characteristic parts of fishes, many species can locomote without them. Furthermore, individuals from which all fins have been removed can swim competently, even though only at low speeds. Their ability for fast starts (acceleration) is impaired markedly. With this impairment

FIN LOCOMOTION

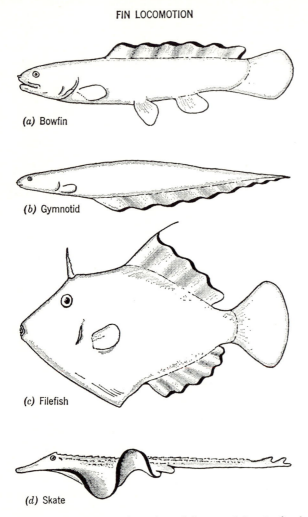

(a) Bowfin

(b) Gymnotid

(c) Filefish

(d) Skate

Fig. 6.2   Locomotion through undulation of longitudinal structures in: (*a*) the bowfin (*Amia calva*); (*b*) a gymnotid "eel" (*Gymnotus*); (*c*) a filefish (*Monacanthus*); and (*d*) a skate (*Raja*). (Source: Breder, 1926).

competence as a predator is reduced and susceptibility as prey is increased. Yet, fins are important in fish locomotion. A great number of fishes are able to move about or make specialized movements solely by use of their fins, even though most may still rely primarily on flexures of the body for general swimming. The caudal fin is of greatest importance for swimming in most fish at low (sustained) speeds and for all fish where high speed (burst) activity is important. A few highly specialized kinds of fishes move exclusively by means of their fins, as, for example, the seahorses (*Hippocampus*) and the trunkfishes (Ostraciidae).

All of the various fins, excepting the pelvics are used in some species as primary organs of propulsion (Fig. 6.2), although not necessarily for the strongest swimming of which the individual is capable. The long-based dorsal of the bowfin (*Amia*) can be undulated to give forward or backward movement to the fish. The anal is similar in elongation and function in the feather-backs (*Notopterus*) and in the electric eel (*Electrophorus*), in both of which it actually extends caudad far enough to fuse with the tail fin. Both the dorsal and the anal fins are moderately long-based in the triggerfishes and filefishes (Balistidae) and undulate to give slow forward or backward progression to the fish, even when the tail fin is at rest. In a few fishes, notably the trunk-fishes (Ostraciidae), the caudal fin is the primary organ of locomotion, having a sculling action. The pectorals, by undulation, are the basic locomotor structure in the skates, stingrays, eagle rays, and mantas (Rajiformes). They are also capable by sculling action of giving movement in such diverse groups as the seahorses (*Hippocampus*), sculpins (Cottidae), and the percid darters (Etheostomatinae).

Both the paired and the median fins are of great use to typical fishes as stabilizing and maneuvering organs (Fig. 6.3). The median fins have an obvious role as keels, helping the individual to remain upright. Because of the intrinsic musculature at the bases of the fins (Figs. 3.10 and 3.11), making possible finely differentiated movements of fin parts, all of the median fins also have value for maneuvering. The pelvics and pectorals too can serve as bilge keels to retard rolling although their principal functions seem to be for such maneuvers as climbing and diving (Fig. 6.4), banking, turning, and stopping in the water. For these actions, the pectorals are the primary organs and the pelvics, the secondary ones, as learned from experimental removal of fins from living fishes.

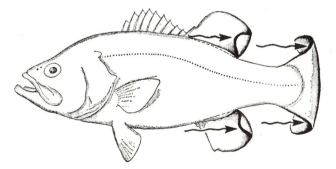

Fig. 6.3 Median fins curved to opposite sides to act as a sea anchor in the largemouth bass (*Micropterus salmoides*). (Based on Breder, 1926).

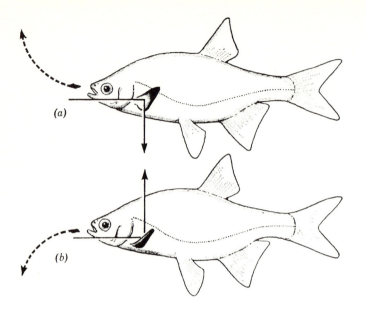

Fig. 6.4   Use of the pectorals to direct climbing (a) or diving (b) in the water by the golden shiner (*Notemigonus crysoleucas*). The curved arrows indicate direction of travel and the angled arrows, vectors of force. (Source: Breder, 1926).

To compare a man-made streamlined device with a natural one, likened to a glider, the body of a fish is the fuselage, the tail fin is the rudder, and the paired fins are the wings, ailerons, flaps, stabilizers, and elevators. The median fins share the steering function of the caudal fin and the gas bladder acts in part as an additional stabilizer.

In the sharks (Squaliformes), the paired fins act to elevate the anterior of the body. The posterior of the body is elevated by lift generated by the heterocercal tail. The lift forces of the pectoral and caudal fins counteract the negative buoyancy of the body lacking a swim bladder. In scombrids, that also lack a gas bladder, the pectoral fins and caudal peduncle generate similar lift forces. In contrast most bony fishes (Osteichthyes) have gas bladders and at equilibrium thus have a specific gravity about equal to water. As a result maintenance of position at any habitual depth offers little problem. Such stability, however, is dynamic and when a fish climbs or dives in the water column the gas bladder compensates to achieve dynamic equilibrium between gas bladder and fins.

The pectorals take on an additional function of serving as brakes. In relatively elongated teleosts, such as salmons (*Oncorhynchus*), the pectorals are situated in front of the center of gravity and posteriorly located (abdominal) pelvic fins keep the tail end of the body from tipping up when the fish stops. In many short, deep-bodied fishes, such as the sunfishes (*Lepomis*), the pectorals are near the center of gravity but the pelvics are in a forward position and counteract the upward thrust of the pectorals when the fish stops.

Although all fins are of importance in stabilizing action, the function is not completely fixed to any one. When one or more fins are removed experimentally (or accidentally), compensatory action of the others still gives reasonable balance. It is only when all fins have been excised that serious impairment of balance becomes evident. Of all fins, the pelvics have been held to have least significance in stabilization. It has been pointed out, however, that fishes with abdominal pelvic fins (soft-rayed fishes generally) tend to have short dorsals and anals whereas those with the pelvics in the thoracic or jugular positions (spiny-rays, generally) often have long dorsals and anals for compensation. The mechanisms of balance involves both the sense of sight and the sense of equilibrium housed in the membranous labyrinth of the inner ear (Chapter 11).

The fins of fishes serve many other functions than locomotion or balance.

MYOMERES AND SWIMMING

When the muscle fibers of a myomere contract they shorten. If the myomeres on one side of a fish are shortened, the head and tail bend toward that side and the myomeres on the opposite side of the body stretch. Alternating serial contractions of the myomeres on the two sides of the body flex the fish, or successive parts of the fish, into a waveform.

The wave resulting from the serial contractions passes smoothly along the body, from head toward tail, because of the overlapping of the muscle segments typical of the gnathostomes. The tail beat frequency (the rate at which waves travel over the body) varies with swimming speed, fish size, and kind of fish. The muscular contractions that rise to a backward-moving propulsive wave result in forward swimming when the wave travels backwards at a faster rate than the derived swimming speed. In the myosepta, the expression of forces such as these is seen in the development of strengthening intermuscular bones in many teleosts. If the body wave moves from tail to head, the fish swims backward. This is a notable capacity of eels and other elongated fishes that have anguilliform locomotion. The lampreys are an exception since they cannot produce reverse waves; their compensation is in great flexibility, enabling with ease the tightest hairpin turn.

The principle of the action of the forces that are applied by a contraction of a myomere may be diagrammatically represented as in Figure 6.5. In this figure, AB represents the midline of the body, with A being toward the head.

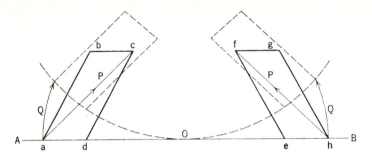

Fig. 6.5 Action of a myomere in bending the body of a fish. For explanation see text.
(Source: Nursall, 1956).

Horizontal sections through the forward and backward flexures of a myomere
are represented respectively by the rhomboids abcd and efgh. A myomere ex-
erts its force to bend the body at these flexures. At the flexures, the myomeres
are attached to the midline of the body at the angles Odc and Oef whereas
the muscle fibers in the myomere are more or less parallel to AB. When the
muscle fibers contract, they pull on the anterior septum of the forward flexure
and on the posterior septum of the backward flexure. P represents the pull of
the muscle fibers on the midline. Muscle contraction draws the flexures
towards one another as the midline bends about the central point, O. Upon
flexion, the myomeres and the midline AB bend to the arc shown by the
broken line. Serial contractions of the myomeres form a smooth wave along
the body as a result of the overlapping and nesting of the myomeres, as already described (Fig. 3.7).

SWIMMING

Swimming has both mechanical and energetic aspects, as summarized in the
following two subsections.

Swimming and Body Movement

Alternating serial flexure of the body muscles is by far the principal means
of forward movement among fishes. Enough variation exists in the outward
manifestations of this kind of muscle action to have led observers to classify
such swimming movements into three broad categories: anguilliform or eel-
like, ostraciform or trunkfish-like, and carangiform or jack-like.

   The anguilliform type of locomotion is serpentine in nature, as seen in a
crawling snake or swimming eel. It is brought about in an eel by sequential,
alternate contractions of the myotomes on each side of the body.

   Ostraciform locomotion is a wig-wag motion, seen especially in the sculling
action of the tail. It is induced by simple alternate contraction of all of the

muscle segments on one side of the body and then on the other. These alternating contractions cause the short tail to swish back and forth like a paddle behind the relatively rigid trunk of the fish. The body moves in a series of short cross arcs in the water as the fish progresses forward. The sculling of a boat by an oar run out over the stern has the same effect.

The third, and most common type, of locomotion is carangiform in which the fish drives itself forward by side-by-side sweeps of the tail region. This is, in a sense, intermediate between the anguilliform and ostraciform types. It is brought about by alternate contractions of myotomes on first one side of the body and then the other, starting at the head and travelling backward.

The vertebrae form a chain of levers hinged to turn only in the horizontal plane. Thus in both carangiform and anguilliform swimming the vertebral column whips or undulates from side to side in contrast to the whales and their relatives where it moves up and down. This throws the body of the fish into short curves, alternating from one side to the other, with the amplitude of these curves increasing toward the tail. The large lateral movements of the large-area caudal fin generate lateral forces that cause the head to move (recoil) also. This is largely resisted by the depth of the body and the high mass anteriorly. One could think of these motions as being a series of small segments out of the continuous anguilliform type. When the forces involved in the swimming of a fish in carangiform fashion are considered, it is seen that as the tail region is bent by a wave of muscular contractions passing backwards over the body the resultant forces of the water on the body drive the fish forward (Fig. 6.6). Anguilliform locomotor forces are simply multiples of the carangiform type with each of the serial undulations providing the force of a single carangiform bend. The cycle of lateral movements of the caudal fin or any other small length of the body has been likened to that of a variable pitch propellor, periodically reversing its direction of rotation and simultaneously altering its pitch.

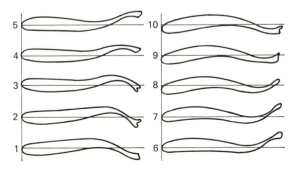

Fig. 6.6 Body shape during a propulsive cycle for subcarangiform motion. Drawn from tracings of the rainbow trout (*Salmo gairdneri*) taken at successive 0.03 second intervals. (Provided by P. W. Webb).

Energy Costs of Swimming

Although knowledge of fish propulsion energetics is incomplete, the energy balance sheet in its essentials requires the measurement of the metabolic energy made available to the locomotory muscles, the proportion of metabolic energy converted to mechanical energy, and the amount of muscle mechanical energy used in propelling the fish. An example of such a balance sheet is given in Figure 6.7 for relatively low-speed swimming in the rainbow trout (*Salmo gairdneri*).

At sustained activity levels, fish remain in equilibrium with their energy supply and thus metabolism is aerobic. Burst or acceleration swimming levels require the release of large amounts of energy in very short periods of time. Such energy is immediately available from ATP (adenosinetriphosphate) in the muscles, which, along with other energy-rich compounds are replenished by anaerobic metabolic pathways (mainly glycolysis).

Two systems of muscle fibers are involved in locomotion, usually in homogeneous blocks of red and white muscle. Red muscle represents some 1 to 20 percent of myotomes in fishes, depending on habitat. Its shortening speed is slow and functions largely aerobically at sustained activity levels. White muscle (twitch muscle) has a higher shortening speed, functions mainly anaerobically, and is used in burst and acceleration swimming.

The propulsive muscles that operate the caudal propeller system generate thrust that is equal to the drag force. Power required for swimming ranges from five times that of an equivalent vehicle such as a submarine to a little more than the foregoing value.

NONSWIMMING LOCOMOTION

Although swimming is the typical form of locomotion for fishes, they also burrow, crawl, leap, soar, and propel themselves by jets. Burrowing is shallow in such fishes as the skates (Rajidae) and flounders (Pleuronectidae and Soleidae) which do little more than to wriggle into the surface of the bottom while throwing the debris over the body to cover it, with the exception of the mouth and eyes. Burrowing is deeper in the mudminnow (*Umbra*) and the Japanese weatherfish (*Misgurnus*), among other fishes, which on alarm literally swim and dive into soft bottom material. The African lungfishes (*Protopterus*) excavate a tubular passage and remain in it to withstand drying while breathing air during estivation. Young lampreys (Petromyzonidae) also tunnel into the bottom throughout the period of their larval life of three years or longer. Only the head protrudes from the passage, but it is withdrawn into the sand or mud when the individual is startled. The sandlances (Ammodytidae) and the sandfishes (Trichodontidae) are consistently found in loose sand to depths of 15 cm. Some marine snake eels (Ophichthidae), at sizes

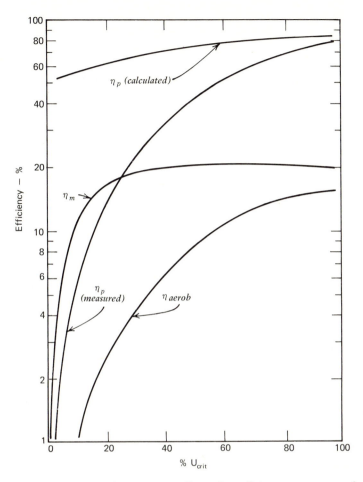

Fig. 6.7 Relations between overall aerobic efficiency, η_{aerob}, muscle efficiency, η_m, and caudal fin propeller efficiency, η_p, as a function of swimming speed expressed as a percentage of the critical swimming speed, for *Salmo gairdneri*. Values for η_p (measured) were calculated from measured oxygen consumption and swimming drag and expected muscle efficiencies. Values for η_p (calculated) were calculated from Lighthill's model. (Based on Webb, 1975).

between 50 and 90 cm., when startled burrow quickly, tail first, to depths of a meter. Their near relatives, the worm eels (Echelidae) are almost as adept at concealment. One of the snake eels (*Pisodonophis*) ascends fresh waters to burrow for food in rice fields of India and has developed specialized extension tubes on the nares that close to keep out the mud when the fish is tunneling.

To crawl on the bottom or on land is usually accomplished by combination of body and fin actions. Bottom crawlers seem either to be eel-shaped like the crevice-creeping morays (Muraenidae) or to have modified pectoral fins like the searobins (Triglidae) with finger-like rays at the leading edge of the enlarged fins. The pectorals have developed more or less arm-like bases to their finger-like tips in other bottom dwellers such as the toadfishes (Batrochoididae), goosefishes (Lophiidae), frogfishes (Antennariidae), and most pronouncedly the batfishes (Ogcocephalidae). The oriental climbing perch (*Anabas*) crawls on land using its pectorals and spiny gill covers and may even climb up on tree roots or other debris along the shore to feed, breathing air while out of the water. The mudskipper (*Periophthalmus*) skitters and jumps on mud flats after its food in Australia, Asia, and Africa. The chisel jaw (*Pantodon*) of Africa is able to traverse considerable distances on land by a rather distinctive creeping on its paired fins. The eel (*Anguilla*), in the course of migrating downstream to the ocean to spawn, has often been found on land, cutting the corners of streams on dewy mornings. Perhaps the most notorious land-walker of all is the walking catfish (*Clarias batrachus*).

Some fishes have literally taken to the air for the very diverse purposes of escaping, preying (as on aerial insects), and perhaps of playing. The leaping of a tarpon (*Megalops*), sailfish (*Istiophorus*), or other fighting game fishes when hooked is really rapid swimming up to and through the surface to hurl the body into the air, like the submarine-launching of an aerial projectile. The sailfish can rush into aerial leaps as great as 40 feet. Both the gizzard shad (*Dorosoma cepedianum*) and the little brook silverside (*Labidesthes sicculus*) represent schooling jumpers; locally the name skipjack has been applied to both of these species in recognition of the habit. Mullets (*Mugil*) and carp (*Cyprinus*) handily escape the nets of shore seiners by jumping over them. Striking jumps, often in the face of strong currents, are made by salmons (*Oncorhynchus* and *Salmo salar*) to surmount falls or other obstacles enroute to spawning grounds in headwater streams. Many of the jumping fishes, including the tarpon, gizzard shad, and others have a peculiar elongation of the last rays of the dorsal fin which may be related to the habit in a way as yet experimentally undetermined.

Extended leaps and adequate velocity can produce soaring flight provided wing-like structures have also been developed. This is not true flight, but rather gliding. In attaining speeds adequate for soaring or gliding the fish swims up through the surface of the water at a rather sharp angle and at a high speed and then soars on extended paired fins. After breaking the water and before gliding, some kinds of flying fishes (Exocoetidae) increase velocity for takeoff by surface propulsion with rapid tail vibration in which only the lower lobe of the caudal has contact with the water. Soaring flight has evolved independently in several groups of fishes including the marine flying fishes proper (Exocoetidae) and the flying gurnards (Dactylopteridae), and the freshwater hatchet-shaped Gasteropelecinae of South America and the chiseljaw,

Pantodon buchholzi of Africa. The pelvics are enlarged apparently to augment the soaring capacity of flying fishes of the genus *Cypselurus*, making veritable biplanes of them when considered along with the enlarged pectorals.

At contraction of the gill chambers water is jetted toward the tail and a forward thrust results. Many fishes at rest continually compensate for this thrust by pectoral fin movement. Jet propulsion seems to constitute an important part of the thrust for speed in take-off from rest, as may be readily observed in predacious fishes such as the pikes (*Esox*).

SPEED OF TRAVEL

Speed of travel in fishes may be thought of as cruising speed (routine activity), maximum sustainable speed, and top, burst speed. Cruising speeds are those involved in ordinary travel. Most information on cruising velocities has been gained by the subsequent recapture of marked fishes at various times and distances from points of release. Obviously such estimates of rate of travel are of necessity lower than actual because information is lacking on detours and stop-overs. Maximum sustainable speeds are those which a fish can maintain for lengthy intervals. They have been measured experimentally for several small fishes by determination of the length of time they can swim in one place when currents of known velocity pass by them. They are higher than cruising speeds but lower than top speeds. Top speed involves maximum thrust for short duration. It can come at take-off from resting position or it can be the sudden lunge of a game fish.

Actual travel speeds vary by species, size of the individual and temperature of the water. Routine travel speeds, in kilometers per hour and miles per hour, have been determined as follows for various species:

FISH	SPEED OF TRAVEL	
	kilometers/hour	miles/hour
blenny (*Zoarces*)	0.31	0.5
goby (*Gobius*)	0.37	0.6
rock gunnel (*Pholis gunnellus*)	0.43	0.7
sprat (*Clupea sprattus*)	0.87	1.4
stickleback (*Spinachia spinachia*)	0.99	1.6
flounder (*Pleuronectes flesus*)	1.49	2.4
eel (*Anguilla rostrata*)	1.61	2.6
plaice (*Pleuronectes platessa*)	1.80	2.9
searobin (*Trigla*)	1.92	3.1
goldfish (*Carassius*)	2.23	3.6
northern pike (*Esox lucius*)	2.92	4.7
brown trout (*Salmo trutta*)	3.29	5.3
mackerel (*Scomber scombrus*)	4.22	6.8

Upstream travel speeds of Pacific salmons have been recorded as averaging between 4.0 and 6.5 kilometers per day (approximately 7 and 10 miles).

In contrast some extreme speeds have been reported from nature. The striped bass (*Morone saxatilis*) and black basses have been logged at 7.5 kilometers per hour (12 mph). The log for mackerels (*Scomber*) is 12.75 kilometers per hour (20.5 mph) and that for the barracuda (*Sphyraena*) is 16.75 kilometers per hour (27 mph), with values near these also observed for salmon and for "blue shark." The flying fishes (Exocoetidae) attained 22 kilometers per hour (35 mph) and the dolphin (*Coryphaena*), 23 (37 mph). The marlin (*Makaira*), tuna (*Thunnus*), bonito (*Sarda*), and albacore (*Thunnus alalunga*) reportedly can hit 25 to 31 kilometers per hour (40 to 50 mph). The sailfish (*Istiophorus*) and the swordfish (*Xiphias*) top off around 37 kilometers per hour (60 mph). These values are experimentally unconfirmed and that of the sailfish, at least may be unrealistic from point of view of horsepower required.

Speeds involving maximum effort on the part of the fish are of extreme importance in the escape from enemies, including seines and other moving fishing devices of humans. Sudden bursts of top speed are vital to predators in the capture of food and to stream-running spawners like the Pacific salmons (*Oncorhynchus*) in leaping over falls or dams half a dozen feet high and a dozen feet in length. For such a feat salmon must leave the water at an estimated minimum velocity of 9 kilometers per hour (14 mph). Maximum sustainable speeds and sudden bursts are, however, the exception rather than the rule in ordinary movement or travel. The salmons have a maximum sustainable speed close to 5 kilometers per hour (8 mph), not 9 as required for a high jump.

MIGRATIONS

From what is known about travel in fishes, it may be concluded that most species have relatively small home ranges to which they restrict most of their movements. However, some fishes are great travelers and seem to be continually on the move. Quite stationary but active nevertheless are many small stream, lake, and inshore marine fishes. Very mobile are pelagic fishes and others that often travel great distances between fresh and marine waters. Included in the latter category are the salmons (*Oncorhynchus* and *Salmo*) which do most of their feeding in the sea but spawn in fresh water, and the eels which do the opposite. In between these two extremes there are variations that grade from one into the other. No really sharp line can be drawn between the uses of the terms movement and migration. Both may take the fish in vertical directions such as from the deeps to the surface waters or up or down slopes. They may also take the fish in horizontal directions such as offshore and onshore or upstream and downstream. In their course they may be within fresh waters (potamodromy or limnodromy), in the ocean (oceanodromy), or they may be between the two (diadromy), running from marine into fresh water (anadromy), or from fresh water into the sea (catadromy).

Travels within streams have been termed potamodromous and are very diverse. They are spectacular among such riverine swimmers as the bocachica (*Prochilodus*) which, with other fishes (Mylinae and *Leporinus*), travels vast distances upstream in river systems of South America at flood stages. Most lake-dwelling suckers (Catostomidae) and lampreys (Petromyzonidae) ascend tributary streams for spawning, as do some populations of the walleye (*Stizostedion*). Stream dwellers among the suckers, lampreys, carps (Cyprinidae), and the trouts (Salmoninae) typically travel upstream for spawning. Limnodromous movements, onshore with oncoming darkness and offshore with oncoming light, characterize such fishes as the trout-perch (*Percopsis*), yellow perch (*Perca flavescens*) and rock bass (*Ambloplites*) among many others. Mass movements in lakes in pursuit of swarms of food organisms have been clearly shown for many species, very dramatically including the white bass (*Morone chrysops*). The lake trout (*Salvelinus namaycush*) and many of the whitefishes and ciscoes (*Coregonus*) move to shoal waters for spawning, as do lakelocked populations of the walleye (*Stizostedion vitreum*).

Marine, oceanodromous travelers are as diverse in their movement habits as freshwater ones. The prevalent direction of travel in the North Temperate Zone is northward as the water warms and southward as it cools. Various herrings including true herrings (*Clupea*) and the menhaden (*Brevoortia*) travel along the coasts in this pattern of direction, as do many other coastal and pelagic marine fishes including mackerels (*Scomber*) and tunas (*Thunnus*).

Diadromous fishes that migrate freely between fresh and marine waters include the anadromous and catadromous kinds previously described. Anadromous fishes are well exemplified by the striped bass (*Morone saxatilis*), Pacific salmons (*Oncorhynchus*), Atlantic salmon (*Salmo salar*), and the marine-run sea lampreys (*Petromyzon*). Of these, the greatest distances have been recorded for Pacific salmons which may spawn hundreds of miles inland after having traversed several thousands of miles at sea during growth. The freshwater eels (*Anguilla*) of North America and Western Europe are a classical example of catadromous fishes. In the autumn adults begin to run seaward from as much as 500 miles inland and from elevations including 7000 feet. In the sea some individuals travel over 3000 miles to find, by remarkable navigation, regions of the Sargasso Sea for spawning. The young drift and swim back toward continental waters in a voyage requiring as much as three years. On arrival they metamorphose from their leptocephalus larval stage and ascend streams to complete growth to sexual maturity over a period of a dozen years or more before reversing their travel. Oceanic orientation of such fishes, followed by subsequent natal stream recognition, as shown by many marked salmons (*Oncorhynchus* and *Salmo salar*), is among the marvels of animal navigation.

Completely free movement between fresh and marine water, not for the purpose of breeding, has been termed amphidromous. Some gobies (*Sicydium*) of tropical rivers drift to the sea as larvae but return to fresh water while still young fry. Sport fishermen of the southeastern United States are well acquainted with the habits of the tarpon (*Megalops*) which sporadically enters freshwater lagoons and quiet rivers. The irregular movements of the Asiatic milkfish (*Chanos*) into and out of fresh water make it amphidromous.

Generally, a migration is a more or less continuous and direct movement from one location to another. Such movement is under the control of the fish, genetically determined, and is influenced by environmental factors. Typically, a migration includes one or more returns to the starting location (homing tendency); such species-specific migration patterns most often involve aggregations of the fish and a sense of direction. Migrations are mostly for spawning or feeding and are generally time oriented, although sometimes they take place when fishes actively flee adverse conditions.

Many factors influence migratory movements, including homing, of fishes; they may be grouped for convenience as physical, chemical, and biological. Physical factors include bottom materials, water depth, pressure (water and atmospheric), current and tide, turbidity, topography, gradient, temperature, and light—intensity, photoperiod, and quality. Among chemical factors are salinity, alkalinity, hydrogen ion concentration, dissolved gases, odors, tastes, and pollutants. Recognized as biological factors of migration are blood pressure, sexual development, phototaxis, social response, predators, competitors, hunger, food, memory, physiological clock, and endocrine state. The foregoing factors interact in various ways to continue, direct and arrest migrations and movements. Throughout, nervous and endocrine mechanisms are important in orientation and timing.

Travel habits of fishes may vary with latitude. Some northern marine sculpins (*Myoxocephalus*) enter fresh waters most often in the cold part of their range whereas the tropical sawfishes (Pristidae) move in the same direction in the warmest parts of their range. In reverse, the threespine stickleback (*Gasterosteus aculeatus*) is marine in the cold waters of the north but to the south it becomes predominantly a freshwater dweller.

SPECIAL REFERENCES ON LOCOMOTION

Bainbridge, R. 1958. The speed of swimming fish as related to size and to the frequency and amplitude of the tail beat. *Jour. Exptl. Biol.*, 35: 109–133.
——. 1958. The locomotion of fish. *New Scientist*, 4: 476–478.

———. 1961. Problems of fish locomotion. *Symp. Zool. Soc. London*, 5: 13–32.

Breder, C. M. 1926. The locomotion of fishes. *Zoologica*, 4 (5): 159–297.

Gerking, S. D. 1959. The restricted movements of fish populations. *Biol. Rev.*, 34: 221–242.

Gero, D. R. 1952. The hydrodynamic aspects of fish propulsion. *Amer. Mus. Novitates*, 1601, 32 p.

Gray, J. 1953. The locomotion of fishes *In* Essays in Marine Biology being the Richard Elmhirst Memorial Lectures. Oliver and Boyd, London and Edinburgh, pp. 1–16.

———. 1957. How fishes swim. *Sci Amer.*, 197 (2): 48–54.

Lighthill, M. J. 1969. Hydromechanics of aquatic animal propulsion. Annual Review of Fluid Mechanics, 1: 413–446.

Nursall, J. R. 1956. The lateral musculature and the swimming of fish. *Proc. Zool. Soc. London*, 126 (1): 127–143.

Webb, P. W. 1974. Pisces (zoology). [In] McGraw Hill Yearbook of Science and Technology, pp. 333–336.

Webb, P. W. 1975. Hydrodynamics and energetics of fish propulsion. *Jour. Fish. Res. Bd. Canada*, 190, 158p.

SPECIAL REFERENCES ON SKELETON

Goodrich, E. S. 1913. Studies on the structure and development of vertebrates. Macmillan and Co., Ltd., London.

Gregory, W. K. 1933. Fish skulls. *Trans. Amer. Phil. Soc.*, 23: 75–481. (Offset reprint, Eric Lundberg Publ., Laurel, Fla., 1959.)

Harrington, R. W., Jr. 1955. The osteology of the American cyprinid fish *Notropis bifrenatus*, with an annotated synonymy of teleost skull bones. *Copeia*, 1955 (4): 267–290.

Hyman, L. H. 1942. Comparative vertebrate anatomy. 2nd Ed., Univ. Chicago Press, Chicago.

Norden, C. R. 1961. Comparative osteology of representative salmonid fishes, with particular reference to the grayling (*Thymallus arcticus*) and its phylogeny. *Jour. Fish. Res. Bd. Canada*, 18 (5): 679–791.

Shufeldt, R. W. 1899. The skeleton of the black bass. *Bull. U.S. Fish Comm.*, 19: 311–320.

7

Blood and Circulation

A closed circulatory system that carries respiratory gases, cell wastes, excretory metabolites, and minerals and nutrients in solution or suspension ties the blood and its vessels closely to the functions of respiration, fluid balance, excretion, and even digestion. Though described in separate chapters, the inter-dependence of the several organ systems ought to be remembered throughout. In fishes, a large area of semipermeable tissue is exposed to the water at the gills and the function of the gill-blood complex is not only gas exchange through this tissue but also the excretion of certain nitrogenous wastes and the elimination or uptake of diffusible minerals.

Excepting the lungfishes (Dipnoi), fishes have a clear-cut single circulation from body to gills and return (Chapter 3; Figs. 3.16 and 3.17). In this single circulatory route, the heart pumps reduced blood that is relatively low in oxygen and high in carbon dioxide. The volume of blood involved is quite low, ranging between 1.5 and 3 percent of the body weight in highest bony fishes (Teleostei) as compared to 6 percent or more in mammals. However, at least one shark, the spiny dogfish (*Squalus acanthias*), has a blood volume equal to 5 percent of its body weight. Blood plasma and cells are renewed in a greater number of organs and systems among fishes than in mammals. Another remarkable feature of the circulatory system of fishes is the passage of blood through a number of special capillary or sinusoid systems within the course of either arterial or venous circulations. Such special systems of capillary arrangement are termed portal systems. They occur in the gills, liver (hepatic portal system), and the kidney (renal portal system). In addition, among physoclistous fishes there is yet another capillary-like network in the rete mirabile, a part of the gas bladder. Teleosts have similar arrangement in the choroid gland of their eyes and in some fast swimmers such as the mackerel sharks (Lamnidae) and the tunas and mackerels (Scombridae) capillary beds of special type occur also in other organs, muscles prominently included, probably promoting more efficient gas exchange between blood and tissues.

HEART AND CIRCULATORY VESSELS

The ground plan of the circulatory system has already been shown, as have its components—the heart, blood and lymph vessels, and fluid tissues including the blood cells.

The Heart

In most fishes the heart is situated immediately posterior to the gills (Fig. 3.17). However, among the highest bony fishes (Teleostei), which have gill covers (operculate condition), the heart is relatively farther forward in the body than among the non-operculate sharks and relatives (Elasmobranchii). In a few fishes posterior displacement of the heart is quite extensive, as in the South American lungfish (*Lepidosiren*) and the Asiatic swampeels (*Monopterus* of the family Synbranchidae). Ordinarily the heart is well protected by the shoulder girdle.

The membranous pericardial sac in which the heart is located is spacious in the sharks and relatives (Elasmobranchii) and more closely adhering in the bony fishes (Osteichthyes). Its lining, the parietal pericardium, joins the visceral pericardium that covers the heart muscle where the main veins enter at the sinus venosus and where the main artery, the truncus arteriosus (ventral aorta), leaves the heart on its way to the gills. Similar coalescence of the parietal and visceral pericardia occurs where the blood vessels of the heart itself, the coronary arteries and veins, enter and leave the heart. The space between parietal pericardium and the visceral pericardium of the heart is a derivative of the cranial portion of the embryonic coelom or body cavity. The pericardial cavity in the adult fish is closed towards the rear, except in the sharks and relatives (Elasmobranchii) and in the sturgeons and relatives (Acipenseriformes) where there may be one or two secondary connections between pericardial and peritoneal cavities.

The heart varies considerably in relative development and size (Figs. 3.16 and 7.1). In connection with the low blood volume and the low blood pressure of fishes, heart size, as expressed in parts per thousand of the body weight is lower in fishes than in other vertebrate groups. Slow-moving and relatively

HEARTS

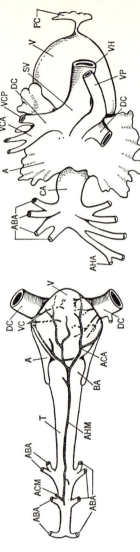

(a) Shark (ventral)

(b) Gar (dorsal)

(c) Trout (ventral)

(d) Lungfish (dorsal)

Fig. 7.1 Variation of proportions among fish hearts and associated blood vessels: (*a*) spotted dogfish (*Scyliorhinus caniculus*), ventral view; (*b*) longnose gar (*Lepisosteus osseus*), dorsal view; (*c*) rainbow trout (*Salmo gairdneri*), ventral view; (*d*) African lungfish (*Protopterus annectens*), dorsal view. A, atrium; ABA, afferent brachial artery; ACA, right coronary artery; ACM, commisural artery; AHA, hyoidean afferent artery; AHM, median hypobranchial artery; BA, bulbus arteriosus; CA, conus arteriosus; DC, duct of Cuvier; PC, pericardium; SV, sinus venosus; T, truncus arteriosus; V, ventricle; VCA, anterior cardinal vein; VCP, posterior cardinal vein; VC, coronary veins; VE, epigastric vein; VH, hepatic vein; VJ, jugular vein; VP, pulmonary vein; VS, subclavian vein. (Sources: (*a*) and (*c*) Grodzinski; (*b*) Rose; (*d*) Goodrich; all in Bronn, 1938).

sedentary fishes have relative heart weights of less than 1 part per thousand of the total body weight whereas the value in fast swimmers such as the mackerels and tunas is 1.2 parts per thousand and the heart of flyingfishes (Exocoetidae) has a relative weight of 2.1 parts per thousand. Fishes differ from land vertebrates inasmuch as their heart-to-body-weight ratios stay relatively constant throughout their lives.

The sinus venosus is a thin-walled sac that collects venous blood; its outline is modified among different fishes because of the variable arrangement of entry for the hepatic veins that bring blood from the liver, and for the ducts of Cuvier that gather the blood from the anterior and posterior cardinal veins. The sinus venosus has a volume that is never more than a fraction of that of the atrium (auricle) into which it opens by a median orifice. In lungfishes (Dipnoi) where this opening is transposed to the right, a sphincter muscle and two valves at the orifice regulate the flow of the blood into the atrium, the first of the two chambers of the heart proper. These sinus valves (sinuatrial valves) may be solely composed of heart lining (endocardial tissue) as in the sturgeons (*Acipenser*) and some sharks (Elasmobranchii), but they are most commonly of both endocardial and heart muscle or myocardial composition.

The atrium occupies the largest part of the dorsal pericardial region and may partially enfold the ventricle (Fig. 7.1, a and d) or have a simple inverted pyramidal form (Fig. 7.1, b and c). Muscle fibers in its walls are arranged to approach the atrial wall toward the atrio-ventricular orifice and thus, upon contraction, to move the incoming blood toward the ventricle. Both the sinus venosus and the atrium are relatively thinwalled in contrast to the ventricle.

The entrance into the ventricle is ellipsoid or circular in outline, lies in the atrial midline, and is surrounded by a flat sphincter. The opening is surrounded by a variable number of pocket-shaped valves (atrioventricular valves) that prevent the return of blood into the atrium. Sharks and their relatives (Elasmobranchii) and most bony fishes (Osteichthyes) have two rows of atrioventricular valves, but some, such as the bowfin (*Amia calva*) and the mrigal (*Cirrhinus mrigala*), have four; the paddlefish (*Polyodon spathula*) has five; and the gars (*Lepisosteus*) and the bichirs (*Polypterus*) have six. However, these valves have not been found in the lungfishes (Dipnoi).

The ventricle is a strong, relatively thickwalled and muscular sac in the ventrocaudad portion of the pericardial sac. Its shape is inverted-pyramidal in the shark-like fishes (Elasmobranchii), with the broad base directed caudad towards the liver. In the lungfishes the ventricle is roundish but in most of the bony fishes (Osteichthyes) it is laterally compressed. Two layers of muscle are common—a superficial, dense cortical layer and a spongy interior one. Codfishes and their relatives (Gadiformes), however, do not have the cortical muscle layers. The outer muscle layer may have circular fibers, as in the sturgeons (*Acipenser*), or it may form whorls, as in the requiem sharks (Carcharinidae), or there may be subdivisions of fibers that cross one another as in many of the highest bony fishes (Teleostei). Among fishes generally, muscle bundles of the spongy layer protrude into the lumen of the heart as trabeculae and form numerous subdivisions (diverticula) of the chamber. Muscle fibers of this layer may run parallel or at right angles to the axis of the heart or may even proceed radially from the surface toward the interior.

The ventricle leads anteriorly into the conus arteriosus, a specialized part of the truncus arteriosus. The conus is composed of heart muscle, like the ventricle, and is largest in the lungfishes (Dipnoi) where it is also slightly bent. It is quite straight in other fishes but varies in length. The conus ranges from relatively long in the gars (*Lepisosteus*) to short and hardly detectable in many of the highest bony fishes (Teleostei) (Fig. 7.2). Related to the length of the conus is the variation in numbers of valves to prevent the backflow of the blood into the ventricle. As many as seventy-four such valves occur among gars, arranged in eight successive rows: two to seven such rows are found in shark-like fishes (Elasmobranchii) and usually one (rarely two) in most bony fishes (Osteichthyes).

The location of pacemakers and nodal tissues in the fish heart is at present obscure. Many tissues of the heart of the hagfish (Myxinidae) exhibit pace-maker potentials. Electrocardiograms of the bullhead shark (*Heterodontus*) and the eel (*Anguilla*) show a V wave that originates from the sinus venosus and precedes the other events of cardiac excitation. In the shark, *Scyllium*, the sinu-atrial valve region contains the pacemaker. Small and young fishes

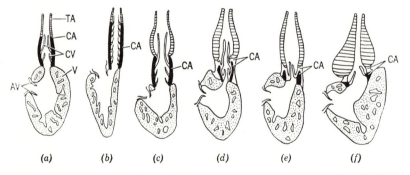

(a) (b) (c) (d) (e) (f)

Fig. 7.2 Variations in relative development of the musculature (in black) in the conus arteriosus and its valves in different fishes: (a) cat shark (*Scyliorhinus catulus*); (b) longnose gar (*Lepisosteus osseus*); (c) bowfin (*Amia calva*); (d) deepsea bonefish (*Ptero-thrissus gissus*); (e) African bonytongue (*Heterotis niloticus*); (f) European redeye minnow (*Leuciscus rutilus*). AV, atrioventricular valve; CA, conus arteriosus; CV, valves in conus; TA, truncus arteriosus; V, ventricle. (Sources: (a), Grodzinski: (b), Stohr; (c), Boas; (d) Senior; (e), Smith; (f) Hoyer; all in Bronn, 1938).

have a more rapid heartbeat than old and/or large specimens; the heartbeat also becomes accelerated with a rise in water temperature and vice versa. A 10-centimeter goldfish at 20°C, for example, has a normal heart rate of around seventy beats per minute. The heart is known to react to stress or noxious stimuli, for example, atropine, by slowing due to impulses from the vagus nerve; this phenomenon has been used in learning experiments with fishes.

The lungfish heart (Fig. 3.16) approaches the amphibian heart in several of its modifications. Apart from changes in the cardinal veins there exists a pulmonary vein that leads into the sinus venosus and has a tubular prolongation into the left half of the partially divided atrium; the right half of the atrium receives nonoxygenated blood from the sinus venosus. The atrial division is due to a constriction and an internal network of muscle and connective tissue fibers. The ventricle also has a division of fibrous connective tissue and cartilage. In the large, muscular conus arterious there is a spiral valve that assures continued division of aerated and non-aerated blood. Even the ventral aorta is still divided so that the nonoxygenated blood, which flows in its dorsal portion, reaches the afferent pulmonary vessel (pulmonary artery) whereas, from the ventral division of the ventral aorta, oxygenated blood flows through the first two gills and is conveyed to the dorsal aorta and thence to the body. The degree of subdivision of the heart differs among the different genera of lungfishes and is correlated with the prevalence of air breathing.

The heart is nourished by two groups of coronary arteries. One group, the anterior coronaries, originates from the hypobranchial vessels (which branch ventrad from the efferent branchial arteries) principally to supply the conus arteriosus and the ventricles. The second group, the posterior coronaries, issues from the coracoid or even the subclavian artery and approaches the heart posteriorly to supply the remainder of the heart wall. Coronary veins are less numerous than the arteries and bring the blood they collect into the sinus venosus or the atrium.

The Gill and Head Vessels (Figs. 7.3 and 7.4)

From the heart, the blood of fishes goes forward into the strong-walled, median truncus arteriosus or ventral aorta, after passing through the valved conus if present. The truncus arteriosus of sharks and relatives (Elasmobranchii) and lungfishes (Dipnoi) appears simply as an elongation of the conus arteriosus. The border between the two structures is marked, however, by a change from striated cardiac muscle in the conus to smooth muscle, characteristic of blood vessels generally, in the truncus. Most bony fishes (Osteichthyes) have a distinct swelling of smooth musculature at the beginning of the truncus arteriosus, which can be seen clearly by inspection with the naked eye; this swelling is the bulbus arteriosus which is large in the highest bony fishes (Teleostei), and smallest in the lowest ones (Chondrostei; Fig. 7.2).

Three to seven (rarely more) pairs of afferent branchial arteries arise from the truncus arteriosus largely to carry the blood upwards into the gills. The afferent branchial arteries curve along the gill arches and the farther away they go from the ventral aorta the more reduced they become in diameter (Fig. 7.3). Enroute they give off arterioles that become capillaries in the gill lamellae where the exchanges of gases and other solutes between blood and outside water take place.

The largest number of seven to fourteen afferent branchial arteries occurs in the lamprey-hagfish group (Cyclostomata) and depends on the number of gill pouches to be supplied. Most shark-like fishes (Elasmobranchii) have four (Fig. 7.3) with the exceptions of the sixgill shark (*Hexanchus*) which has five and the sevengill shark (*Notorynchus*) which has six. Lungfishes (Dipnoi) have four or five pairs of afferent branchials; the other bony fishes generally have four (Fig. 7.3), although six gill arches are laid down in the embryos of these groups. In a number of fishes such as the chimaeras (*Chimaera*), cods (*Gadus*), certain gobies (*Gobius*), puffers (Tetraodontidae), and mudskippers (*Periophthalmus)* the arteries leading to the gills arise separately from the main trunk of the ventral aorta. However, in most sharks the anterior-most vessel to arise from the truncus gives rise both to the first afferent branchial artery and to the hyoid artery. Most bony fishes show a common root for gill arteries IV and III but in some, such as the carp (*Cyprinus carpio*) and other minnows (Cyprinidae), gill afferents II and I have a common origin.

A branch of the afferent artery leads into each gill filament where it breaks up into capillaries and lacunae in the lamellae of the filament. The aerated blood from the filament collects laterally into an efferent loop around each internal branchial aperture. The loops collect into the efferent gill arteries that merge to become the dorsal aorta. Thus afferent gill arteries carry to the gills venous blood that is oxygen-poor, and carbon-dioxide-rich. The flow is greatly slowed in the vast capillary bed of the gills. The blood flow in the gill lamellae is such that reoxygenation and removal of CO_2 occur gradually from their aboral to oral edges, in order to make optimal use of the flow of the water that bathes the gills. Drop in blood pressure on the collecting side of the branchial vessels begins in the lamellae of the gills. Dorsal aortic diastolic pressures have been recorded as the following percentages of ventral aortic blood pressure: skate (*Raja binoculata*), 15; dogfish (*Squalus suckleyi*), 26; trout (*Salmo gairdneri*) and carp (*Cyprinus carpio*), between 50 and 60. Blood pressure and flow in the dorsal aorta of the cod (*Gadus morhua*) is pulsatile even though blood flows through the capillaries of the gills before entering this vessel.

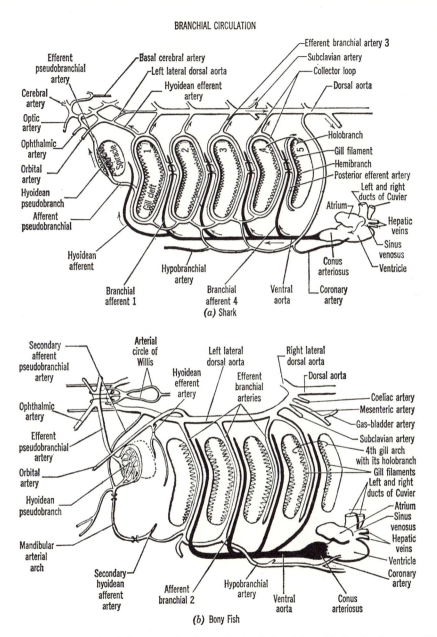

Fig. 7.3   Generalized diagram of gill circulation (left side) in (*a*) the shark group (Elasmo-branchii) and (*b*) the bony-fish group (Osteichthyes) as shown in a cod, *Gadus*. (Based on Goodrich, 1930).

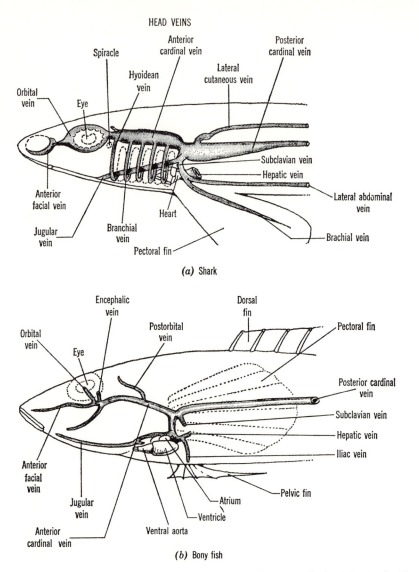

HEAD VEINS

(a) Shark

(b) Bony fish

Fig. 7.4 Lateral view of head veins of: *(a)* the spotted dogfish shark (*Scyliorhinus caniculus*) and *(b)* a bony fish, a goby (*Gobius panizzae*) with branchial veins not shown. (Based on Grodzinski in Bronn, 1938).

Efferent branchial arteries may be single, as among the highest bony fishes (Teleostei), or paired, one for each hemibranch of the gill arch, as in the sharks, rays, and skates (Elasmobranchii), with the exception of the anterior, hyoid arch that has one efferent branchial artery only. Intermediate conditions prevail in the lungfishes (Dipnoi), the paddlefish (*Polyodon*), the sturgeons (*Acipenser*), and the gars (*Lepisosteus*).

In the lamprey and shark groups of fishes, the efferent branchial vessels rise dorsally one by one to join the median aorta. The first efferent branchial artery largely provides the blood supply for brain and head regions although some arterial vessels, principally the hyoid artery, arise from the first afferent gill vessel. In higher bony fishes (Actinopterygii), the efferent branchial arteries of each side rise to collect into a trunk vessel on each side. The result is that over the gill region the dorsal aorta is paired and does not become a median vessel until just behind the gill region.

A special arrangement is found in the blood supply to the spiracle of sharks, rays, and skates (Elasmobranchii) and to its homologous structure in the bony fishes (Actinopterygii), the hyoidean pseudobranch. Both are supplied with oxygenated blood from a mandibular artery or its derivative and in bony fishes an efferent pseudobranchial vessel leads the blood that has passed the pseudobranchial capillaries through the ophthalmic artery to the choroid gland of the eye. Thus the vascular connections of the pseudobranch suggest for it a nonrespiratory primary function in keeping with its other anatomical and physiological properties (Chapter 8). The pseudobranch may contain receptors sensitive to oxygen and carbon dioxide tension.

The veins of the dorsal head region (Fig. 7.4) begin as facial, orbital, postorbital and cerebral veins and join into the paired anterior cardinals that lead into the common cardinal or duct of Cuvier and from there, into the heart. The jugular vein collects venous blood from the lower head region and is tributary to the common cardinal vein in bony fishes (Actinopterygii). In sharks and relatives (Elasmobranchii), the jugular vein unites with the subclavian vein from the pectoral fins and shoulder before merging with the duct of Cuvier. The elasmobranchs have several veins with relatively larger inner space (lumen) than have bony fishes; the sharks, rays, and skates also have several venous sinusoid pockets of sluggish blood flow, such as the orbital sinus, and also complex venous anastomoses such as in the chondrocranial and hypobranchial regions.

The Main Blood Vessels of the Body (Fig. 7.5)

As regards size, the blood vessels in the body of a fish are principally of two kinds: (*a*) the major arteries and veins, which follow the long axis of the body; (*b*) branches of the major vessels, which lead to and from the skin, the skeleton, and the musculature, and which also provide circulation for the brain, the spinal cord, and the intestinal organs.

The dorsal aorta is the main route of blood transport from the gills to the body; it is unpaired behind the gills in all fishes. In the trunk region, the

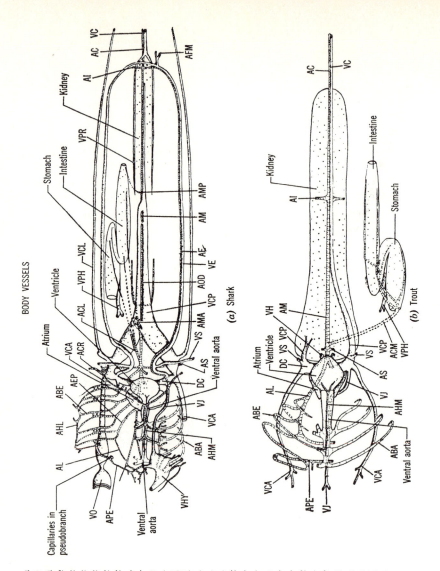

BODY VESSELS

(a) Shark

(b) Trout

Fig. 7.5 Comparison of main body vessels (ventral view) of (a) the antarctic smooth dogfish (*Mustelus antarcticus*) and (b) the rainbow trout (*Salmo gairdneri*): ABA, afferent branchial artery; ABE, efferent branchial artery; AC, caudal artery; ACL, coeliac artery; ACM, coeliacomesenteric artery; ACR, coracoid artery; AE, epigastric artery; AEP, epibranchial artery; AFM, femoral artery; AHL, lateral hypobranchial artery; AHM, medial hypobranchial artery; AI, iliac artery; AL, lateral aorta; AM, median aorta; AMA, anterior mesenteric artery; AMP, posterior mesenteric artery; AOD, oviducal artery; APE, efferent pseudobranchial artery; AS, subclavian artery; DC, duct of Cuvier; VC, caudal vein; VCA, anterior cardinal vein; VCL, lateral cutaneous vein; VCP, posterior cardinal vein; VE, epigastric vein; VH, hepatic vein; VHY, hyoidean venous sinus; VJ, jugular vein; VO, orbital venous sinus; VPH, hepatic portal vein; VPR, renal portal vein; VS, subclavian vein. (Trout based on Grodzinski in Bronn, 1938; shark based on Parker in Bronn, 1938).

206

dorsal aorta lies directly beneath the vertebral column of jawed fishes (gnathostomes) or under the notochord of jawless fishes (Cyclostomata). After giving off a series of dorsal, ventral and lateral segmental vessels (see below) the aorta becomes smaller and becomes known as the caudal artery as it traverses the hemal canal of the tail vertebrae. The hemal canal is a rigid and incollapsible tunnel that serves to shelter the dorsal aorta and caudal vein from the waves of muscular tension that would otherwise bear on them as the flexures of swimming pass down the trunk. In the tail region, the caudal vein doubles back along the course of the caudal artery and returns blood toward the heart. In the sharks tail pumps have been described for *Heterodontus*, in which the pumping is dependent on tail movement, and *Mustelus*, in which waves of muscle contraction squeeze venous reservoirs in the tail of the resting shark.

The numerous branch vessels that arise from the aorta and the corresponding veins are arranged more or less to match the body segments. For each segment of the embryo there arises a pair of arteries that divide into dorsal, lateral and ventral branches; uniquely in lampreys and hagfishes (Cyclostomata), these arteries have valves to prevent backflow of blood towards the aorta. Modifications of the embryonic segmental arrangements have occurred by the time a fish is an adult. Nevertheless, the dorsal segmental branches still essentially supply the dorsal musculature and the dorsal fin(s) but have branches intruding between the muscle bundles (myomeric cones) of the epaxial musculature of the sides. The lateral segmental arteries follow the dividing line (horizontal skeletogenous septum) between epaxial and hypaxial regions. The embryonic ventral segmental arteries supply muscle and fin elements in the tail region but also become asegmental and highly modified as certain vessels of the intestinal tract, mainly the coeliaco-mesenteric system with its branches, and the iliac arteries (Figs. 7.5 and 7.6).

The venous systems in many fishes (Cyclostomata; Elasmobranchii) have elaborate sinuses; the system in the hagfishes (*Myxine*) also has secondary "hearts" in the form of contractile bulbs along the way, as in veins of the tails. In addition, venous blood returning to the heart flows through sinus-like areas of the hepatic and renal portal systems. The renal portal veins that arise from the caudal vein of the embryo break up in the developing kidney into a network of increasingly smaller capillaries which then collect into the posterior cardinal or postcardinal veins. The postcardinal of each side terminates in the corresponding common cardinal sinus or duct of Cuvier of that side, which in turn passes into the sinus venosus. In the shark (*Heterodontus*) there is a gradient in venous blood pressure along the lateral cutaneous vein (Fig. 7.5), from a pressure that is positive and steady at the peripheral end to a pressure that is negative and fluctuating in time with the heart beat at the central end.

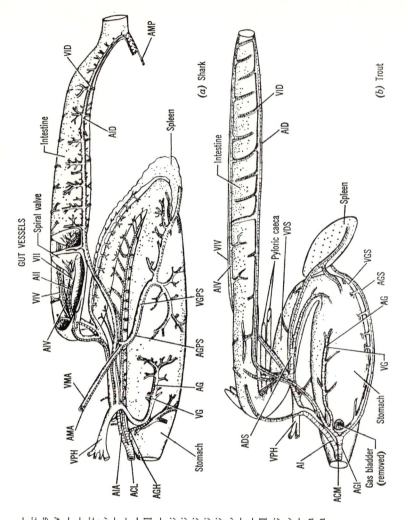

GUT VESSELS

(a) Shark

(b) Trout

Fig. 7.6 Blood vessels of the digestive tract of (a) the antarctic smooth dogfish shark (*Mustelus antarcticus*) and (b) the rainbow trout (*Salmo gairdneri*): ACL, coeliac artery; ACM, coeliacomesenteric artery; ADS, duodenosplenic artery; AG gastric artery; AGH, gastrohepatic artery; AGI, gastrointestinal artery; AGPS, gastropancreatosplenic artery; AGS, gastrosplenic artery; AI, intestinal artery; AID, dorsal intestinal artery; AII, intraintestinal artery; AIA, anterior intestinal artery; AIV, ventral intestinal artery; AMA, anterior mesenteric artery; AMP, posterior mesenteric artery; VDS, duodenosplenic vein; VG, gastric vein; VGPS, gastropancreatosplenic vein; VGS, gastrosplenic vein; VID, dorsal intestinal vein; VII, intraintestinal vein; VIV, ventral intestinal vein; VMA, anterior mesenteric vein; VPH, hepatic portal vein. (Trout based on Grodzinski, 1938; shark based on Parker in Bronn, 1938).

208

A curious feature exists in the circulation of the intestine of lampreys (Petromyzonidae) where the posterior part of the digestive tract is reached by vessels that have an artery placed in the lumen of a vein. Furthermore, the blood from the intestine returns to the heart not through an hepatic portal system but reaches the cardinal veins through a suprarenal venous sinus.

The veins from the stomach and the intestine of jawed fishes collect into the hepatic portal vein to reach the liver. In the liver, the portal vein gives rise to a myriad of sinusoids that expedite transfer of dissolved food substances to the liver tissue. The blood is then collected into hepatic veins which enter into the sinus venosus. Most fishes have two hepatic veins, but minnows (Cyprinidae) and cods (*Gadus*) have three, whereas such fishes as the mudskippers (*Periophthalmus*), the gars (*Lepisosteus*), and the bowfin (*Amia*) have one; in the South American lungfish (*Lepidosiren*), the posterior vena cava receives the blood from several small hepatic veins.

The Lymphatic System

In fishes, as in the other vertebrates, lymph is collected from all parts of the body by a system of paired and unpaired ducts and sinuses that finally return it to the main blood stream (Fig. 7.7) Unlike higher vertebrates, fishes lack lymph nodes. The lymphatic systems of lampreys and hagfishes (Cyclostomata) are characterized by more numerous and more diffuse connections with the blood vascular systems than exist in other groups of fishes. Because of this close connection, the vessels have been described as a hemolymph system. The lampreys and hagfishes have a spacious abdominal lymph sinus into which enter the lymph vessels of the kidneys and gonads. The sinus

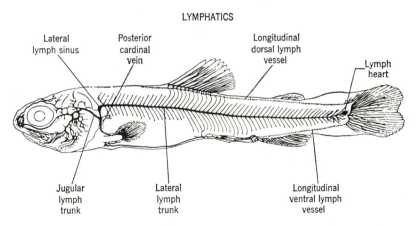

Fig. 7.7 Lateral view of superficial lymph vessels and the lymph heart of a young brown trout (*Salmo trutta*). (Based on Hoyer in Bronn, 1938).

has several openings into the cardinal veins which are, however, provided with valves so as to allow flow of the lymph into the vein and prevent a back-flow of venous blood into the lymph sinus. Superficial and deep lymph sinuses occur in the head region of lampreys (Petromyzonidae) where a lymphatic peribranchial sinus has valved connections with the jugular veins.

In the sharks and relatives (Elasmobranchii), the lymphatic system consists of lymph vessels rather than sinuses and lacks the contractile lymph "hearts" of cyclostomes and bony fishes (Osteichthyes). Subvertebral lymph trunks situated in the hemal canal of the tail vertebrae collect the lymphatic fluids from the tail region and then merge into the abdominal lymph ducts which form a dense network of vessels with the renal and gonadal lymphatic systems. Lymph-collecting vessels from the segmental musculature and the intestinal organs also flow into the subvertebral lymph trunks which in turn open into the cardinal sinus approximately at the level of the origin of the subclavian artery from the aorta. The subvertebral lymph trunks also extend into the head where they collect the lymph from cranial and branchial regions.

The lymphatic system of fishes is derived from the venous rather than from the arterial part of the blood vascular system, and sharks and their relatives (Elasmobranchii) as well as primitive bony fishes (Chondrostei and Holostei) progressively illustrate its increasing development, definition, and complexity. In the highest bony fishes (Teleostei), the lymphatics attain a well-defined arrangement comparable to that of terrestrial vertebrates. The teleosts have widely branching subcutaneous lymphatics; lymph from the head is collected through branching sinuses and flows into a subscapular sinus in the pectoral region where it is joined by fluid from the three main lymph ducts of the body—the dorsal, lateral, and ventral subcutaneous lymph trunks (Fig. 7.7). Neural, ventral, and hemal submuscular lymph trunks collect the tissue fluids from the body musculature whereas the lymph ducts of the visceral organs are divided into a superficial and a deep system. The deep visceral lymphatics collect the fat absorbed from the intestinal mucosa (chyle) into a coeliaco-mesenteric lymph trunk that connects to the remaining lymphatics, probably through the subvertebral trunk. The gas bladder, the gall bladder, the ab-dominal portion of the kidneys, and other organs of the body cavity drain their lymph into the paired pararenal lymph vessels that end in the pericardial sinus.

The lymph of rayfin bony fishes (Actinopterygii) finally flows into the main blood stream through the anterior (cephalic) lymph sinus, which opens into the cardinal veins as shown in the congers (*Conger*) and freshwater eels (*Anguilla*). The orifice connecting blood and lymph systems may also lie in the jugular veins as they do in some morays (*Muraena*) and in pikes (*Esox*) or in the posterior cardinal veins as in some members (*Salmo*) of the salmon family (Fig. 7.7). A false lymph heart that propels the lymph by relying on gill movements has been found in the cephalic sinus of the congers (*Conger*) and the freshwater eels (*Anguilla*), but a true lymph heart with valves and cardiac muscle-like contractile fibers occurs in the tail of both *Anguilla* and *Salmo*. The true lymph hearts appear as small flat blisters ventral to the last (penultimate) vertebra on the hypurals and are covered with muscle and skin. They communicate with the body lymph vessels and the caudal vein (Fig. 7.7) and are double-chambered and valved; they are thought to further venous flow but not to be vital to it.

BLOOD, TISSUE FLUIDS, AND BLOOD-FORMING ORGANS

Fish blood consists of fluid plasma and contains particulate entities that are largely blood cells or their remains. Lymph, the part of the plasma that perfuses the blood vessels to bathe the tissues, is mostly fluid but a few of the cellular components of blood may be found in it; red blood cells (erythrocytes) occur more frequently in the lymph of fishes, than in the lymph of higher vertebrates. In hagfishes (Myxinidae), the sinus spaces may be divided into red and white lymphatics, the former having a relatively high content of red blood cells.

As in humans, different blood types also exist in fish. Some resemble the Rh system of humans and the B system of cattle.

Fish Plasma

Plasma is a clear liquid in which are found, apart from the blood cells, dissolved minerals, absorbed products of digestion, waste products of the tisues, special secretions, enzymes, antibodies, and dissolved gases. Species differences include differences in sedimentation coefficients of principal plasma proteins. For the cod (*Gadus callarias*) the electrolyte composition of the blood in mmoles/liter is as follows: sodium (Na), 180; potassium (K), 4.9; magnesium (Mg), 3.8; calcium (Ca), 5.0; chloride (Cl), 158; and phosphate (PO_4), 3.1. The concentrations of sodium and chloride are generally lower in freshwater teleosts. Blood of the sharks (Squaliformes) has a higher Mg content but a weaker alkaline reserve than blood of higher bony fishes (Actinopterygii).

The solute content of a solution is indicated by the freezing point depression (Δ), which is also a measure of the osmotic pressure. With increased osmotic pressure there is an increased tendency for water to diffuse across a semi-permeable membrane to dilute the solution. The Δ's of fish plasma range from 0.5°C for freshwater bony fishes, through 1.0°C for the few freshwater sharks and relatives (Elasmobranchii), and between 0.6° and

1.0°C for marine bony fishes to reach a peak of 2.17°C in marine elasmo-branchs. The value of Δ for sea water is 2.08°C (Chapter 9).

Fish have low levels of plasma proteins compared to higher vertebrates. The principal blood plasma proteins of fish are albumin (controls osmotic pressure), lipoproteins (transport lipids), globulins (bind heme), ceruloplasmin (binds copper), fibrinogen (blood clotting), and iodurophorine (unique to fish, binds inorganic iodine). The relatively low levels of fibrinogen and pro-thrombin-like proteins do not correlate with the known rapid clotting time of fish blood. The blood of rainbow trout (*Salmo gairdneri*), which is capable of living at temperatures slightly above 0°C, clots rapidly at low temperatures. By contrast, the blood of carp (*Cyprinus carpio*), a warmwater fish, does not clot rapidly until the temperature reaches about 15°C.

Antarctic fishes living at ambient water temperatures as low as −1.9°C resist freezing partly because of the presence of a unique group of glyco-proteins in their serum. The only amino acids present in these antifreeze glycoproteins are alanine and threonine in a ratio of 2:1. The molecular weight of antifreeze glycoproteins ranges from 2600 to 33,000. Apparently these antifreeze glycoproteins lower the freezing point of fish blood by interfering with the growth of ice crystals.

A number of enzymes have been isolated from fish plasma, among them lipase and carbonic anhydrase which occur more abundantly in marine than in freshwater fishes. The serum of certain teleosts, including the freshwater eels (*Anguilla*), some catfishes (Siluridae), and tunas (*Thunnus*), has been shown to be toxic when injected into the mammalian bloodstream. Serum of the freshwater eel destroys blood cells by hemolysis and influences the tonus and the permeability of blood vessels in mammals.

Red Blood Cells

Nucleated, yellow-red erythrocytes have been found in the blood of all fishes (Fig. 7.8), except in three small Antarctic species that have a low, cold-adjusted metabolism and inhabit oxygen-rich waters, and in the ribbon-like leptocephalus larvae of eels (*Anguilla*) and certain deepsea fishes. These exceptional forms can apparently perform their gaseous exchanges by diffu-sion. Mature red blood cells of fishes are oval, tiny, and range in long diam-eter from 7 microns in many fishes such as in certain wrasses (*Crenilabrus*) to 36 microns in the African lungfish (*Protopterus*); a micron is 0.001 of a millimeter and the round human erythrocyte has a diameter of 7.9 microns. Enucleate red blood cells are rare but the proportion of immature, roundish red blood cells varies considerably in fishes. Blood of the sharks (Squali-formes) like that of eels (Anguillidae) and mackerels (Scombridae) may contain as much as 20 percent of immature cells, whereas blood of other

BLOOD CELLS

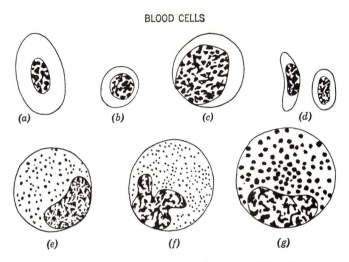

Fig. 7.8 Blood cells of fishes: (*a*) erythrocyte; (*b*) lymphocyte; (*c*) monocyte; (*d*) throm-bocytes; (*e*) acidophilic granulocyte; (*f*) neutrophilic granulocyte; (*g*) basophilic granu-locyte. (*d*) is from a shark (*Scyllium canicula*); all others are from a carp (*Cyprinus*). (Based on Maximow in Bronn, 1938).

active bony fishes may have a far lower percentage as in the remoras (*Echeneis*) and pipefishes (Syngnathus).

Oxygen transport in blood depends on the iron compound, hemoglobin, the respiratory pigment of the blood. The hemoglobin content of whole fish blood varies with the number of erythrocytes present; expressed as per-centage of dry weight of erythrocytes we find values of 14 to 19 percent in the smooth dogfish shark (*Mustelus canis*) and 37 to 79 percent in a number of marine and freshwater teleosts. The oxygen carrying capacity (Chapter 8), at 95 percent saturation, iron content of the blood, and number of red blood cells in fishes often go hand in hand and vary with life history stage, habits, and environmental conditions (Table 7.1). This has been well shown by Grodzinski (1938) from whose work Table 7.1 is taken.

Table 7.1. Comparison of Certain Characteristics of Red Blood Cells of Fishes

Species	Oxygen Retention Capacity (percent-age of blood by volume)	Number of Erythrocytes (per mm^3)	Fe Content (mg per 100 cc)
Goosefish (*Lophius*) (sedentary)	5	867,000	13.4
Mackerel (*Scomber*) (active)	16	3,000,000	37.1

Circadian variations in the number of erythrocytes also occur as shown in a blenny (*Blennius*) and a wrasse (*Crenilabrus*). Fishes respond to "shock" with a lowering in number of red blood cells, among other reactions. However, they can rapidly increase the number of young erythrocytes in the blood when subjected to blood losses, a reaction somewhat comparable to the so-called high altitude effect in mammals.

White Blood Cells (Fig. 7.8)

Apart from erythrocytes with their respiratory pigment, fish blood contains several types of colorless or white cells (leucocytes), all of which are ovoid to spheroid in shape. Whereas red blood cells range from 20,000 to 3,000,000 per cubic millimeter of blood, white cells vary between 20,000 and 150,000 per cubic millimeter in different groups of fishes. Among the white cells there are granulocytes that may make up between 4 and 40 percent of all white cells. Their average diameter is about 10 microns but ranges from 24 to 33 microns in the African lungfish (*Protopterus*). Granulocytes are subdivided, according to their staining reactions into neutrophils, which are most common; acidophils (eosinophils); and basophils, which are rare in fish blood. Some elasmobranchs possess a fourth type of granulocyte, the heterophil. Also present are agranular lymphocytes and monocytes as well as oval, usually smaller, thrombocytes. The agranular leucocytes are the most numerous white cell components in fish blood. The monocytes serve a macrophage function. Lymphocytes appear to differentiate into two populations, one concerned with the production of antibodies and the other with cellular immunity. Thrombocytes account for about half of all leucocytes in fish and are involved in blood clotting.

Blood Formation

In warmblooded vertebrates formation of blood cells is restricted to bone marrow, the spleen, and the lymph nodes; in fishes and amphibians many more organs take part in the manufacture of blood cells (hematopoiesis).

The blood vessels, and from them also blood cells, are differentiated early in the embryology of fishes. In the adult, blood cells may still be formed from the lining of blood vessels, but other more distinct blood-cell forming centers have emerged. In the lampreys and hagfishes (Cyclostomata), all of the kinds of blood cells are formed in the diffuse spleen found in the submucosa of the digestive tract; further diffuse hematopoietic sites probably also lie in various blood vessels. Erythrocytes and granulocytes are formed in the protovertebral arch dorsal to the notochord.

In the gnathostomatous fishes there is a distinct spleen; it is more or less divided into a red, outer cortex and white inner pulpa, the medulla. The spleen manufactures erythrocytes and thrombocytes in its cortical zone and lymphocytes and some granulocytes in its medullary region. In higher bony fishes (Actinopterygii) red blood cells are also destroyed in the spleen; it is not known whether other organs also function in blood decomposition or how blood destruction comes about in the jawless fishes (Agnatha) or in the sharks and rays (Elasmobranchii). Thrombocytes are formed in the mesonephric kidney of fishes (Chapter 9) and granulocytes come from the submucosa of the digestive tract, the liver, the gonads, and the mesonephric kidney. Beneath the submucosa of the esophagus, sharks, rays, and chimaeras (Chondrichthyes) have flat, yellow patches of loose connective tissue filled with white blood cells which are manufactured there; if the spleen is removed, this structure, the organ of Leydig, also takes over erythrocyte production. In the Chondrichthyes and the lungfishes (Dipnoi), the spiral valve of the intestine produces several white blood cell types. In the sturgeons (*Acipenser*), the paddlefish (*Polyodon*), and the South American lungfish (*Lepidosiren*) the heart is surrounded by reddish-brown, lobular tissue of spongy character where both lymphocytes and granulocytes are produced. Cellular elements in the cranial cartilages of some sharks (Squaliformes) and of the chimaeras (Chimaeridae) and in the large headbones of gars (*Lepisosteus*) and the bowfin (*Amia*) also produce all types of blood cells, thereby anticipating the hematopoietic function of the bone marrow of higher vertebrates.

Blood and Diseases

In fishes, as in humans and other vertebrates, the circulatory system plays an important role in combating infectious diseases. Antibodies against specific antigens are probably formed in the liver, kidney, spleen, and thymus except in lampreys (Petromyzonidae), which do not have clearly defined spleen or thymus. The immunoglobulin of lampreys is a unique molecule among other immunoglobulins so far described from vertebrates. It has a tetrameric structure, lacks interchain covalent bonding, lacks heavy and light chain structure, and has a prominent alpha helix. Fish in general make phage neutralizing and agglutinating antibodies following antigenic stimulation. Only two species of sharks and two holosteans, the bowfin (*Amia calva*) and the Florida gar (*Lepisosteus platyrhincus*), have been demonstrated to be capable of precipitin production. Antitoxins and lysins are also known to exist in fish blood.

Individual differences in disease resistance also exist with one individual fish being more resistant against certain bacteria than another. Such differences are based on genetically induced variations in the antibody producing system and probably also on other, overall, slight differences in individual body chemistry. Thus, disease-resistant strains of several species of trout

(Salmoninae) have been bred but it is also possible to inoculate especially valuable fishes against specific diseases. Weak (attenuated) strains of live bacteria can be used or, as is done more frequently, repeated injections are given of sera containing dead bacteria and their antigens. The fish so treated then acquire temporary passive immunity.

Another aspect of infection is the change that occurs in the blood picture, so useful in diagnosing certain diseases in man. In infected fishes there are changes in the hemoglobin content of the blood as well as in the numbers of white and red blood corpuscles. For example, blood of the carp (*Cyprinus carpio*) with a common sore-disease due to the bacterium *Pseudomonas punctata*, may show the following changes: lymphocytes fall in number, from 92.0 to 49.4 percent of all white blood cells; monocytes rise from 5.7 to 38 percent, and polymorphonuclear leucocytes increase from the normal of 2.3 to 12.6 percent; the number of erythrocytes falls from more than two million per cubic millimeter of blood to less than one million and the hemoglobin content of the blood is reduced accordingly.

BODY TEMPERATURE

Fishes, with amphibans and reptiles, are cold-blooded poikilothermous vertebrates. The body temperature adjusts passively to that of the surrounding water. Rapid warming and cooling can be lethal to some fishes, especially delicate coldwater species such as the trouts (*Salmo*) and chars (*Salvelinus*). Difficulties in the adjustment to temperature changes are due, not only to inherited low tolerance to thermal change but, especially in relation to a temperature rise, to metabolic, or to respiratory stress.

Small to moderate changes in environmental temperature spread rapidly through the entire fish due to positive or negative heat transfer through skin capillaries and the large capillary bed of the gills. Young fishes, with their higher gill to body surface ratio adjust more rapidly than older ones.

Limited control of body temperature exists in some fishes. The perches (*Perca*), sunfishes (*Lepomis*), and some catfishes (*Ictalurus*) react to external frightening stimuli with a slight increase in body temperature.

Tunas and lamnid sharks can maintain muscle temperatures well above ambient. The large bluefin tuna (*Thunnus thynnus*) can maintain a red muscle temperature of between 26° and 32°C, while the ambient temperature varies between 6° and 30°C. The red muscle, which is the warmest part of the muscle, is supplied with energy by oxidation of fats and is used for sustained swimming. White muscle is used in burst swimming of short duration, and energy is supplied by anaerobic glycolysis. The ability of tuna to thermoregulate is related to the presence of a rete mirabile or "wondernet" formed by arteries and veins arising from a pair of cutaneous vessels and supplying the red muscles. The retia act as countercurrent heat exchangers in such a way that venous blood warms arterial blood entering the muscle and reduces heat loss. The retia form a thermal heat barrier retaining metabolic heat in the tissues and preventing its loss in the gills.

SPECIAL REFERENCES ON BLOOD AND CIRCULATION

Black, E. C. 1940. The transport of oxygen by the blood of freshwater fish. *Biol. Bull.,* 79: 215–229.

Carey, F. G., and J. M. Teal. 1969. Regulation of body temperature by the bluefin tuna. *Comp. Biochem. Physiol.,* 28: 205–213.

Catton, W. T. 1951. Blood cell formation in certain teleost fishes. *Blood, Jour. Hematol.,* 6: 39–60.

Good, R. A., J. Finstad, and G. W. Litman. 1972. Immunology. The biology of lampreys, 2: 405–432. Academic Press, New York.

Grodzinski, Z. 1938. Das Blutgefaessystem. *In* Bronn's Klassen und Ordnungen des Tierreichs. Band 6, Abt. 1, Buch 2, Teil 2, Leiferung, 1: 1–77.

Holmes, W. N., and E. M. Donaldson. 1969. The body compartments and the distribution of electrolytes. *In* Fish Physiology, Academic Press, New York. 1: 1–89.

Hoyer, H. 1938. Das Lymphgefaessystem. *In* Bronn's Klassen und Ordnungen des Tierreichs. Band 6, Abt. 1, Buch 2, Teil 2, Lieferung, 1: 78–101.

Randall, D. J. 1970. The circulatory system. *In* Fish Physiology, Academic Press, New York. 4: 133–172.

Root, R. W. 1931. The respiratory function of the blood of marine fishes. *Biol. Bull.,* 61: 427–456.

Satchell, G. H. 1971. Circulation in fishes. Cambridge University Press. London.

Skramlik, E. von. 1935. Ueber den Kreislauf bei den Fischen. *Ergeb. d. Biol.,* 2: 1–130.

8

Respiration

One of the fundamental needs of a fish, like other animals, is to have an adequate supply of oxygen in the tissues so oxidation can occur and provide the necessary energy for life. The success of a fish depends upon its ability to obtain oxygen from the external environment by means of vascularized gills, lungs, or skin; to transport this oxygen to the tissues; and to unload the oxygen to the tissues. In like manner, the carbon dioxide, which results from cellular oxidation of carbon-containing compounds, must be transported in the blood and eliminated at the gills or other respiratory structures. In some scaleless fishes, gas exchange with the water also takes place through the skin and in the embryos of fishes various tissues serve as temporary breathing structures. Among these structures the yolk sac is relatively highly vascularized early and has respiratory functions. After hatching, transitory vascularization of developing pectoral fins may assist the developing gills. Temporary opercular vascularization for gas exchange may also occur, as in the bowfin (*Amia*). Several adaptations also exist for taking oxygen directly from the air. Included are modifications of the gills, the mouth cavity, the intestine, and the gas bladder.

The gross anatomy of the gills in the major groups of fishes has been described previously. Emphasis in this chapter is given to the function of the gills in respiration, to the various respiratory modifications of the gills of fishes, and to the pneumatic evaginations of the digestive tract. The function of fish blood in respiratory gas exchange will also be treated briefly.

STRUCTURE AND FUNCTION OF GILLS

Gills in Lampreys and Hagfishes

The seven paired gill sacs or branchial pouches of the sea lamprey (*Petromyzon marinus*) are representative of the gill condition in all lampreys (Petromyzonidae). The gill sacs open toward the lumen of the alimentary tract. Each is divided from the next by a thin diaphragm that adheres to the body wall. Except for the inner edge of the diaphragm, the inside of each gill sac is covered with radially arranged gill filaments which have small secondary cross folds on them to enlarge the respiratory surface. The branchial pouches, supported by a cartilaginous branchial basket of lattice-work structure (Chapter 3; Fig. 3.13), communicate with the exterior through more or less pronounced, epithelium-lined branchial atria that open to the exterior through the gill pores or external branchial apertures. Between each branchial atrium and its external gill pore there is a short, oblique, posteriorly directed branchial canal.

Lampreys rarely use the suctorial mouth for inspiration, even when the mouth is not attached to an object or a host fish. Fifty to seventy contractions of the gill pouches have been counted per minute in the sea lamprey (*Petromyzon*) attached to a prey, but have been found to rise to a hundred and twenty or even to two hundred contractions per minute in the rapidly swimming river lamprey (*Lampetra fluviatilis*). Respiratory water enters the pouches by tidal flow but is expelled through contraction of branchial compressor muscles attached to the branchial basket and the divisions between pouches; sphincters surround each gill pore. Water intake is also aided by enlargement and contraction of the self-contained nasal sac (nasopharyngeal pouch) which works like a hydraulic system. Conversely, the alternate filling and emptying of the gill pouches also serve to promote water exchange in the nasal sac.

The "suction cup" type of attachment afforded by the suctorial mouth (buccal funnel) of lampreys implies that the pressure inequalities due to breathing are not translated forward into the mouth cavity. When the rasping tongue is at work, the buccal funnel is closed posteriorly by the semi-annularis muscle and the inside openings of the gill pouches are protected by the velum. Such an almost complete separation of pathways is unique among gill-breathing verebrates. Between the gill pouches there are septa that contain venous blood sinuses, cartilaginous supports from the branchial basket, and muscles. These interbranchial septa receive an afferent artery each from the truncus arteriosus. The afferent artery divides into anterior and posterior branchial-pouch arteries and further into the arteries of the filaments. Filamental arteries spread into capillaries and lacunae in the gill lamellae where gas exchange occurs. Oxygenated blood leaves the lamprey gills, headed toward the dorsal aorta, by a system of vessels arranged in the septa parallel to the afferent pouch arteries. The physiology of gas exchange in cyclostomes is essentially similar to that in higher fishes.

The hagfishes (Myxiniformes) have two distinct breathing habits. When an individual is not feeding, it may lie buried in the mud, except for the anterior part of its head. When the animal is in this position, water reaches the gill pouches from the nasopharyngeal cavity. When the mouth is buried in food, respiratory water flows alternately in and out of the gill pouches through the esophageocutaneous duct. This duct opens externally behind the last gill pouch.

Gills in Sharks and Rays (Fig. 8.1)

Five, more rarely six or seven, external gill slits exist ventrally in a series on each side in rays and their relatives (Rajiformes), but laterally in sharks (Squaliformes). An anterior addition to this series of openings in both the sharks and rays (Elasmobranchii) is the spiracle, corresponding to a vestigial primitive first gill slit. Internal to the spiracle, gill lamellae may persist as a hyoidean pseudobranch that is supplied with oxygenated blood and has an arrangement of cells which suggests that the pseudobranch of the elasmo-branchs may not have a respiratory function but may be secretory as in bony fishes (Osteichthyes) or may take part in blood-cell formation. In sharks, respiratory water typically enters through the mouth, but in the bottom-dwelling rays the spiracle admits most of the water that flows subsequently over the gills.

Well-developed septa with cartilaginous supports and individual gill-arch muscles characterize each holobranch; the oral and aboral sides of the septa each carry a hemibranch composed of the gill tissue proper (Fig. 8.1.). Both primary and secondary gill filaments are present in each hemibranch. The distal ends of the primary gill filaments are detached from the septum (Fig. 8.1) so that two hemibranchs in apposition may form an effective barrier that forces the water to penetrate between all filaments when it seeks its exit as a result of the suction and pressure of the respiratory movements. The respiratory cycle of sharks and rays can be divided into three consecutive phases. In the first phase, coraco-hyoid and coraco-branchial muscles contract to widen the angle enclosed by the gill arches and to enlarge the oropharyngeal cavity, water enters by suction through the mouth cavity and/or spiracle, and during this phase the gill flaps are held to the skin by exterior water pressure; thus the external gill slits are closed. In the second phase, abductors of the lower jaw and gill arches relax, but adductor muscles (interarcual adductors) between the upper and lower portions of each gill

GILL CIRCULATION

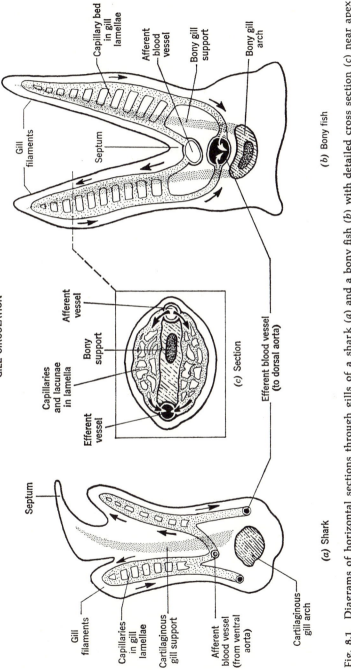

Gill
filaments

Capillary bed
in gill lamellae

Afferent
blood
vessel

Bony gill
support

Bony gill
arch

Septum

(b) Bony fish

Capillaries
and lacunae
in lamella

Afferent
vessel

Bony
support

Efferent
vessel

(c) Section

Efferent blood vessel
(to dorsal aorta)

Septum

Gill
filaments

Capillaries
in gill
lamellae

Cartilaginous
gill support

Afferent
blood vessel
(from ventral
aorta)

Cartilaginous
gill arch

(a) Shark

Fig. 8.1 Diagrams of horizontal sections through gills of a shark (a) and a bony fish (b) with detailed cross section (c) near apex of (b). Arrows indicate direction of blood flow.

221

arch contract and the mouth cavity begins to function as a pressure pump. While the oral valve prevents forward flow out of the mouth, the water is directed backward toward the internal gill clefts. Contraction of the inter-arcual adductors bulges the oral portion of the interseptal spaces; hydro-static pressure at the inner gill surfaces is then reduced and water is drawn into the gill cavities, which are still closed toward the outside. In the third phase, the interarcual adductors relax, another set of muscles contracts to narrow the internal gill clefts, and the water is forced through the gill lamellae. Then the flaps at the external gill clefts open passively and allow the water to flow to the ouside. However, many sharks, for example the mackerel sharks (Lamnidae), do not show such pronounced breathing move-ments and can only take in sufficient respiratory water while swimming; they suffocate readily when immobilized by capture or other causes.

Gills in Bony Fishes (Fig. 8.2)

Bony fishes in general share the same basic arrangement of a single external branchial aperture, on each side of the head, beneath a gill-covering oper-culum. However, loss of the spiracle, reduction of gill slits to four, and a gradually deepening indentation between the two hemibranchs of each holo-branch (Figs. 8.1 and 8.2) form a transition from lower (Chondrostei) to higher (Teleostei) rayfin bony fishes. Also, the forked efferent branchial

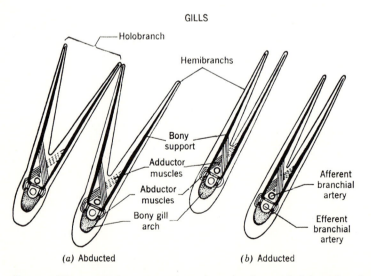

Fig. 8.2 Diagrammatic cross sections through adjacent holobranchs of a bony fish show-ing supporting and muscular elements which enable (a) abduction and (b) adduction. (Based on Bitjel in Grassé, 1958).

artery of each aortic arch of sharks, which to some extent persists in the chondrosteans, is replaced by a single efferent vessel in the teleosts. In all bony fishes, muscles move the bases of the hemibranchs so that the pressure in, and possibly also the flow of water from, the oral cavity is regulated (Fig. 8.2). Strong adductions of the two hemibranchs of a holobranch toward one another occur in "coughing," a violent sweeping of water over the gill lamellae to free them from accumulated detritus.

Although lacking a spiracle, rayfin bony fishes (Actinopterygii) retain a hyoidean pseudobranch (Fig. 8.3) that is free in some fishes but skin-covered in others. This portion of the gill region is the first to develop in the embryo and probably has an early respiratory function. In the adults of most bony fishes, however, the pseudobranch receives oxygenated blood through a secondarily established blood supply from the aorta and related vessels and also has a direct vascular connection to the choroid gland of the eye. Pseudobranchs are absent in the eels and their relatives (Anguilliformes), the elephantfishes (Mormyridae), the catfishes (Siluroidei), and featherbacks (Notopteridae), and other specialized groups in which the choroid gland of the eye is also missing.

Each filament of a pseudobranch has a thin cartilaginous supporting rod and blood vessels in its core, around which are arranged layers of pseudobranch cells and sinusoidal blood spaces. The large, granular pseudobranch cells, which dominate the epithelium, are covered at the outer edges by squamous pavement epithelial cells (Fig. 8.5). The blood space is supported by pilaster cells, whose thin extensions form the endothelial lining of the blood space. Within the pseudobranch cells are large numbers of mitochondria and tubular membrane arrays in close proximity to them. These

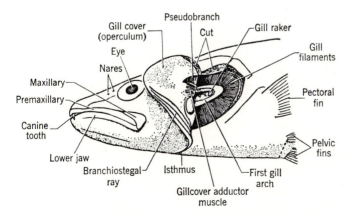

Fig. 8.3 Position of the pseudobranch as shown in the European walleye (*Stizostedion lucioperca*), family Percidae. (Source: Rauther in Bronn, 1940).

tubules are extensions of the plasma membrane and display secretory activity. The tubules may be the site of production of carbonic anhydrase, which facilitates the formation of carbonic acid from carbon dioxide and water.

Because of direct vascular connection with the eye, the high intraocular oxygen tension, the similarity of rete mirable-like capillary arrangement in the choroid gland, and the gas-secreting complex of the gas bladder which also contains much carbonic anhydrase, the pseudobranch is suspected to be instrumental in the metabolic gas exchange of the retina. The retina is the most oxygen-demanding tissue of the body. The pseudobranch may also have a functional relation, through the production of carbonic anhydrase, to the filling of the gas bladder; killifishes (*Fundulus*) are unable at least to refill the organ when their pseudobranchs are removed. Furthermore, certain deepsea fishes (*Gonostoma bathyphilum, Aphyonus gelatinosus*) with fat-instead of gas-filled gas bladders lack pseudobranchs although their close relatives with proper gas bladders have them. Because of its direct vascular connection with the choroid gland on the eyeball, the pseudobranch has also been implicated in the regulation of intraocular pressure.

Each gill filament bears many subdivisions or lamellae (Fig. 8.4) that are the main seat of gas exchange. In some fishes, such as the eels (Anguilliformes), acidophilic secretory cells occur at the bases of the lamellae. The free edges of the lamellae are extremely thin, covered with epithelium, and contain a vast network of capillaries supported by pilaster cells (Fig. 8.5). The relative number and size of the lamellae determine the respiratory area of the gills. The respiratory area varies greatly with the habits of fishes. Fast-swimming pelagic marine fishes such as mackerels (*Scomber*) and menhadens (*Brevoortia*) have not only more gill lamellae per millimeter of gill filament but also possess more than five times the gill area (in square millimeters per gram of body weight) found in sedentary fishes such as the toadfishes (*Opsanus*).

The uptake of oxygen from the respiratory water is furthered not only by the subdivision of gill filaments into lamellae but also by the direction of blood and water circulation. These circulations are really a counter-current system where the oxygen gradient of the water, flowing from the oral to the aboral side of the gills, decreases as the blood in the lamellae flows from aboral lamellar afferent to oral lamellar efferent blood vessels (Fig. 8.4.). The counterflow of blood and respiratory water maintains an even diffusion gradient across the gills for oxygen to enter and carbon dioxide to leave the fish. The efficiency of this counter-current system in the tench (*Tinca*) gives a mean utilization of 51 percent of the oxygen content of the water whereas experimental reversal of the direction of flow over the gills, from gill slit to mouth, gives a utilization of only 9 percent.

The aquatic respiratory cycle of operculate fishes (Osteichthyes) grossly

LAMELLAE ON GILL FILAMENTS

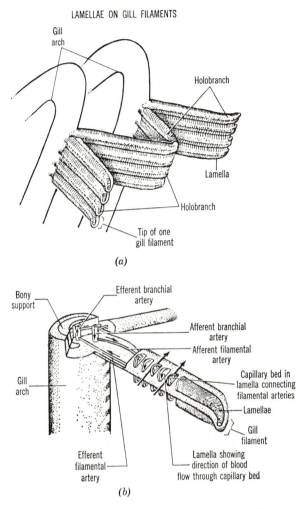

(a)

(b)

Fig. 8.4   Diagram (a) of several abducted gill filaments on two adjacent gill arches, and (b) the lamellae on a single enlarged gill filament. Small arrows show direction of blood movement and large arrows, water movement. (Based on van Dam, 1938).

resembles that of sharks and rays (Elasmobranchii) inasmuch as both suction and pressure propel the water through the gills (Fig. 8.6).

At the beginning of inspiration, just after the gill covers have closed forcefully, the mouth is opened while various muscles contract. Included in these muscles are the sternohyoid and the elevator of the palatine arc. At the same time, the branchiostegal rays are spread and lowered and the mouth cavity is enlarged to create negative water pressure in it. Water is thus

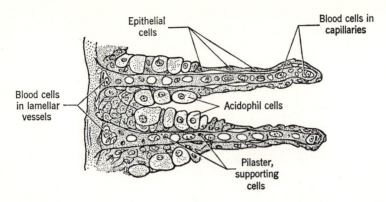

Fig. 8.5 Section through two adjacent lamellae of a gill filament. (Source; Keys and Willmer in Grassé, 1958).

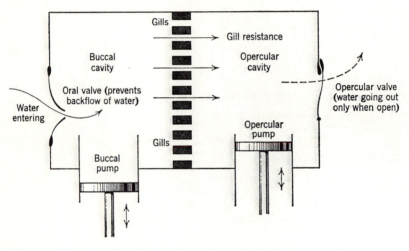

Fig. 8.6 Diagram of mechanism for gill ventilation by means of two pumps; the stage shown is shortly after the commencement of inspiration. (Based on Hughes, 1960).

drawn into the mouth and after a slight time lag the space between the gills and the operculum is enlarged as the gill covers are abducted anteriorly although the opercular skin flaps are still closed posteriorly by the outside water pressure. A pressure deficit is now also created in the gill cavity and the water flows over the gills. Then the buccal and opercular cavities begin to be reduced while the oral valves prevent the flow of water out of the mouth and the mouth cavity begins to function as a pressure-pump instead of a suction-pump. The operculum, with the opercular flaps still closed, has now reached its furthest state of abduction and water is accumulating outside the gills. At this point the opercula are quickly brought toward the body, the gill flaps open and the water is expelled, being prevented from flowing backwards by excess pressure in the buccal cavity as compared to the epibranchial cavity.

The foregoing basic respiratory process is modified according to different life habits. Fast swimmers including the mackerels (Scombridae), and the several trouts and salmons (Salmoninae) may leave their mouth and gill flaps open to bathe their gills with the stream of water displaced by swimming. In general, the gill cavities of swift-swimmers are smaller than those of sedentary fishes. Groundfishes, such as the flounders (Pleuronectidae) and the goosefishes (Lophiidae), have an enlarged and highly distensible opercular cavity; in them, the mouth does not open widely during inspiration but breathing movements are slow and deep. The propelling and friction-reducing effects of the expelled water may, by its jet action, at one time assist in locomotion (Chapter 6) or at other times counteract braking and reversing movements, especially among bottom-dwelling fishes which are prone to swim backwards more than others.

Forceful expelling of water through the mouth by using the breathing mechanism as a pressure pump with reversal of the normal direction of flow occurs in the bonefish (*Albula vulpes*) and goatfishes (Mullidae). When the fish finds a worm or mollusk on its sandy feeding grounds, the prey will be uncovered for feeding by a jet of water from the mouth through the rapid adduction of gill covers. Among the bottom dwellers of coral reefs, the morays (Muraenidae) leave the mouth open almost continuously while breathing. The opercular pumping apparatus is small and the branchiostegal mechanism greatly reduced and the respiratory water is mainly propelled by buccal pumping. Flow of water out of the mouth in the morays is prevented by means of small sphincters at the internal gill slits. The internal gill slits lead to a large flesh-covered gill chamber that ends in a very small external branchial aperture.

The rigid skeleton of such fishes as the trunkfishes (Ostraciidae) and the puffers (Tetraodontidae) restricts respiratory action of the gill cover. However, these fishes have a compensatory high rate of exchange of respiratory water brought about by rapid breathing (up to a hundred and eighty breathing movements per minute).

Fishes that make temporary oral attachment to stones of torrential streams, such as the loach-like Asiatic hillstream catfishes (Sisoridae and Akysidae), have solved the problem of water intake while attached either by suspending inhalation or by developing grooves protected by barbels that allow water intake with only a slight reduction in the force of suction. Ventilation is accomplished almost solely by opercular action. An extreme adaptation to breathing in torrential streams is found in the gyrinochelid (*Gyrinocheilus*) and in a catfish of the Andes (*Arges*). These fishes can take in and expel respiratory water through an inhalant slit in the horizontally divided gillcover while staying attached to the substrate with their suctorial mouths (Chapter 5). Larvae of some fishes have transitory "external gills" that protrude from the external gill slits or opercular region (Fig. 8.7); examples include the larvae of the South American lungfish (*Lepidosiren*) and the bichirs or reedfishes (*Polypterus*). In the reedfishes the blood supply of the larval external gills comes from the hyoid artery and the single gill fold is supported by a cartilaginous element; the entire structure is lost when the true gills take over at metamorphosis into the adult body form.

FISH BLOOD AS A GAS CARRIER

Oxygen diffuses very slowly from one liquid into another. Thus fishes, like other vertebrates, have evolved in their red blood cells a gas-carrying device of high efficiency. This device enables a fish to take up in one unit volume of blood the oxygen contained in 15 to 25 times the same volume of water. The red blood cells account for 99 percent of this uptake; the volume of oxygen carried in the plasma amounts to less than 1 percent of the total.

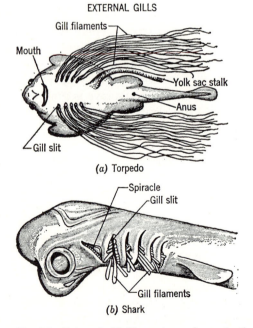

Fig. 8.7   External gill filaments as shown in the embryonic young of two elasmobranchs: (a) torpedo (*Torpedo marmorata*), ventral view; (b) spiny dogfish (*Squalus acanthias*), lateral view. (Based on Bolk, Göppert, Kallius, and Lubosch, 1937).

Hemoglobin is the respiratory pigment of fishes and other vertebrates and is in the red blood cells. An iron atom lies at the center of each group of atoms that form the pigment called heme, which gives blood its red color and its ability to combine with oxygen. Each heme group is enfolded in one of two or four chains of amino acid units that collectively constitute the protein part of the molecule, called globin. The hemoglobins of most vertebrates have molecular weights near 65,000. The range of oxygen carrying capacities is illustrated in Table 7.1; the highest of which is 16 volume percent as in a mackerel (*Scomber*) per 100 cubic centimeters of blood. In comparison, mammals average about 20 volume percent. It appears that habits and habitat, such as a bottom-dwelling versus a pelagic existence, have led to lower and higher hemoglobin contents of blood respectively in the examples in Table 7.1. This adaptation parallels those in respiratory anatomy, such as the gill surface per unit of body weight and the size of the gill cavity.

Inasmuch as oxygen (O_2) is taken up, transported, and released by the red blood cells, we may speak of the processes of "loading" and "unloading" of oxygen and of the respective tensions at which these processes occur. To have adequate measures for comparison, two stages are chosen: (1) T_l or T_{sat}—the loading tension of blood; that partial pressure of O_2 at which hemoglobin of a particular species is 95 percent saturated with oxygen; (2) T_u or $T_{\frac{1}{2}sat}$—the unloading tension of blood; that partial pressure of O_2 at which the hemoglobin is 50 percent saturated, or, in other words, the oxygen tension at which half the hemoglobin of the blood is in the oxygenated state and half is in the unoxygenated state. The half-saturation tension is a measure of the affinity of hemoglobin for oxygen. If hemoglobin has a low $T_{\frac{1}{2}sat}$ it has a high affinity and vice versa.

The oxygen dissociation curve (Fig. 8.8) describes the equilibrium of oxygen with hemoglobin. The curve may be hyperbolic, as shown for the eel (*Anguilla*) or sigmoid, as for humans. The shape of the equilibrium curve is influenced by the degree of interaction of the four polypeptide chains and their heme groups. The three-dimensional structure of the chains determines which amino acid residues will be at the surface of the molecules and available for subunit aggregation. Lack of interaction between hemes leads to a hyperbolic curve, whereas with cooperativity between hemes the curve tends to be sigmoidal. A hyperbolic curve with high oxygen affinity is characteristic of fish that can live in water with a low oxygen concentration, for example, the eel. The hemoglobin of the eel becomes saturated at much lower oxygen tensions than that of mammals whereas other fishes, such as the Atlantic mackerel (*Scomber scombrus*), resemble mammals fairly closely in this regard and require high oxygen tensions to achieve saturation. The difference between the T_{sat} and the $T_{\frac{1}{2}sat}$ determines the total amount of oxygen delivered to the tissues. Blood described by a sigmoid curve is able to deliver more

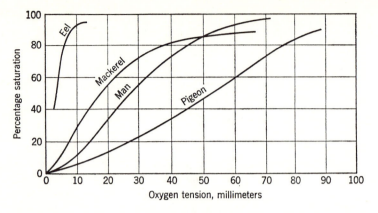

Fig. 8.8 Comparison of hemoglobin saturation curves as a function of oxygen pressure. (Based on Prosser and Brown, 1961).

oxygen to the tissues than when described by a hyperbolic curve. Thus, it is easy to see why an active fish like the mackerel has evolved hemoglobin described by a sigmoid dissociation curve.

As the partial pressure of carbon dioxide increases (often written as P_{CO_2} as in Fig. 8.8), higher oxygen tension is required to reach T_l, and T_u is lowered accordingly. This phenomenon, called, after its discoverer, the Bohr effect, is often more pronounced in fishes than in other vertebrates and facilitates the unloading of oxygen to tissue cells where the CO_2 tension is relatively high. The degree of protonation of carboxy and d-amino groups of the amino acids in the polypeptide chains is determined by both oxygenation and binding of CO_2 to these groups. Great differences in the Bohr effect have been found in different species (Fig. 8.9). The mackerel (*Scomber scombrus*) is a fish of the high seas that never encounters low levels of O_2 and lives under uniform but low CO_2 tensions. Its blood is very susceptible to small changes in carbon dioxide, whereas blood of the carp (*Cyprinus carpio*) and especially the bullhead catfishes (*Ictalurus*), which often live in stagnant waters, is relatively unaffected by changes in carbon dioxide levels. An increase in hydrogen (H) ion concentration (that is, lowering of values on the pH acidity-alkalinity scale) has similar effects to an increase in CO_2 tension on the oxygen-carrying capacity of hemoglobin. The shape of the oxygen dissociation curve of tuna hemoglobin changes with pH; cooperativity of the heme groups is lost as the pH drops from 9 to 6. When high CO_2 tensions prevail, even partial pressures of oxygen as extreme as 100 atmospheres cannot completely saturate the blood (the so-called Root effect). An increase in temperature raises the partial pressure of oxygen required to saturate the

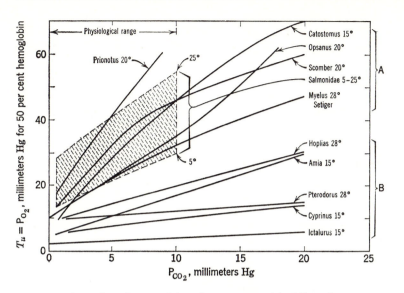

Fig. 8.9   The Bohr effect in fishes from oxygen-rich (A) and oxygen-poor waters (B). (Source: Fry in Brown, 1957).

blood; in addition, the absolute oxygen carrying capacity of fish blood is somewhat lower at higher than at lower temperatures.

Metabolism may be defined as the chemical changes in living cells by which energy is provided for vital processes and activities and by which new material is assimilated to repair the waste. The standard metabolic rate of fish is approximately equivalent to the basal metabolic rate measured in clinical studies of humans. The oxygen consumption rate of fish is a measure of their metabolism.

Oxygen consumption can be controlled several ways: (1) by the intensity of oxidative metabolism at the cellular level; (2) by the ventilation rate, which controls movement of water over the gills and hence the diffusion gradient across the gills; (3) by internal convection, that is the velocity of blood circulation and volume of blood brought to the gills; and (4) by long-term adjustments in size of respiratory exchange surfaces or the oxygen affinity of hemoglobin.

Oxygen-consumption rate is independent of the external oxygen concentration down to a critical oxygen concentration. Below the critical oxygen concentration or tension, oxygen uptake is dependent upon oxygen concentration down to the incipient lethal level below which the fish cannot survive indefinitely. The critical oxygen tension, T_c, varies with species. For the mummichog (*Fundulus heteroclitus*), T_c is 16 mm Hg whereas for the puffer (*Sphoeroides maculatus*) T_c is 100 mm Hg.

Oxygen uptake at rest varies with species—for the shark *Scyliorhinus* at

15°C oxygen uptake runs 54 cc oxygen/kg wet weight-hour, and for the California killifish (*Fundulus parvipinnis*) at 18–21°C, 130–230 cc oxygen/kg wet weight-hour. In general, oxygen consumption increases with rise in temperature up to some critical value, beyond which deleterious effects become evident and the rate falls off rapidly. Seasonal acclimation takes place in some fishes; hence, a Temperate Zone fish would have a higher rate of oxygen uptake at 15°C in the winter than in the summer.

The rate of oxygen consumption is ordinarily lower in the larger individuals of a given species. The relationship between oxygen consumption and body weight is an exponential function,

$$R = kW^x$$

where R = oxygen consumption in volume of oxygen consumed per unit time, k = a constant, W = body weight, and x = an exponent ranging in value from 0.6 to 1.0.

Other factors that may influence oxygen consumption include activity, age, reproductive state, nutrition, disease, and intrinsic regulatory mechanisms (nervous and hormonal control).

Increases in carbon dioxide tensions or decreases in pH lead to decreased oxygen consumption, but increased ventilation rate. Not only do the oxygen-transport facilities of the blood become less efficient at higher CO_2 tensions (Bohr effect) but warming of the water further leads to increased respiration because the tissues demand more oxygen at higher than at lower temperatures. This "vicious" circle of respiratory inflation (the harder the fish breathes the higher the cost in terms of oxygen requirements) may lead to fish mortalities due to asphyxiation. Especially at sites where human utilization of water contributes to high temperatures and high concentrations of carbon dioxide, such as near thermal and sewage effluents, die-offs of fishes may occur above the absolute lower lethal concentrations of dissolved oxygen.

Carbon dioxide is considerably more soluble in water than O_2 as shown by the fact that, at 15°C, 1 liter of CO_2 can dissolve in 1 liter of water. Thus the low amount of free carbon dioxide in natural waters favors waste gas elimination at the gills by diffusion.

Carbon dioxide in the venous blood of fishes is carried primarily as bi-carbonates but also in solution in the plasma. Carbamino hemoglobin carries about 7 percent of the carbon dioxide in venous blood of humans. Although carbamino hemoglobin may be present in some fishes (e.g., trout, *Salmo*), indirect evidence rules out the presence of carbamino hemoglobin in dogfish (*Mustelus*) and in carp (*Cyprinus*). The change of bicarbonates into CO_2 and water is catalyzed by the enzyme carbonic anhydrase, found in the acidophil cells of the gills, in red blood cells, and in other tissues.

ADAPTATIONS FOR AIR BREATHING AMONG FISHES

Fishes have evolved many respiratory adaptations for air breathing. Obligate or habitual air breathers occur in more than two dozen genera of some twenty diverse families of living fishes.

It is generally held that air breathing in fishes evolved in association with environmental oxygen deficiency; however, the Australian lungfish (*Neoceratodus*), the holostean fishes *Amia* and *Lepisosteus*, and many teleosts breathe air even though there is sufficient dissolved oxygen. Internal oxygen deficiencies resulting from higher metabolic activities may have also provided selective pressure for development of air breathing. Structural adaptations are of two main kinds, those not involving the gas bladder, and those in which the gas bladder serves as a lung.

Structures Other Than Lungs

Many tropical freshwater fishes and some brackishwater ones as well assume a temporary air breathing habit that enables them to leave the water in short periods or makes it possible for them to tolerate oxygen depletion in the water.

Listed below are air breathing adaptations that do not involve the gas bladder as a lung. Morphologically unmodified gills can secrete a protective slime cover that still permits gas diffusion in the Asio-african mastacembelid spiny eels (*Mastacembelus*). The interior wall of operculum and gill chamber is folded and highly vascularized in the mudskippers (*Periophthalmus*). Diverticula of the mouth and pharyngeal cavities (Fig. 8.10) develop in these fishes and in the snakeheads (*Channa*). The diverticula carry folded respiratory epithelium with a rich blood supply, usually from the afferent gill circulation.

The intestinal tract proper is modified into a thinwalled stomach that helps in respiration in such armored catfishes as the Loricariidae. In the loaches (Cobitidae) the middle and posterior portions of the intestine serve both digestive and respiratory functions. In some of the cobitid loaches the intestinal digestive phase alternates with a respiratory one at short intervals with all blood-forming and homeostatic structures taking part in adjustment to the phases. In others of these loaches the digestive tract is nonrespiratory during the winter and only assists in breathing under stagnant summer conditions.

The lining of the branchial cavity is extended into a long sac reaching the tail regions, with access to the gill cavity between the second and third gill arches in such a fish as *Saccobranchus*, a member of the Asian catfish family Saccobranchidae (Fig. 8.10). The gills are fused to form a valve over the ori-

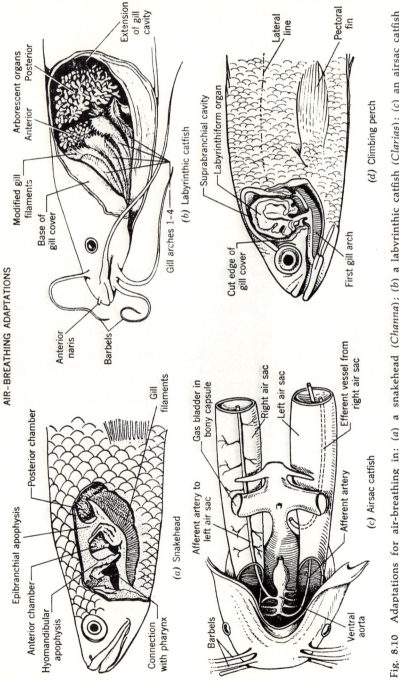

AIR-BREATHING ADAPTATIONS

Extension of gill cavity

Arborescent organs
Anterior   Posterior

Modified gill filaments

Base of gill cover

Gill arches 1-4

(b) Labyrinthic catfish

Anterior naris

Barbels

Lateral line

Pectoral fin

Suprabranchial cavity

Labyrinthiform organ

Cut edge of gill cover

First gill arch

(d) Climbing perch

Epibranchial apophysis

Posterior chamber

Anterior chamber

Hyomandibular apophysis

Gill filaments

Connection with pharynx

(a) Snakehead

Gas bladder in bony capsule

Right air sac

Left air sac

Afferent artery to left air sac

Efferent vessel from right air sac

Afferent artery

Barbels

Ventral aorta

(c) Airsac catfish

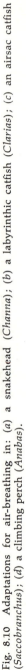

Fig. 8.10 Adaptations for air-breathing in: (a) a snakehead (*Channa*); (b) a labyrinthic catfish (*Clarias*); (c) an airsac catfish (*Saccobranchus*); (d) a climbing perch (*Anabas*).

234

fice of the air sacs. Blood is supplied from the branchial afferent artery V, hence the structure is not homologous to the lungs of amphibians, in spite of a certain superficial resemblance to them. Another instance of gill cavity enlargement is found in the Clariidae in *Clarias* and related catfish genera where an opening between the second and third gill arches leads to two sacs with highly vascular arborescent, bush- or shrub-like extensions on cartilaginous supports that originate from gills II and IV. The sac walls function in aerial respiration and the entire structure receives its blood supply from all four afferent branchial arteries; oxygenated blood is returned to the corresponding efferent branchial vessels (Fig. 8.10). The climbing perches (Anabantidae) have a slightly different air-breathing structure called a labyrinth inasmuch as it too is arborescent (branching). It is lined with folds of respiratory epithelium and is derived from the first gill arch. The labyrinth is situated in a moist pocket that takes up much of the preopercular dorsolateral head region (Fig. 8.10).

Most fishes with air-breathing tissues such as the foregoing can spend several hours out of water and may be seen creeping on dry land by means of their pectoral fin and/or their movable, spiny opercular covers. For example, a snakehead (*Channa*) can live out of water for more than 24 hours and be restored to normal life even after the skin has become dry. Some species must have access to atmospheric air even when water is accessible for breathing; whether their aquatic gills are insufficient or whether they must use their air-breathing tissue to rid themselves of carbon dioxide is not known.

Lungs

Although the gas-bladder lungs of the lungfishes (Dipnoi) are respiratory like the lungs of higher vertebrates, the homologies of the organs are not clear. However, most comparative anatomists believe that the lungs of the lungfishes and the gas bladder of higher fishes are related in evolution. In addition to respiration, the gas bladder has developed many functions in the diversification of fishes, such as gravity adjustment, hydrostatic equilibrium, sound production, and assistance in sound reception.

Only in the lungfishes (Dipnoi) has the gas bladder evolved structurally to resemble a lung as it exists in other vertebrates (Fig. 8.11). Internal septa, ridges, and pillars divide the air spaces into smaller compartments opening into a median cavity. Further subdivisions of these compartments by progressively smaller reticulated septa terminate in alveoli-like pockets richly covered with blood vessels. The short pneumatic duct and the lung parenchyma are richly supplied with smooth muscle, which is important for the mechanics of breathing and may exert additional influence in distributing the

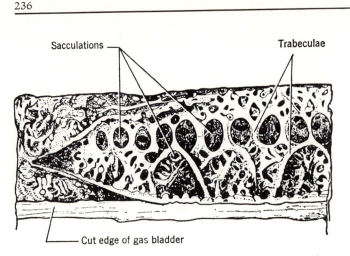

Sacculations

Trabeculae

Cut edge of gas bladder

Fig. 8.11 Sacculation as shown in a portion of the wall of the gas bladder in an African lungfish (*Protopterus*). (Based on Spencer in Bronn, 1940).

air inside the lung. The fine structure of the lung, as revealed by electron microscopy, does not differ basically from that in higher vertebrates.

Like the lungs of all other vertebrates, the lungs of the lungfishes originate as a median ventral evagination of the embryonic foregut. The Australian lungfish (*Neoceratodus*) has one lung that lies dorsal to the gut but maintains a pneumatic duct that opens into the ventral wall of the gut. In the African lungfishes (*Protopterus*) and in the South American lungfish (*Lepidosiren*) the lung is paired and bilobed but is ventral to the gut.

The blood supply to the lungs comes from the last efferent branchial artery (the fused third branchial artery in *Protopterus* and the fourth in the other lungfishes), and blood aerated in the lungs returns to the heart directly through a pulmonary vein into the left side of the atrium (Fig. 3.16). A "spiral" valve also divides the conus arteriosus and partially separates oxygenated and unoxygenated blood in its distribution to the gill vessels.

The African lungfish (*Protopterus annectens*) estivates in a cocoon of hardened slime that is brown and parchment-like. The animal rests coiled in the cocoon with the mouth upwards, its lips connected by a funnel to an air passage from the surface. An estivating lungfish commonly stays in its cocoon surrounded by hardened earth for the three to four months of the dry season, but encasement for over a year has been recorded. During estivation, metabolism is reduced and oxygen consumption falls accordingly. When the rains begin, the tube leading downward to the mouth of the fish fills with water, air can no longer reach the lungs, and the animal is "awakened" by incipient asphyxiation. The cocoon is ruptured subsequently and as the ground begins to be softened by the rain the fish can emerge and use its gills for the rest of the year till the next dry season. Presumably the South American lungfish estivates similarly but the Australian species survives only when in water. It neither burrows nor forms a cocoon.

THE GAS BLADDER

The gas bladder, less precisely designated as the "swim bladder" or the "air bladder," is characteristic of true fishes and reaches its fullest development as a hydrostatic organ among the spiny-rayed teleosts (Acanthopterygii). The gas bladder functions also as an accessory breathing organ, as a sound producer, and as a resonator in sound perception. There remains the possibility that the original function of the gas bladder was that of a lung. In certain deepsea mouthfish relatives (e.g., the gonostomatid *Yarrella*) the bladder may also serve as a fat-storage organ of possible buoyant function. In its weight-regulating (hydrostatic) function, gas secretion is accomplished through a special structure in the bladder wall, the gas-secreting complex with its rete mirabile or "wondernet" of circulatory vessels and a gas gland. In many rayfinfishes (Actinopterygii) gas resorption is also performed by a special structure, the oval organ situated in the posterior portion of the gas bladder, instead of through blood vessels over the entire bladder wall.

The Gas Bladder as a Respiratory Organ

Many physostomous fishes use the gas bladder as a temporary or supplementary organ of respiration. The giant redfish (*Arapaima gigas*) of the Amazon is one of the largest freshwater bony fishes and attains some 5 meters (16 feet) and 200 kilograms (441 pounds). It lives in swamps and has the gas bladder particularly well developed for aerial respiration (Fig. 8.11). The gars (*Lepisosteus*), the bowfin (*Amia calva*), some mormyrids (Mormyridae), and the feathertail (*Notopterus*) are additional examples of facultative air-breathers with a highly sacculated or alveolar lining in the gas bladder. These fishes can survive in water devoid of oxygen if enabled to swallow air which then can pass into the gas bladder via the pneumatic duct (and in some can pass out of the gas bladder by an opening at the anus—Fig. 8.12). The relative importance of air breathing by the bowfin which lives in Temperate regions of North America depends also on the temperature of the water. When water temperatures reach 20 to 25°C, the fish regularly ascend for air even in well-aerated water. Since the gas bladder in physostomous fishes always contains more carbon dioxide than the air, it has been suggested that elimination of this waste gas is performed there also, but the extent of this function is not known.

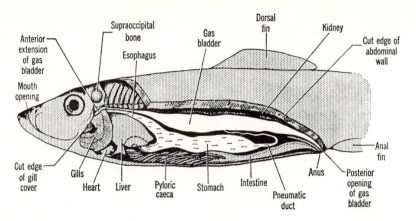

Fig. 8.12 Extension of gas bladder to regions of inner ear and anus in a herring (*Clupea*), as in several clupeids. (Based on Maier and Scheuring in Bronn, 1940).

The gas bladder is supplied with blood from the dorsal aorta as in the gars (*Lepisosteus*) or from the branchial vessels as in the bowfin (*Amia*). However, in most fishes the blood comes from the coeliacomesenteric artery; a dual supply of blood vessels is also common. Blood returns to the heart from the gas bladder by draining into one of the postcardinal veins, as in the gars, or through paired veins that become a common vessel before entering the sinus venosus, as in the bowfin.

The mudminnows (*Umbra*), facultative airbreathers of sometime stagnant waters, have both alveoli and gas absorbing and secreting organs in their gas bladders.

The Gas Bladder in Sound Reception

The density of fish flesh is so nearly equal to that of sea water that the animal would be virtually "transparent" to sound if it were not for bone and the gas bladder that provide acoustical discontinuities based on their difference of densities from that of water; both bone and gas bladder may act, therefore, as sound conductors or resonators. In certain groups of fishes the gas bladder may extend into the region of the inner ear and variations in pressure due to sound waves may be transmitted directly to the perilymph. Thus, in the cods (Gadidae) and the porgies (Sparidae), an anterior extension of the gas bladder touches the headbones near the sacculus of the inner ear. In the herrings (Clupeidae) extensions of the gas bladder grow into cartilaginous capsules (prootic and pterotic bullae) in close apposition to the perilymph spaces of the superior and inferior portions of the inner ear (Fig. 8.12); a similar condition is found in the cichlid *Etroplus*.

Transmission to the inner ear of changes in volume of the gas bladder is accomplished in a very specialized manner in the Order Cypriniformes (=Ostariophysi). A large number of predominantly freshwater families are included here. They have in common a peculiar set of small paired bones or ossicles that constitute the Weberian apparatus (Fig. 3.25) which connect the gas bladder with the ear. These ossicles are derived from the apophyses of anterior vertebrae; the hindmost of them, the tripus, touches the anterior wall of the gas bladder and is connected with a ligament to the next bone, the intercalare, or, when this is missing, to the scaphium which, in turn, is attached to the minute claustrum. The claustrum of each side touches a membranous window of the sinus impar that lies in the basioccipital bone of the head and is an extension of the perilymph system of the inner ear. If the claustrum is absent as in the gymnotids (Gymnotidae), the scaphium touches the membranous window of the sinus impar instead. Volume changes of the gas bladder cause the Weberian ossicles to move in such a manner that pressure changes are transmitted to the perilymph and thence to the sensory cells of the inferior portion of the labyrinth which is the seat of sound reception. In some species, however, the gas bladder is encased in a capsule of bone or connective tissue; in these fishes rhythmic compression of the gas bladder momentarily lessens rather than augments the pressure on the perilymph.

Among the cypriniform fishes, the shape and size of the gas bladder varies with the family. In the minnows (Cyprinidae), for example, the gas bladder is two-chambered anteroposteriorly with a constriction between the chambers formed by a sphincter-like accumulation of the smooth muscle on both portions of the bladder near the constriction. In predominantly bottom dwelling cypriniform groups such as the loaches and their relatives (Homalopteridae, Gastromyzonidae, and Cobitidae), the posterior portion of the gas bladder almost disappears and the anterior portion lies in a bony capsule and is cushioned by a jelly-like fluid. In some catfishes (Sisoridae, for example), only an anterior bladder chamber remains which may be encapsulated and displaced toward the outside so as to be in touch with the skin through a lateral opening in the capsule. The mechanical arrangement here as a sound registering structure reminds one of an old fashioned, gramophone pickup of the vibrating-membrane type.

The cypriniform fishes have a wider range of sound perception and better sound discrimination than fishes that lack the Weberian apparatus. Elimination of the gas bladder in the minnows (Cyprinidae) greatly reduces the auditory range. In fishes that do not have the Weberian ossicles, bony elements of the pectoral girdle or the skull may abut the gas bladder. In the squirrelfishes (Holocentrus) and the triggerfishes (Balistes), these bony connectives probably play a role in sound conduction.

Many fishes are thought to be sensitive to changes in atmospheric pressure. A loach (*Misgurnus fossilis*) is sensitive to such changes. However, it has not yet been demonstrated whether such small but gradual pressure changes are perceived through the gas bladder, the Weberian ossicles, and the inner ear or whether the fish reacts to changes in other variables which usually attend a rise or fall in barometric pressure such as rain or the cessation of rain, and changes in dissolved oxygen. For a long time there has been a controversy as to whether or not the gas bladder and the inner ear together make a fish sensitive to small but rapid changes in water pressure. However, several species of minnows (Cyprinidae) can detect rapid changes as small as a few centimeters of water pressure. The participation of the inner ear in sensing this kind of pressure change is ruled out by eliminating the organ through a delicate operation; thus the gas bladder wall is implicated as the sensor.

The Gas Bladder in Sound Production

The concept that fishes are mute has been dispelled by hydrophone recordings at many depths where many different species of fishes and some of their invertebrate food organisms have been shown to produce sound. Only a few hundred out of the more than 20,000 fish species have so far been clearly identified as sound producers. Three general types of sonic mechanisms are present in fishes: (1) *stridulatory*, produced by friction of teeth (e.g. grunts, Pomadasyidae), fin spines, or bones; (2) *hydrodynamic*, resulting from swimming movements, especially from rapid changes in direction or velocity; and (3) *gas bladder*. The gas bladder is directly involved in making sounds in such fishes as the drums (Sciaenidae), the grenadiers (Melanonidae), and the gurnards (Triglidae) where striated muscles originating on the dorsal body wall and inserted on the gas bladder wall vibrate (Fig. 8.13). In the toadfishes (Batrachoididae) striated muscles in the bladder itself produce rapid volume changes resulting in audible signals. The sound-producing muscular gas bladder of the toadfishes is so self-contained that it can be taken out and made to produce sound by electrical stimulation. Sounds produced by the gas bladder usually have low frequencies, whereas sounds made by the teeth or other bony parts tends to be higher-pitched.

In most species where the gas bladder emits sounds either through its own special modification or through adjacent structures, the sounds are significant in breeding behavior or in the defense of territories. However, sounds of problematical significance can be elicited (mostly under stress) from many more species of fishes than have been shown to have elaborate breeding or territorial behavior. Thus, for instance, highly excited minnows (Cyprinidae), loaches (Cobitidae), and eels (Anguillidae) produce high-pitched sounds by release of air from the bladder through the pneumatic duct. Herrings

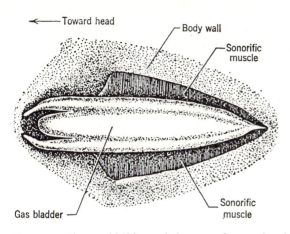

Fig. 8.13   The gas bladder and the sonorific muscles that vibrate it for sound production in the spotted sea trout (*Cynoscion nebulosus*), a member of the drum family (*Sciaenidae*). (Source: Tower in Bolk, Göppert, Kallius, and Lubosch, 1937).

(Clupeidae) in nets have been heard to expel gas from their gas bladders through a duct that ends at the anus (Fig. 8.12). The circumstances under which these sounds are produced implicates them as warning sounds. In the deepsea fishes, species-specific sounds may be unusually significant because light other than bioluminescence is absent.

The Gas Bladder as a Hydrostatic Organ

The density of fish flesh of about 1.076 is greater than that of water which is approximately 1.0005 for fresh, and 1.026 for sea water. In order to be completely weightless and to expend a minimum of energy in maintaining position, the fish may lay down fats and oils in muscle and liver or it may use a gas inclusion to reduce its overall weight. The gas bladder of bony fishes (Osteichthyes) is such an organ and has, as one of its functions, that of bringing the overall density of a fish closely to that of the surrounding water. Lampreys and hagfishes (Cyclostomata), the sharks, rays, and their relatives (Elasmobranchii), and the living coelacanth lobefin (*Latimeria*) do not have gas bladders. The sharks and rays are suspected to regulate body buoyancy by means of adjusting the "water ballast" contained in the body cavity and operated through their abdominal pores. The gas bladder among fishes may make up between 4 and 11 percent of the body volume, 4 to 6 percent for marine species and 7 to 11 percent for freshwater ones; the difference being accounted for by the density differences of the two media.

   The division of fishes into physostomous (bladder with opening) and physoclistous (bladder closed) has a functional as well as a morphological basis. There is a gradual, not an abrupt, change from one to the other condition as concerns gas secreting and resorbing structures and in the adults of many physostomous species the gas bladder loses the pnuematic duct that connected it to the outside in the juveniles or young. This condition is called paraphysoclistous and is found, for example, in the mid-depth lanternfishes (Myctophidae). Nevertheless the soft-rayed fishes (Malacopterygii) are prevalently physostomous and the spiny-rayed ones (Acanthopterygii), physoclistous. The truly physoclistous teleosts, with perch-like fishes (Perciformes) as the typical example, adjust the pressure in their gas bladders entirely through secretion or resorption of gases from or to the blood.

The position of the gas bladder in relation to the center of gravity of a fish is important in swimming and maintaining its position (Chapter 6). The yellow perch *(Perca flavescens)* for example has the center of gravity below the gas bladder, an arrangement that tends to keep the fish effortlessly in the normal swimming position.

The upside-down catfish *(Synodontis batensoda)* uses a displaced gas bladder to insure the unusual swimming position of the species.

The Filling and Emptying of the Gas Bladder. Although in most natural waters and the arterial blood of fishes, the partial pressures of oxygen and nitrogen are about 0.2 and 0.8 atmospheres, respectively, the partial pressures of oxygen and nitrogen in the bladder may be 100 and 20 atmospheres, respectively. This ability of the bladder to concentrate O_2 and N_2 some 500 and 30 times, respectively, is the unique property of the organ.

The gas bladder in most physostomes must be filled by gulping air at the time the yolk sac is lost. Although these fishes are capable of gas secretion and absorption by means of the blood supply when they are adult, they are unable to fill the gas bladder initially without access to the atmosphere; examples include the trouts and salmons (Salmoninae). Many physoclists have pneumatic ducts in larval life and also must gulp atmospheric air for the first filling of the gas bladder; examples include the sticklebacks (Gasterosteus), the guppy (Lebistes), and the seahorses (Hippocampus). Fossil evidence in the lobefins (Crossopterygii), the larval pneumatic duct among physoclists, and the prevalence of the true physoclistous condition among many of the highest teleosts have been interpreted to mean that early fishes were physostomes rather than physoclists.

Any deepsea forms that have retained a functional gas bladder and presumably spend their entire lives in deep water, such as the grenadiers (Melanonidae) of the abyss, must have evolved a different mechanism for the initial filling of the gas bladder unless they have an early life stage in the upper waters accessible to air, like most lanternfishes (Myctophidae).

Adjustment to varying hydrostatic conditions such as would be encountered when a fish with a gas bladder undertakes vertical migrations implies the existence of structures that increase or reduce gas bladder pressure and volume according to the needs of the animal. Fish are able to vary the gas content in such a way that the gas volume is almost constant regardless of the hydrostatic pressure. Boyle's law, which establishes that the volume of a gas changes inversely with pressure, naturally also applies to the gas bladder. Since gas secretion and absorption are physiological processes, there are limits to the speed with which fishes with gas bladders can change their position in depth. The European perch (*Perca fluviatilis*) for instance, at 20 meters (66 feet) can, without discomfort rise, and sound rapidly only between 16 and 24 meters, that is, the limit of quick vertical movement for this species is 20 percent of the total depth either up or down from its own level. The much more spectacular nightly vertical migrations to the ocean surface of the lanternfishes (Myctophidae) and other mid-depth species, encompassing 400 meters or more, are accomplished by the pronouncedly enlarged and highly efficient gas secreting and resorbing structures of their gas bladders.

Another familiar occurrence illustrates the limitations that a closed gas bladder places on the vertical movements of a fish. Anyone who has hooked a fish from a depth of more than a few meters will have noticed the everted stomach when it is hauled into the boat. The explanation is that the gas bladder expanded rapidly on the way up, far more rapidly than the capacity of the gas-resorbing mechanism allowed, and a portion of the digestive tract was pushed out of the mouth of the fish.

The transfer to the lumen of the gas bladder of the gases carried in the blood is accomplished through highly vascular regions on the wall of the bladder. Parts of the bladder wall together with special blood vessels make up the two parts of this gas-secreting complex (Fig. 8.14); these two are (1) the gas gland and (2) rete mirabile. The gas gland is a region of bladder epithelium, which may be one-layered, folded, or composed of stratified epithelium several cell layers thick. Here, small blood vessels from the underlying rete mirabile reach the epithelial cells. The rete mirabile, or "wonder-net," is composed of many capillaries and venules placed perpendicularly to the bladder wall. In physoclist bladders the rete and the secretory epithelium have positions resembling the stem and hat of a mushroom. The blood is supplied from the dorsal aorta through a branch of the coeliaco-mesenteric artery and is carried off by a branch of the renal portal vein. In an eel (*Anguilla*) over 100,000 arterioles and a slightly smaller number of venules give the rete a total surface of over 2 square meters where blood vessels are opposed to one another. The arterial blood to the bladder and the venous blood from the bladder are in intimate diffusional contact with each other. Systems of this kind, called countercurrent multipliers, have the capacity of building up concentration differences of various substances from one end of

GAS BLADDER

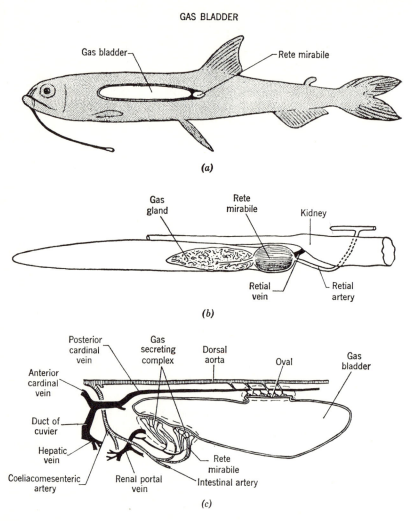

Fig. 8.14   The gas bladder: (a) position of bladder and rete mirabile in a deepsea snaggle-tooth (*Astronesthes*); (b) details of the gas bladder in *Astronesthes*; (c) generalized blood supply of the gas bladder in physoclistous bony fishes. (a and b based on Marshall, 1960, and c, on Goodrich, 1930).

the organ to the other (Fig. 8.15). In the rete, it is mainly oxygen which is built up, along the capillaries, to very high tensions at the bladder wall.

Experiments where an isotope of oxygen was traced from the outside water into the gas bladder show that the rete is capable of concentrating oxygen in the epithelium of the gas gland at far higher tensions than those that per-sist in the water around the fish. The epithelial cells of the gas gland then

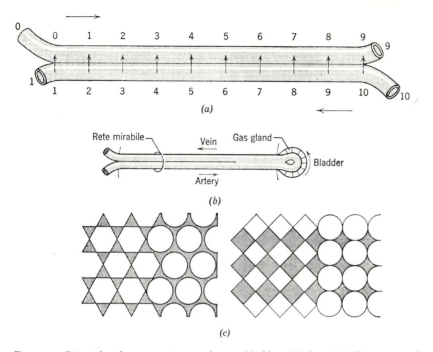

Fig. 8.15 Principle of gas secretion in the gas bladder of fishes. (*a*) Illustration of counter current principle. (*b*) One counter current capillary in the rete mirabile. (*c*) Arrangement of blood vessels in rete mirabile. (Cross section). (Source; Scholander, 1957).

transfer the dissolved gases from the capillaries into the gas bladder. The transfer process rests on the intracellular formation of minute gas bubbles that enter the bladder. In addition, cells of the gas gland are impermeable to dissolved gases although they can secrete bubbles; they can therefore act as a barrier to gas diffusion, preventing gas movement out of the gas bladder into capillaries of the epithelium and the rete.

For example the whitefish (*Coregonus*) from deep water has almost pure nitrogen in its gas bladder. Such gas concentrations that differ radically from those of the blood speak further for active cellular gas secretion from the gland into the gas bladder rather than for accumulation of gas by dissociation of oxygen from arterial blood in the rete capillaries.

Fishes from deeper waters in general fill the gas bladder mostly with oxygen; as searobins (*Trigla*) are lowered from shallow into deeper water the percentage of oxygen in their gas bladders increases, from 16 percent at 1 meter to 50 percent at 8 meters; at 175 meters, 87 percent oxygen occurs in the gas bladder of a conger eel (*Conger*).

Resorption of gas from the gas bladder is accomplished in a number of ways:

a. Through diffusion into blood vessels all over the gas-bladder wall, outside those of the gas-secreting complex, as in the sauries (Scombresocidae) and the killifishes (Cyprinodontidae).

   b. Through a specialized resorbent capillary network, sometimes connected with the gas-secreting complex (Fig. 8.16), in such a manner that it is expanded and fully active when the fish rises. At that time the gas gland is collapsed to allow for rapid resorption of gas and subsequent release thereof through the gills. When the fish descends the gas gland expands and the resorbent capillary network collapses so as to enlarge the area of contact of the gas gland for rapid secretion of gases into the bladder, as shown in the hatchetfishes (Sternoptychidae).

   c. Usually, but not invariably, to the rear of single-chambered gas bladders or in the posterior portion of two-chambered ones there is in the bladder wall a thinned region where a complex of capillaries is separated

GAS BLADDER STRUCTURES

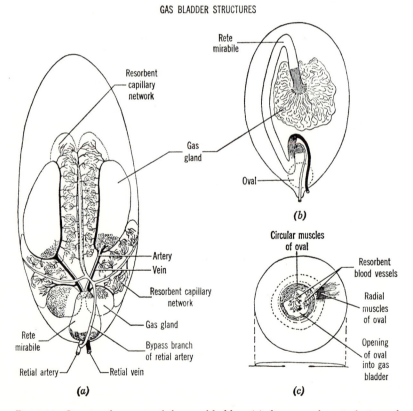

Fig. 8.16   Structural aspects of the gas bladder: (*a*) diagram of ventral view of gas bladder in a deepsea mouthfish (*Maurolicus*) with detail of blood supply; veins are black, arteries, white; (*b*) the gas gland in a deepsea berycoid (*Melamphaes*) as seen in ventral view and (*c*) the oval, in detail. (Source: Marshall, 1960).

from the bladder lumen only by a one-cell layer of epithelium. This area of blood vessels which come from the intercostal arteries and drain into the cardinal veins is called the oval organ because it is usually surrounded with a sphincter that controls the rate of gas resorption by expanding or contracting an oval orifice to the region of the capillaries (Fig. 8.16). The oval organ is typically found in the cods and their relatives (Gadiformes) and in spiny-rayed fishes (Acanthopterygii). Embryologic evidence points to a homology of the pneumatic duct of physostomes with the oval organ of physoclists.

The gas bladder is innervated by branches from the vagus nerve and from the coeliac ganglia. Nerve ends have been observed in the reabsorbent region, the oval, the rete, and in the secretory epithelium. Near the heart pole of the rete there are numerous large ganglion cells. The muscular layer of the gas bladder wall is also well supplied by nerves. The functioning of this nerve supply is not yet understood completely. It appears that reabsorption of gas is stimulated by catecholamines, whereas deposition is increased by the absence of such stimuli.

The Gas Bladder in the Distribution and Ecology of Fishes.   The gas bladder has, throughout the evolution of fishes, assumed the various functions which were previously discussed, but in present-day fishes the most wide-spread function of the organ is to regulate buoyancy. Gas inclusion and weight reduction may in some habitats be a liability rather than an asset. For example, fishes of swift streams must expend extra energy in maintaining their preferred position, if they are of equal density with the water instead of being heavier. Many stream darters (Etheostomatinae) have lost the gas bladder whereas a lacustrine member of the same percid subfamily, the logperch (*Percina caprodes*), retains the organ. The loaches of southern Asia (Cobitidae) that inhabit rapid streams in the hills as compared to those in lakes have lost only the posterior portion of the tripartite gas bladder typical of the group; the anterior two chambers that are connected with the inner ear persist in all loaches and are instrumental in hearing (Chapter 11). These relations suggest that evolutionary adaptation to strong current repeatedly and independently has led to the loss of the gas bladder.

Perhaps the most striking examples for the loss of the gas bladder on the one hand and its modifications, including extreme development, on the other, come from fishes in the deepwater zones of the oceans. Though sampled only here and there, in consequence of the difficulties of deepsea trawling, there are indications that small mesopelagic fishes between 200 and 1000 meters are indeed extremely numerous. They are in fact so numerous that they are largely responsible for producing the sound scattering layers in the ocean that can be tracked by sonar. The sound scattering layer, like the fishes, is found closer to the surface at night and at far greater depths during the day. The echo has the characteristics of a reflection from small gas bubbles and it is believed that it is the gas bladders of the fishes that reflect the sound waves.

The best known representatives of the mesopelagic species-complex of fishes are the lanternfishes (Myctophidae); they make long nightly vertical migrations to the upper waters of the ocean in order to feed on plankton. Their predators, such as the deepsea viperfishes (*Chauliodus*) and deepsea swallowers (*Chiasmodon*), naturally also undertake extensive vertical migrations. Many mesopelagic species have highly efficient and greatly enlarged gas secreting complexes and resorbent capillary networks or else oval organs. In fact, the gas bladders of mesopelagic fishes show better than those of any other group that the rete mirabile and the gas gland are two distinct parts of the gas-secreting complex. The relative extent of the gas secreting complex on the gas bladder of such very active surface swimmers as the flyingfishes (Exocoetidae) is one-tenth the area of the bladder surface but in the lanternfishes (Myctophidae) and other mid-depth groups it is far greater in extent. It is these enlarged structures for gas exchange that make possible the extensive migrations of mid-depth fishes, often extending over the range from 3 to 400 meters, upward at night and down again in the morning.

In certain fishes, such as the Melamphaeidae, from the lower mesopelagic zone (500 meters and below), gas bladders become filled with fat; they probably do not migrate as extensively as some other groups such as the lanternfishes (Myctophidae) and though fat is lighter than fish muscle or bone, a fat filled space of gas-bladder size cannot greatly contribute to lighten the fish as a whole.

As one proceeds farther down in the ocean, to the bathypelagic zone between 1000 and 4000 meters, one encounters less favorable conditions for life. Food is sparse and consequently so are the fishes themselves. The latter show the typical adaptations to unfavorable trophic (feeding) conditions. The gas bladder disappears, bone is reduced, and the muscles are flabby and watery in consistency. Bathyscaphe observations show the animals to be lethargic and to move about but little as compared to the highly active, darting habits of fishes of the mesopelagic zone. Due to the pressure of the overlying water column at these great depths, gases have become noticeably denser than at the surface and a gas bladder would have to be relatively large to be of value unless excessive pressures could be generated in it; the filling and maintenance of pressure in such a gas bladder would require much work and therefore abundant food as a source of energy. Evolutionary adaptation has taken another path and lightened the bone and the flesh of fish rather than provide for an inclusion of gas.

Finally, on arriving at the ocean bottom in the bathyal, abyssal, and hadal realms, anywhere between 2000 and 5000 meters or beyond, there are again zones of food concentration, this time in the form of benthic invertebrates and fishes that ultimately live on the accumulated products of the surface zone. Among the fishes, gas bladders reappear and are found in 50 percent of the limited number of species so far known. The fishes uniformly also lay down heavier bone and have firmer flesh than those which swim above them. The blood vessels of the rete mirabile of deepwater benthic species are the longest known, in relation to the size of the gas bladder. The elongation of this portion of the gas-secreting complex is not surprising if one remembers that the bladder-filling function relies on a countercurrent principle, increasing in efficiency as the length of the opposing vessels also increases. Conjectures are that a gas bladder assists many of these deepest water, bottom fishes in maintaining themselves very slightly over the bottom as they search for food.

Physoclistous fishes predominate in the marine environment where even mid-depth families such as the Argentinidae and Microstomidae, relatives of freshwater physostomes such as the salmons (Salmonidae), have gradually lost the original connection between gas bladder and esophagus. Many physoclists are found among the freshwater fishes but physostomes with well developed pneumatic ducts predominate in the flowing-water portion of the freshwater fish fauna.

SPECIAL REFERENCES ON RESPIRATION

Black, E. C. 1951. Respiration in fishes. *Univ. Toronto Stud. Biol. Ser.*, 59 (Publ. Ont. Fish. Res. Lab. 71): 91–111.

Dendy, L. A., R. L. Deter, and C. W. Philpott. 1973. Localization of Na^+, K^+-ATPase and other enzymes in teleost pseudobranch. I. Biochemical characterization of subcellular fractions. *J. Cell Biol.*, 57: 675–688.

Dendy, L. A., C. W. Philpott, and R. L. Deter. 1973. Localization of Na^+, K^+-ATPase and other enzymes in teleost pseudobranch. II. Morphological characterization of intact pseudobranch, subcellular fractions, and plasma membrane substructure. *Jour. Cell. Biol.*, 57: 689–703.

Fry, F. E. J. 1947. Effects of the environment on animal activity. *Univ. Toronto Stud. Biol. Ser.* 55 (Publ. Ont. Fish. Res. Lab. 68): 1–62.

Hughes, G. M. 1960. A comparative study of gill ventilation in marine teleosts. *Jour. Exptl. Biol.*, 37 (1): 28–45.

Hughes, G. M., and M. Morgan. 1973. The structure of fish gills in relation to their respiratory function. *Biol. Rev.*, 48: 419–475.

Johansen, K. 1970. Air breathing in fishes. *In* Fish Physiology, Academic Press, New York. 4: 361–411.

Jones, F. R. H., and N. B. Marshall. 1953. The structure and function of the teleostean swimbladder. *Biol. Rev. Cambridge Philos. Soc.*, 28: 16–83.

Leiner, M. 1938. Die Physiologie der Fischatmung. Akademische Verlagsgellschaft, Leipzig.

Maren, T. H. 1967. Carbonic anhydrase: chemistry, physiology, and inhibition. *Physiol. Rev.*, 47: 595–781.

Randall, D. J. 1970. Gas exchange in fish. *In* Fish Physiology, Academic Press, New York. 4: 253–292.

Steen, J. B. 1970. The swim bladder as a hydrostatic organ. *In* Fish Physiology, Academic Press, New York. 4: 413–443.

van Dam, L. 1938. On the utilization of oxygen and regulation of breathing in some aquatic animals. Drukkerij Vollharding Groeningen, Netherlands.

Wittenberg, J., and B. A. Wittenberg. 1961. The secretion of oxygen into the swim-bladder of fish. II. The simultaneous transport of carbon monoxide and oxygen. *Jour. Gen. Physiol.* 44 (3): 527–542.

Woskoboinikoff, M. 1932. Der Apparat der Kiemenatmung bei Fischen. *Zool. Jahrb. Abt. Anat. u Ontog. Tiere,* 55: 315–488.

9

Excretion and
Osmotic Regulation

OSMOREGULATORY AND EXCRETORY ORGANS
(FIGS. 3.15 AND 3.18)

Vertebrates eliminate some metabolic wastes through the gut and the skin, but most are eliminated through special excretory organs, the kidneys. In elimination fishes and other aquatic animals have a particular problem in that their gills and oral membranes are permeable both to water and salts. In the ocean the salinity of the water is more concentrated than that of the body fluids of the fish, and water is drawn out, but salts tend to diffuse inward; hence marine fishes drink sea water. In contrast, in fresh water, fishes lose salts and take up water through the gills because their internal salt concentration is greater than that of their surroundings (Fig. 9.1).

Many nitrogenous wastes of fishes pass through the kidneys that also assist in water-salt balance (homeostasis) by the excretion or retention of certain minerals. The gills also take a prominent part in waste excretion, eliminating mainly ammonia.

The typical fish kidney is made up of many individual units or nephrons, each consisting of a renal corpuscle (Malpighian body; Fig. 9.3) and a kidney tubule. The tubules join in collecting ducts that finally lead to the outside through the mesonephric duct with its various terminal modifications. The Malpighian body is made up of a glomerulus, a blood vessel tightly coiled

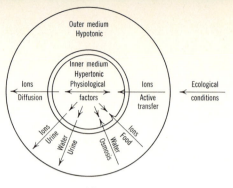

(a)

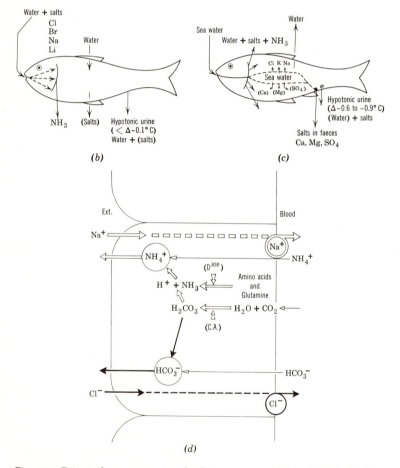

(b)

(c)

(d)

Fig. 9.1 Principal processes involved in osmoregulation in (a) freshwater fishes, (b) inflow and outflow of salts and water in a freshwater fish, and (c) osmotic regulation in marine bony fishes. (Sources: (a), Wikgren, 1953; (b) and (c), Baldwin, in Brown, 1957). (d) Schematic representation of ionic exchanges in the branchial cell of the goldfish, *Carassius auratus*. D[ase], deamidation and deamination enzymes; (C. A.) carbonic anhydrase; active transport of ions requiring energy. (Source: Maetz and Garcia Romeu, 1964).

with afferent and efferent arterioles and encapsulated by thin kidney cells (Bowman's capsule). Some teleosts have kidneys with frequent small glomeruli or a few medium-sized glomeruli; others show a tendency for the glomeruli to become smaller, less frequent and poorly vascularized as in the horned sculpins (*Myoxocephalus*), whereas still others have very few glomeruli, and these are nonfunctional in their adult stages; for example, the goosefish (*Lophius*) loses the connection between glomeruli and kidney tubules when the fish becomes mature. Finally there are a fair number of marine fishes, including Antarctic bony fish, which are lacking glomeruli entirely (for instance, the toadfishes, *Opsanus*). The glomerulus and the capsule together act as ultrafilters where blood pressure brings about nonselective dialysis of molecules up to a molecular weight of about 70,000; thus, serum proteins are retained in the blood. The excretory fluid undergoes alteration on its way through the tubules where glucose, various minerals, other solutes, and in some cases (sculpins, Cottidae) water are reabsorbed into the blood by an energy-requiring process. Presence of antifreeze glycoproteins and aglomerulism of the kidneys probably played important roles in permitting Antarctic fish to adapt to near freezing water temperatures. Since these small molecular weight glycoproteins are not filtered into the formative urine, the fish avoid the energy expenditures that would be necessary for their tubular reabsorption. Control of filtration and reabsorption takes place through hormone action; in fishes the adrenal cortex, thyroid, suprarenal bodies, gonads, hypothalamus, and perhaps the pituitary are involved in this regulation (Chapter 11).

Emphasis here will be placed on functional aspects of nitrogen excretion and related processes; freshwater, marine, and diadromous (migrating to and from the sea) fishes are treated separately—as are sharks, skates, and rays (Elasmobranchii)—because they have developed a high urea tolerance through which they achieve osmotic homeostasis differently from other fishes.

Freshwater Lampreys and Bony Fishes

The osmotic pressure of body fluids depends on their mineral and organic compound content. In all freshwater fishes this pressure is higher than that of the surrounding water and there is a tendency for water to diffuse into the animal wherever there is a water-permeable membrane, that is, the gills, the oral membranes, and the intestinal surface. The skin and its mucus greatly reduce water permeability although small amounts of water do enter a fish through the skin. Freezing point depression, Δ, of the blood of freshwater lampreys (some Petromyzonidae) is $-0.38°$ to $-0.46°C$, and that of freshwater teleosts varies around $-0.57°C$ when Δ of the outside water is virtually zero.

To cope with the steady inflow of water resulting from the differences in tonicity between the internal and external media, freshwater fishes produce a copious and highly dilute urine that is hypotonic with regard to the fish. Lampreys in fresh water have a urine flow that normally varies between 15 and 36 percent of the body weight per day. The comparable measurements for freshwater bony fishes (Osteichthyes) range between 5 and 12 percent. The urine has a freezing point depression (Δ) of around $-0.025°C$ for freshwater fishes; higher values of Δ are rare, although $-0.09°C$ has been recorded for an African lungfish (*Protopterus*). The main work of the kidney in freshwater fishes, therefore, lies in water excretion; certain nitrogenous compounds, usually amounting to only a fraction of the total excreted nitrogen, also pass to the exterior by way of the kidney and its associated ducts. The kidney further functions in the retention of sugars and other vital solutes. Some salt losses occur since fishes in fresh water are hypertonic to the outside. These losses are made up by selective absorption of salt through the gills against the natural diffusion gradient.

Kidneys of Freshwater Fishes. Larvae of the lampreys (Petromyzonidae) have functional pronephric kidneys until they are 12 to 15 millimeters long. The nephrostomes have funnels that open into the pericardial coelom. The composition of body fluids is determined by permeability of the body surface and by the developing gill mechanism for the regulation of ionic equilibrium. Coelomic fluids are swept into the nephrostome, at the site of the glomus, the circulatory component of the pronephric tubule; substances to be conserved are reabsorbed into the blood and the remaining aqueous filtrate passes by duct to the urogenital sinus (Fig. 9.2). In metamorphosis of the larvae, the mesonephric kidney takes over and the pronephric openings into the coelom disappear. Arterial blood is filtered through the glomerulus without selective retention of solutes, while the convoluted tubule is instrumental in reabsorbing salts and other substances (Fig. 9.3).

With some exceptions, such as the bowfin (*Amia*) and the gars (*Lepisosteus*) which have nephrostomes opening into the anterior coelom (Fig. 9.4), adult freshwater rayfin fishes (Actinopterygii) have a mesonephric kidney closed

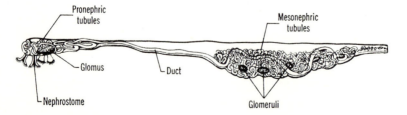

Fig. 9.2 Kidney of a lamprey (*Petromyzon*) larva of 22 millimeters showing relationship of pronephric and mesonephric portions. (Based on Wheeler in Bolk, Göppert, Kallius, and Lubosch, 1938).

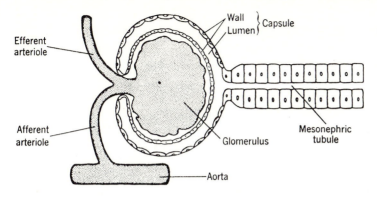

Fig. 9.3 Diagram of a renal corpuscle (glomerulus plus capsule) at the end of a uriniferous tubule as in a fish mesonephros. (Source: Smith, 1953)

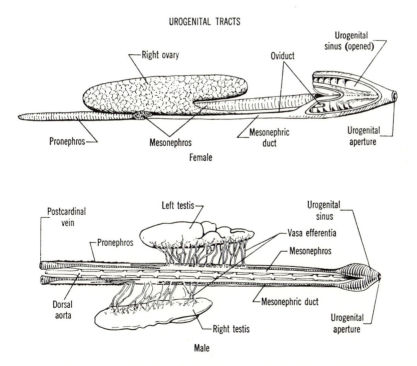

Fig. 9.4 Diagram of relationships of kidney, gonads, urinary ducts, urogenital sinus, and blood vessels in a male and female holostean fish, gar (*Lepisosteus*). (Based on Pfeiffer (male) and Balfour and Parker (female) from Bolk, Göppert, Kallius, and Lubosch, 1938).

off from the body cavity. Here again excess water is filtered from the blood through glomeruli, and salt and sugars are reabsorbed into the bloodstream through the epithelium of the kidney tubules and returned into the surrounding capillaries.

The typical kidney of freshwater rayfin fishes (Actinopterygii) has many glomeruli that are larger than those of the typical marine rayfins with glomerular kidneys; the blood supply is also well developed. The kidney of freshwater fishes is often larger, in relation to the body weight, than that of marine fishes and measurements testify to the passage of great amounts of water through the organ. The number of glomeruli in one kidney from a number of freshwater species often exceeds 10,000 whereas many truly marine species have far fewer of them. These characteristics are related to salt content of the environment and therefore to the volume of urine excreted, as shown by a comparison of kidneys of salmons (*Oncorhynchus*) retained in freshwater to those of the same age in the sea; the former have significantly more glomeruli than the latter. The most telling comparison can be made, however, with the diameter of the glomerulus itself; in several freshwater species this measure ranges from 48 to 104 microns (mean 71), whereas in several marine species it ranges from 27 to 94 microns (mean 48). Some extreme glomerular variations exist; a tropical river-dwelling pipefish (*Microphis boaja*) has none and neither do any of the goosefishes (Lophiidae) that also invade fresh water. The tubules that drain the Bowman's capsules of freshwater fishes have an open neck segment, proximal and distal convoluted portions and, in some forms, an initial collecting region (Fig. 9.5).

The fish kidney has a dual blood supply (Fig. 3.17; Chapter 7)—the renal artery and the renal portal veins. The renal artery sends blood to the glomeruli where the high arterial blood pressure helps to produce the glomerular filtrate. The renal portal veins are connected to the capillary network around

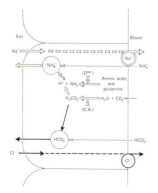

Fig. 9.5 Schematic representation of the kidney unit (nephron) of different fish groups. (Source: Prosser, 1961).

the kidney tubules and their blood joins that coming from the capillary bed within the glomerulus; all blood leaves the kidney region through the post-cardinal veins. The kidneys of lampreys and hagfishes (Cyclostomata), however, lack renal portal circulation.

The urine of freshwater fishes contains creatine, creatinine, unidentified nitrogenous compounds some of which may be amino acids, and a little urea and ammonia. Urinary nitrogen amounts to between 2 and 25 percent of the total nitrogen excreted by freshwater fishes. The bulk passes out through the gills as ammonia (56 percent or more in the carp, *Cyprinus*). The remainder is probably urea and other simple compounds of nitrogen that also leave through the gills. The urine volume, as was already mentioned, depends largely on the water permeability of body surface, gills, and oral lining. Not all fishes have an equally impermeable mucus; water will diffuse twenty times as fast into a lamprey (*Lampetra*) as it does into an eel (*Anguilla*), but only twice as fast into a goldfish (*Carassius*); the urine volumes vary accordingly.

Most important for the outward diffusion of substances from the body of a fish is the ratio of gill to body surface; small fishes, with a high metabolic demand, have relatively larger gill surfaces than their bigger species mates. Pronounced differences in gill-surface to body-surface ratios also exist among species and are reflected in the urine volume produced per unit body weight and unit time. The most unfavorable gill to body surface ratio in any of the fishes is that of the lampreys (Petromyzonidae).

The passive intake of water through gills and mouth and consequently also urine production are temperature dependent; lampreys (*Petromyzon*) kept at 2° to 3°C "bailed out" surplus water through urine production at a rate of 60 milliliters per kilogram of body weight per day, whereas, at 18°C, they passed 500 milliliters per kilogram per day. The extreme range of these values is between 6 and 50 percent of the body weight, a copious urine flow indeed if we consider that freshwater bony fishes (Osteichthyes) rarely accumulate in a day more than 20 percent of their body weight as urine and terrestrial mammals, 1.5 percent.

In connection with copious urine secretion, several freshwater bony fishes have developed a sometimes pronounced posterior evagination of the mesonephric ducts that serves for urine storage; the structure is frequently referred to as a urinary bladder (Fig. 3.15).

Some salts, especially chlorides, are also lost in the urine and through the mucus by fishes. Gills and mouth are yet another site of chloride loss, estimated to equal all others of the body. The ratio of urinary to total chloride losses varies with different species, being 1:100 in a lamprey (*Petromyzon*) and certain salmons (Salmoninae), and 1:35 in the carp (*Cyprinus*), pointing to reabsorption of filtered chlorine in the kidney tubules. Total chloride losses also vary considerably; the lamprey (*Petromyzon*) loses about seven times as much as the carp (*Cyprinus*) when chloride passage out of these fishes is compared per 100 grams of tissue per hour.

The Role of the Gills of Freshwater Fishes in Replacing Salt Losses. Osmotic regulation in freshwater animals is necessary for the maintenance of a fixed internal salt concentration, an important component of homeostasis. Salt loses vary according to species; the Atlantic salmon (*Salmo salar*) for instance, may, in fresh water, lose up to 17 percent of its body chlorides in a day by diffusion whereas the goldfish (*Carassius auratus*) loses only 5 percent. Fishes can also fast for considerable periods and freshwater species rarely, if ever, drink water although some, such as the goldfish, have been shown to swallow small amounts, but not even the carp (*Cyprinus*), a fish most resistant to salt loss, can live for more than a few days without external salt supply.

The gills and the oral membranes are the sites of active ion absorption which is necessary to replace salt losses or to supplement the minerals of the food. The following ions have been shown to be absorbed by these structures: lithium (Li^+), sodium (Na^+), cobalt (Co^{++}), strontium (Sr^{++}), calcium (Ca^{++}), chlorine (Cl^-), bromine (Br^-), acid phosphate (HPO_4^-) and sulfate ($SO_4^=$). Each of these ions has its own absorption threshold but quantitative data on their uptake are meager except for Cl^-. The lamprey (*Petromyzon*), carp (*Cyprinus*), and the roach (*Leuciscus*, family Cyprinidae) take up chloride ions from water with less than 0.05 millimoles Cl per liter (2 ppm); other freshwater fishes, as for example, the perch (*Perca*), require higher chloride concentrations than this before active uptake through the gills takes place. Not only are there species-specific differences with regard to the uptake of chloride and other ions, but it is also highly probable that such differences extend to physiological races, depending on the chemical characteristics of the water in which they occur. The rate of absorption corresponds to the rate of losses by diffusion; the lamprey (*Petromyzon*) absorbs 90 micromoles of Cl per 100 grams of body tissue per hour, the Atlantic salmon (*Salmo salar*), 30, and several minnows (Cyprinidae), between 4 and 30. These values show that active replacement of chlorides by absorption through the gills can considerably exceed passive losses. Many minerals are undoubtedly replaced through the food in sufficient amounts, but some may have replacement mechanisms similar to that ascertained for chlorides.

A mechanism for the uptake of Na^+ and Cl^- by gills of a freshwater fish is illustrated in Fig. 9.1d. Hydrogen ions (H^+), produced from the reaction catalyzed by carbonic anhydrase, react with ammonia (NH_3^+) to produce ammonium (NH_4^+). Most ammonia apparently results from deamidation and oxidative deamination in the gills but a trace of ammonia will be in the blood as a result of these reactions in the liver. Ammonium ions diffuse passively out of the gill cells. In exchange, an equivalent amount of positively charged sodium (Na^+) is carried through the gills by active transport, an energy requiring process. Bicarbonate (HCO_3^-) from the carbonic anhydrase reaction diffuses out of the gill cells and chloride ions (Cl^-) are transported actively through the gills. The net result is a sodium-ammonium exchange and a chloride-bicarbonate exchange. The trademarks of chloride cells involved in active transport of salts are an abundance of mitochondria and a highly developed labyrinth of agranular membranes (endoplasmic reticulum). The autonomic centers of the brain in the brainstem and thalamus exert a powerful influence on salt uptake because fishes with cut spinal cords show a severe reduction in salt- and water-regulatory capacity. Endocrine control of salt and water balance in fishes has also been shown. The gill epithelium and that of the proximal convoluted kidney tubules show certain physiological similarities inasmuch as such agents as the mercury (Hg) ion upset the Na and Cl uptake in both organs, probably through impairment of the enzyme succinic dehydrogenase.

There are great differences among freshwater fishes with regard to salt tolerance so that one may arrange them into groups as stenohaline (relatively intolerant to salinity changes), the freshwater fishes generally, or euryhaline (salt-change tolerant). Among the latter can be ranged the many diadromous species, such as the eel (*Anguilla*), the Atlantic salmon (*Salmo salar*), and also certain sticklebacks (*Gasterosteus*) and killifishes (*Fundulus*). Among stenohaline fishes there is variation in the capacity to adjust to a higher than normal salinity; the carp (*Cyprinus*) and the goldfish (*Carassius*) can tolerate salinities up to 17 parts per thousand, equivalent to a freezing-point depression (Δ) of $-0.9°C$ and also equivalent to a concentration of salts considerably higher than that in their own bodies. Concentrations of both tissue salt and urinary salt rise under such conditions and urine flow is reduced accordingly. The slower such stressful external conditions are imposed on the fish, the greater its tolerance of change. Capabilities of freshwater fishes to adjust to abnormally high concentrations of nontoxic salts also depend on species-specific factors such as gill to body surface ratio, gill histology, neurosecretory and/or hormonal control of membrane permeability as well as oxygen and and temperature levels. Racially determined physiological variations within species also exist in this regulatory function.

Marine Fishes

In contrast to the freshwater environment, marine fishes live in a medium that is hypertonic to the body fluids and tissues and thus they tend to lose water and gain salts through their osmotic membranes (Figs. 9.1c and 9.6). To counteract the water losses, marine fishes drink sea water and thus increase the salt content of the body fluids. Whereas dehydration is prevented

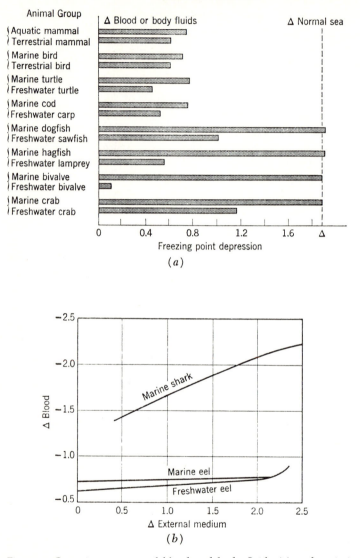

Fig. 9.6 Osmotic pressures of blood and body fluids (a) and variation in blood concentration of three fishes in waters of different concentrations (b). The marine shark is a cat shark (*Scyliorhinus*), the marine eel is a conger (*Conger*), and the freshwater (euryhaline) eel is the European species of *Anguilla*. (Source: Nicol, 1960).

by this process, excess salts must be eliminated. This homeostatic mechanism is peculiarly circuitous and energy consuming.

Kidney Function and Water Balance in Marine Fishes.   Since marine fishes are forced by osmotic conditions to conserve water, urine volume is greatly reduced compared to that of freshwater species. As little as 3 milliliters of urine per kilogram of body weight per 24 hours have been measured, with a tonicity comparable to that of the blood. The kidney tubules are apparently capable of water retention, as known for marine sculpins (Cottidae), because the glomerular filtrate is commonly five times as voluminous as the amount of urine finally excreted. Structurally the kidneys of marine fishes vary greatly; the hagfishes (*Myxine*) have a primitive mesonephros with the pronephros persisting in the adult. The function of the pronephros is not clear. It may serve in some ionic regulatory capacity, but critical experiments have not been performed to examine this possibility. As in freshwater species of lampreys (Petromyzonidae), the circulation to the glomerular and tubular portion of the nephron (kidney unit) is solely arterial, no venous renal portal circulation exists. Some killifishes (*Fundulus*), sculpins (*Cottus*), and the mudskipper (*Periophthalmus*) have been reported as having functional pronephroi but the kidney of rayfin fishes (Actinopterygii) in general is truly mesonephric.

Up to 90 percent of the nitrogenous wastes of marine fishes may be eliminated through the gills mostly as ammonia and as small amounts of urea, but the urine also contains traces of the above compounds. Ammonia appears to be formed from nontoxic precursors at the site of excretion because it is toxic to tissues and only a trace of it occurs in the blood. The urine of marine bony fishes (Osteichthyes) contains creatine, creatinine, some unidentified N compounds, and trimethylamine oxide (TMO). The problem of the metabolic origin of TMO in fishes is unresolved. The osmoregulatory advantage of its presence in the body fluids of marine forms is clear however. Possible explanation for the much higher levels of TMO in marine as compared to freshwater species could include the higher TMO content in the diets of marine fishes, the synthesis of TMO by intestinal microorganisms, and differences in retention capabilities of gills and kidney as well as the possibility of some endogenous synthesis of TMO.

Salt Balance in Marine Fishes (Fig. 9.6).   Among marine cyclostomes, the hagfishes (Myxinidae) occupy a peculiar position inasmuch as their blood has an osmotic concentration that approaches that of sea water, though with a lower chlorine content; urea concentrations in the blood have also been found to vary, most likely depending on whether the hagfish had fed on a bony fish (Osteichthyes) or a member of the shark group (Elasmobranchii) prior to analysis. Being nearly or quite iso-osmotic to their surroundings, the hagfishes do not have to drink water and use whatever water they derive from their food for the formation of their own urine. The slime produced by hagfish skin is high in Mg^{++}, Ca^{++}, and K^+ and may serve as a primitive mechanism for secretion of these cations. Marine bony fishes eliminate their surplus salts, which come mainly from the food and sea water swallowed, through the gills and the gut; only traces leave through the urine. Univalent ions, especially the chloride surplus, pass from intestine to blood stream and leave through the gills whereas divalent ions, for example, magnesium (Mg^{++}) and calcium (Ca^{++}), remain in the intestine where the alkaline nature of the fluids promotes their combination with oxides and hydroxides to form insoluble compounds which are then passed out with the feces. Chloride or salt cells in marine fishes eliminate excess chlorine ions, whereas they absorb them in freshwater fishes. These cells have been found on the bases of the gill lamellae as in an eel (*Anguilla*) as well as on the oral membranes, as in certain killifishes (*Fundulus*). There is uncertainty about the participation in ion excretion and absorption of the entire oral epithelial surface as contrasted with the work of special secreting cells.

Among marine as among freshwater bony fishes, there are some species that are more, and others which are less stenohaline. Adjustment to higher- and lower-than-normal salt concentrations can be made. Mudskippers (*Periophthalmus*) often meet with 45 parts per thousand salinity in isolated tidepools and many marine species can gradually adjust to living in brackish or even almost fresh water; direct transfer over the entire salinity range from 35 parts per thousand to 0 parts per thousand often fails even in otherwise tolerant species such as the milkfish (*Chanos*). An important factor in adjustment to lowered salinity by marine fishes is the presence of a high concentration of calcium ions. Since calcium lowers cell permeability to both salts and water, the slippery dick (*Halichoeres bivittatus*), a reef-dwelling wrasse (Labridae), can be kept healthy and will feed in a mixture of less than $\frac{1}{10}$ salt to $\frac{9}{10}$ fresh water filtered through calciferous coral sand.

Many experiments on osmoregulation and excretion have been complicated by temporary laboratory diuresis. The condition is caused by handling or otherwise upsetting the fish and manifests itself by impaired gill and kidney function, a temporarily increased urine flow, and loss through the kidneys of such ions as magnesium.

Diadromous Fishes

When an eel (*Anguilla*) reaches the freshwater habitat on its upstream migration it faces problems of salt depletion and overhydration as compared to the tendency toward dehydration and salt excess in the ocean. A young salmon (*Oncorhynchus*) migrating downstream to its ocean habitat comes up against these conditions in reverse. It follows then, that both anadromous and catadromous fishes must be versatile in their osmotic adjustment, have glomerular kidneys that can adjust to differences in urine volumes due to different salinities, and possess gills and oral membranes capable to cope both with the uptake and the secretion of certain ions against existing diffusion gradients. Experiments with the eel (*Anguilla*) suggested that the chloride cells can function both as secreting and absorbing units. It should be noted that diadromous fishes are adjusted either to the freshwater or to the saltwater phase of the life cycles and that such adjustments are primarily dependent on genetically determined physiological changes. The sea lamprey (*Petromyzon*), for example, after having entered fresh water in order to spawn cannot adjust to salinities higher than 17 parts per thousand, and young salmons (*Oncorhynchus*; *Salmo salar*) cannot successfully enter the sea before their chloride cells are well developed. The ease of transition depends on anatomical peculiarities such as the gill-to-body surface ratio, well illustrated in different races of stickleback (*Gasterosteus*), which show differential rates of adjustment to fresh water on their spawning migrations along the coasts of western Europe.

Changes in endocrine activity are usually simultaneous with, or precede, changes in salt- and water-balance mechanisms; the pituitary, the thyroid, and the gonads are primarily concerned with changes in physiological adjustment prior to and during migration. An increase in thyroid activity has been reported for salmon (*Oncorhynchus*) on downstream migration, possibly to facilitate the energy-demanding process of salt excretion in sea water, among other functions. Pituitary and gonadal changes often lead to appetitive behavior, such as the preference for fresh water by maturing sticklebacks (*Gasterosteus*) and are therefore important in the timing of migrations.

Sharks, Rays, and Skates

Sharks, rays, and skates, the Elasmobranchii, so adjust their internal osmotic pressure that no or little water passes through their permeable membranes; they can thus avoid some of the osmotic stresses of other fishes. Blood and tissue fluids of marine elasmobranchs have a freezing point depression, Δ, of more than $-2.0°C$, a value slightly higher than that for sea water. Elasmobranch blood is thus hypertonic with regard to the outside water. This hypertonicity is in part due to a high chloride-ion content, higher than is found in the blood of rayfin bony fishes (Actinopterygii). However between 42 and 55 percent of the osmotically active blood solutes are nitrogenous compounds, mostly urea. The tissues of marine elasmobranchs have a high content of urea, as much as 2 to 2.5 percent. Little or no urea is lost in shark gills, a feature paralleled only by the kidney tubules of mammals among all other vertebrate organs. In the elasmobranchs there is a slight water intake through the gill and mouth membranes but the rest of the body surface is relatively impermeable to water. Consequently only a scant flow of hypotonic, urea-containing urine is produced. Representative urine-flow measurements for sharks range from 2 to 24 milliliters per kilogram of body weight per 24 hours, but the volume of glomerular filtrate is about 80 milliliters per kilogram per day; water reabsorption takes place in the proximal convoluted tubules. Trimethylamine oxide (TMO) makes up from 7 to 12 percent or more of that part of osmotic pressure in sharks, rays, and skates which is caused by organic compounds. TMO appears to be important in elasmobranch osmoregulation because, like urea, it is reabsorbed from the glomerular filtrate in the renal tubules.

The water regimen of marine elasmobranchs is regulated first by water intake through the gills so that the blood becomes diluted and urine flow increases. Consequently, blood and tissue urea become lowered and urine flow slows down. This negative feedback can always provide the animals with sufficient fresh water to meet their urinary and other metabolic water requirements. Salts are excreted primarily by the kidneys and the rectal gland (Fig. 3.15a). The volume of urine and of rectal-gland fluid is nearly equal; however, the composition of the two fluids differs in that the latter is essentially a hypertonic sodium chloride solution with little Mg^{++} compared to urine. Minor amounts of Na^+ are excreted by the gills of elasmobranchs. Some reabsorption of salts occurs in the kidney tubules. The young and the few freshwater species of elasmobranchs have special osmotic problems.

Sharks, skates, and rays either bear live young or enclose their eggs in horny urea-containing, impermeable egg cases. In either situation, development takes place in a high-urea environment and the young are born with the mechanisms of urea retention. The serum has a Δ of about $-1.0°C$, and the salts resemble in percentage composition those of freshwater bony fishes (Actinopterygii) with a blood Δ of about $-0.57°C$. The differences of $-0.43°C$ or more result from urea in blood and tissues and it follows that freshwater elasmobranchs must have highly glomerular kidneys and a copious flow of hypotonic urine. Nitrogen is mainly excreted as ammonia through the gills. Urea is also an excretory product of freshwater elasmobranchs; about a third of it passes through the kidneys and its ducts but the bulk leaves through the gills. Chloride losses through diffusion and urine are relatively high but it is not known if freshwater members of the shark group can take up this ion from the water by gill absorption.

ENDOCRINE CONTROL OF EXCRETION AND OSMOREGULATION

The volume of urine output and salt balance is regulated in fishes, as well as in other vertebrates, by endocrine secretions.

Hormones may influence the kidney either by an increase or decrease of blood pressure that alters the filtering rate into the capsule of the renal corpuscle, and thus the amount of fluid excretion. Hormones may also affect renal excretion by specific action on tubule cells to change permeability and reabsorption rates of specific substances. In fishes, where gills and kidneys share the osmoregulatory process, hormones also influence filtration or absorption processes at the gills.

In man and in amphibians, posterior pituitary extracts contain fractions that directly influence urine output by constricting the afferent glomerular arterioles. Experiments with eels (*Anguilla*) show no such effect on the water balance although hypophysectomized killifishes (*Fundulus*) can only be kept healthy in fresh water by injection of pituitary extracts. It is not known to what extent this treatment influences water- as opposed to electrolyte-balance.

Information on the effects of fish endocrine organs and substances on mineral balance is somewhat more extensive than that on urine output, although still very far from adequate. Adrenal cortical hormones that influence sodium and chlorine equilibria in higher vertebrates also play a role in the regulation of gill and kidney functions of fishes. Blood of trouts (*Salvelinus; Salmo*) contains hydrocortisone and corticosterone as well as other cortical hormones and related compounds. Corticosteroids administered to trout lessen sodium elimination through the kidneys and increase its excretion through the gills either by a change in the permeability of the gill epithelium or by affecting the rate of its uptake. In marine fishes, which live in a sodium-rich medium, such a mechanism of sodium conservation in the kidneys is not needed, although a control of gill permeability to the movement of sodium ions would be beneficial. Evidence of such action in marine species is, so far, scanty and not conclusive.

Other endocrine tissues (Table 11.3 and 11.4) in the fish body, implicated to function in mineral balance include: hypothalamus, thyroid, suprarenal bodies, gonads, and urohypophysis.

Neurosecretory cells in the preoptic nucleus of the hypothalamus lose their secretions rapidly in several marine fishes, including dragonets (*Callionymus*), sand lances (*Ammodytes*), and mullets (Mugilidae) when the fishes are temporarily placed in hypertonic sea water. An hour after return to normal sea water, hypothalamic neurosecretory cells are refilled; there follows also a recovery of secretory cells in the neurohypophysis.

The strictly aquatic larvae of the mudskipper (*Periophthalmus*) have a less active thyroid than the adults which lead semi-aquatic lives. Thyroid injections of this fish may induce it to leave the water for lengthened periods. It appears that the thyroid hormone, apart from influencing other changes during metamorphosis, affects the water and mineral balance of fishes, not only in adaptation to temporary life outside the water but also in changes from fresh to salt water such as occur in many migratory fish species. Salmon smolt (downstream migrant juveniles of *Salmo salar*) develop chloride-secreting cells coincident with or perhaps even as a consequence of thyroid hyperactivity that occurs in their development and at a time when they begin their seaward migrations. Detailed investigations on the mode of action of thyroid secretions on osmoregulatory tissues in fishes have not yet been done.

Experiments with perfused gills of eels (*Anguilla*) reveal that adrenalin, the hormone produced in the suprarenal gland(s), has a strong vasodilatory effect on gill vessels and reduces or stops chloride secretion, which normally takes place there. However, it is not known whether adrenalin also influences chloride exchange in the normal, intact animal.

There are differences between male and female fishes of one species in blood calcium and chloride levels and in the water content of their tissues as in the cods (*Gadus*), the puffers (*Tetraodon*), and salmon (*Salmo*). These may be due to gonadal hormones since it is known from mammalian examples that female sex hormones especially exert a hydrating effect by means of sodium retention.

Finally, acetylcholinestarase, the enzyme that destroys the neurohumoral substance acetylcholine, influences the sodium uptake in frog skin and also has the same function in the anal papillae of dipterous insect larvae. In view of the wide distribution of the substance and its comparable effects in widely differing animal groups one may expect that acetylcholinesterase could also play a role in the sodium balance of fishes.

SPECIAL REFERENCES ON EXCRETION AND OSMOREGULATION

Black, V. S. 1951. Osmotic regulation in teleost fishes. *Univ. Toronto Stud. Biol. Ser.*, 59 (Publ. Ont. Fish Res. Lab., 71): 53–89.

Conte, F. P. 1969. Salt secretion. *In* Fish Physiology, 1: 241–292. Academic Press, New York.

Dobbs, G. H., III, Y. Lin, and A. L. DeVries, 1974. Aglomerularism in antarctic fish. *Science*, 185: 793–794.

Fontaine, M. 1956. The hormonal control of water and salt-electrolyte metabolism in fishes. *Mem. Soc. Endocrinol.* (Cambridge University Press), 5: 69–82.

Forster, R. P., and L. Goldstein. 1969. Formation of excretory products. *In* Fish Physiology, 1: 313–350. Academic Press, New York.

Hickman, C. P., Jr., and B. F. Trump. 1969. The kidney. *In* Fish Physiology, 1: 91–239. Academic Press, New York.

Krogh, A. 1937. Osmotic regulation in freshwater fishes by active absorption of chloride ions. *Zeitschr. Vergl. Physiol.*, 24: 256–266.

Marshall, E. K. 1934. The comparative physiology of the kidney in relation to theories of renal secretion. *Physiol. Revs.*, 14: 133–159.

Maetz, J., and F. Garcia Romeu. 1964. The mechanism of sodium and chloride uptake by the gills of a fresh-water fish, *Carassius auratus*. II. Evidence for NH_4^+/Na^+ and HCO_3^-/Cl^- exchanges. *Jour. Gen. Physiol.*, 47: 1209–1227.

Pantin, C. F. A. 1931. The origin of the composition of the body fluids in animals. *Biol. Revs.*, 6: 459–482.

Robertson, J. D. 1974. Osmotic and ionic regulation in cyclostomes. *In* Chemical Zoology, 8: 149–193. Academic Press, New York.

Shewan, J. M. 1951. The biochemistry of fish. *Biochemical Soc. Symposia* (Cambridge, England), 6: 28–48.

Smith, H. W. 1932. Water regulation and its evolution in fishes. *Quart. Rev. Biol.*, 7: 1–26.

———. 1936. The retention and physiological role of urea in the Elasmobranchii. *Biol. Revs.*, 11: 49–82.

———. 1953. From fish to philosopher. Little, Brown and Company, Boston.

Wikgren, Bo-J. 1953. Osmotic regulation in some aquatic animals with special reference to the influence of temperature. *Acta Zool. Fennica*, 71: 1–102.

10

Reproduction

Reproduction (Fig. 10.1) is the process by which species are perpetuated and by which, in combination with genetic change, characteristics for new species first appear.

TYPES OF REPRODUCTION

At least three types of reproduction are possible: bisexual, hermaphroditic, and parthenogenetic. In bisexual reproduction, which is the prevalent kind, sperms and eggs develop in separate male and female individuals. In hermaphroditism (a type of intersexuality), both sexes are in one individual and, as among certain serranids and a dozen or more other families, self-fertilization or true functional hermaphroditism exists. Such synchronous hermaphroditism, from an evolutionary point of view, could be the most advantageous form of reproduction.

Hermaphroditic sex glands are known for several species, including some trout relatives (salmonoids), perches (*Perca*), walleyes (*Stizostedion*), darters (*Etheostoma*), and some of the black basses (*Micropterus*). Some seabasses (Serranidae) are protandric hermaphrodites, being males at first, females later.

Parthenogenesis is the development of young without fertilization, and a condition that has been called parthenogenesis (more properly gynogenesis) occurs in a tropical fish, the live-bearing Amazon molly (*Poecilia formosa*), and is also known in *Poeciliopsis*. Mating with a male is required, but the

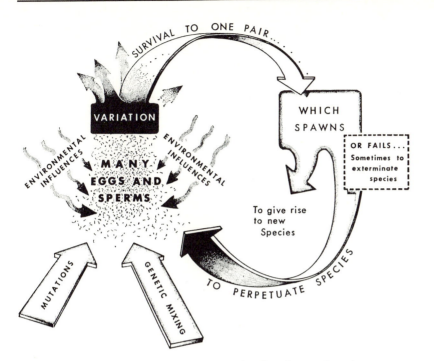

Fig. 10.1   Role of reproduction in the survival and evolution of species.

sperm serves only one of its two functions, that of inciting the egg to develop; it does not take any part in heredity. The resultant young are always females, with no trace of paternal characters.

THE REPRODUCTIVE SYSTEM

The reproductive function in fishes is primarily the job of the reproductive system. Previously described are the components of the system, the sex glands or gonads, ovaries in the female and testes in the male, and their ducts. The endocrine system plays an important regulatory role in reproduction.

SPERMATOZOA AND THEIR FORMATION

The process by which sperm cells are formed in the testes is called spermatogenesis or spermiogenesis. It is in this process that the materials determining the characteristics which the offspring will inherit from its parents are readied for union with like materials in the egg cell. Not only do the spermatozoa or sperms of different fishes contain different hereditary materials, they differ in shape as well (Fig. 10.2a). It is also in this process that sperm cells gain a whip-like tail which makes them motile and with which they make their way ultimately to and into eggs for fertilization. Furthermore, during development spermatozoa attain the differences in shape that they exhibit as species characteristics. In order to insure fertilization, each

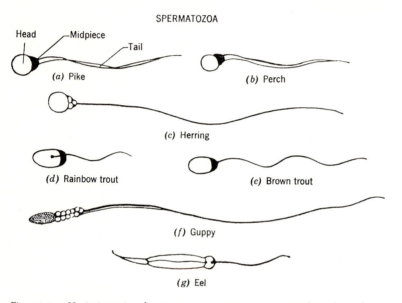

Fig. 10.2a   Variations in shape among spermatozoa (greatly enlarged) of certain bony fishes: (a) northern pike (*Esox lucius*); (b) European perch (*Perca fluviatilis*); (c) Atlantic herring (*Clupea harengus*); (d) rainbow trout (*Salmo gairdneri*); (e) brown trout (*Salmo trutta*); (f) guppy (*Lebistes reticulatus*); and (g) European eel (*Anguilla anguilla*). (Based on Tuzet, Ballowitz, Fontaine, and Vaupel in Grassé, 1958).

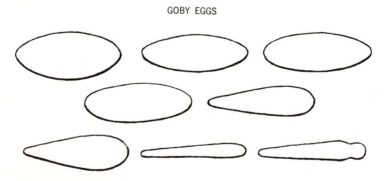

Fib. 10.2b   Nonspherical fish eggs (greatly enlarged) as shown in selected gobies (Gobiidae). (Source: Breder, 1943).

male fish, like the male of other vertebrate species, produces huge numbers of sperm cells so tiny that a million of them can be held in a droplet. The spermatozoa, plus secretions of the sperm ducts, compose the milt that the male fish exudes at spawning (like the semen of man). The sperms are inactive and immobile until the secretion of the sperm ducts has been encountered. Once they have become active and have started to move by the whipping motions of their tails they will soon perish unless an egg is encountered for fertilization. The life span of sperms varies considerably according to species and according to substrate into which they are deposited. If the sperm cells are deposited in water, they live a much shorter time than if they are deposited in a female. If water has nearly the same salt content as the fish body fluid, the sperm cells live longer than if the medium is either higher or lower in effective salt content than body fluids.

Viability of sperm cells is also affected by temperature; in general, they live longer in lower temperatures than in higher ones and, just as human sperms may be stored under refrigeration for artificial insemination, it is possible to store the milt of fishes. The spermatozoa of fishes, like those of man, bull, fowl, and frog, have been successfully preserved by freezing—for example, carp (*Cyprinus*), Atlantic herring (*Clupea*), and salmon (*Oncorhynchus*) sperms have remained viable after freezing.

EGGS AND THEIR FORMATION

Oogenesis is the process of egg development in ovaries parallel to that of sperm manufacture in the testes. The result is not only the largest of the cells to be found in fishes (mostly from 0.5 to 5.0 millimeters in diameter, with few to golf-ball size and larger) but also a cell endowed with the peculiar quality to produce after fertilization a new individual of the species. During oogenesis the surrounding epithelial cells (granulosa) provide the developing egg cell with relatively large quantities of stored food material in the form of granular yolk (protein) and fat commonly in the form of oil droplets.

Fecundity is a general term used to describe the number of eggs produced. The fecundity of an individual female varies according to many factors including her age, size (Fig. 10.3), species, and conditions (food availability, water temperature, season).

Fecundity appears to bear some broad relationship to the care or nurture accorded to the eggs. Thus, the ocean sunfish (*Mola mola*) may produce huge numbers of eggs but they are merely broadcast in the open sea water by the female while swimming with males. A 54-pound fish of this kind is said to produce 28,000,000 or more eggs in a single season. The cod (*Gadus*), which is also a pelagic open-water spawner, produces as many as 9,000,000 eggs per

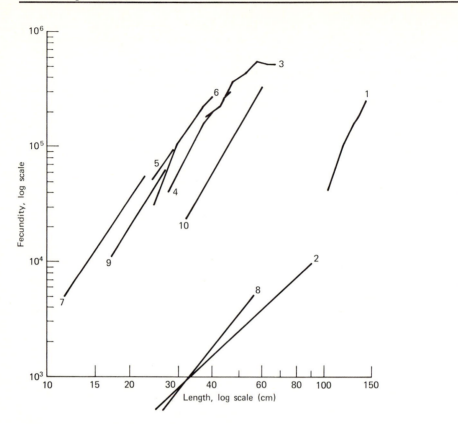

Fig. 10.3 The relationship between fecundity and length within a species. 1. *Acipenser stellatus* (Nikolsky, 1963); 2. *Salmo salar* (Pope et al., 1961); 3. *Cyprinus carpio* (Nikolsky, 1963); 4. *Pleuronectes platessa* (Clyde) (Bagenal, 1966); 5. *Clupea harengus* (northern North Sea) (Baxter, 1959); 6. *Melanogramus aeglefinus* (see Parrish, 1956); 7. *Osmerus eperlanus* (Lillelund, 1961); 8. *Salvelinus fontinalis* (Vladykov, 1956); 9. *Sardinops caerulea* (MacGregor, 1957); and 10. *Sebastes marinus* (Faroe Island) (Raitt and Hall, 1967). (Source: Blaxter, 1969).

female per season. Such species accord no care to their eggs, other than to emit them in the vicinity or in the company of males depositing sperms. Compare with this the sticklebacks (*Culaea* and *Gasterosteus*) which make elaborate nests and then provide parental care to the developing fish. Here the usual number of eggs ranges between thirty and a hundred. In guppies (*Lebistes*), which bear their young alive and thus accord an extreme amount of care (that is, internal incubation), the usual brood number is less than two dozen. Some fishes even develop at one time, only a single egg, or one in each ovary.

The eggs of most fishes have shells secreted around them. The egg cell itself has only a tender membrane when ready to leave the ovary. As the eggs of oviparous species pass through the oviducts, they go through glands that secrete the shells about them. In fishes that accord care to their young, either by having the parents stay on guard or by burying the eggs in gravel, the shells are often more tender and less limy or horny than they are in eggs exposed to waves, currents, or other hazards. In many skates and some sharks, for example, the shells are horny-hard and almost fingernail-like in strength. Such eggs resist damage, even drying, to such an extent that the eggs, when washed up on shore, may lie there for some time and still hatch when returned to the sea. An egg of the little skate (*Raja erinacea*) was once mailed without padding from the New Jersey coast to our Ann Arbor laboratories in a common letter envelope. It was out of the water in transit for several days and traveled some seven hundred miles. Yet it hatched successfully and became a pet for some months.

Fish eggs have many other interesting characteristics than their covering. Some are buoyant (pelagic) and have a specific gravity about like that of fresh water. Although most marine food fishes, for example, cod (*Gadus*), have buoyant eggs, some marine species, for example, herring (*Clupea harengus*), have nonbuoyant eggs that are attached to the substrate. Most stream fishes have eggs that are heavy and sink (demersal), with a specific gravity greater than that of fresh water. Perhaps this is an adaptation that insures these eggs against being swept downstream at the time of deposition, which may take place in gravel beds in rather swift current in sites such as those chosen by stream chars (*Salvelinus*), the trouts and Atlantic salmon of the genus *Salmo*, and the species of Pacific salmons (*Oncorhynchus*).

Ova also vary as to adhesive stickiness. Most buoyant eggs are nonadhesive and do not normally stick to anything for holdfast purposes. Most eggs that sink, however, are at least temporarily adhesive. Trout and salmon (Salmoninae) eggs are temporarily adhesive during the process of water hardening in the early minutes after deposition while they are absorbing water at a great rate. One can almost visualize the eggs as "sucking in" water at such a rate that they stick to things, things stick to them, or they adhere to one another. In this process, the eggs of stream-spawning trouts and salmons pick up little particles of sand that help sink them to the bottom and anchor them there for covering and development. The stickiness of the eggs of other fishes such as the walleye (*Stizostedion*) serves the purpose of keeping the eggs from settling deep into bottom materials and becoming suffocated. Sticky, demersal eggs adhere to sticks and debris on the bottom without falling down into the deeper, softer materials that may be present. Adhesiveness can be due temporarily to the process of water hardening, or more permanently to a pasty, mucoid deposit on the egg that remains sticky throughout the whole developmental period.

Eggs also vary as to shape (Fig. 10.2b), but those of most species of fish are spherical. The spherical shape seems to be the natural, basic contour of most animal eggs. In many anchovies of the herringlike family Engraulidae, the outline of the egg is elliptical, occasionally spherical. But in the minnow genus *Rhodeus*, the European bitterling, and in various other fishes including the cichlids, the eggs are oval in outline (ovoid). In the gobies they range from ovoid to teardrop-shape.

The eggs of fishes also differ considerably according to the appendages that they bear. In the hagfish (*Myxine*), for example, the egg is an elongate ovoid, with a number of tendrils at each end. Apparently, these tendrils anchor the eggs to vegetation or other material. In the American smelt (*Osmerus mordax*), the egg has a low stalk which is adhesive and becomes attached to the stony bottoms of streams where the species most often spawns. In the brook silverside (*Labidesthes sicculus*), there is a single elongate filament that serves first for temporary flotation and then for attachment. The eggs of some skates (Rajidae) and cat sharks (Scyliorhinidae) have tendrils that come out of each of the four corners of the more or less box-shaped case. Most flyingfishes (Exocoetidae) have eggs with numerous filaments, much like long hairs, that come out from the surface of the spherical egg. However, the eggs of a few flyingfishes have no filaments, some have the filaments confined to the two poles of the egg, and others, to one pole.

Quite as spermatozoa exhibit different viability in different situations, so do eggs have various survival features. Interestingly, not all mature eggs are laid in any one period of spawning. Some remain in the ovary and are normally resorbed. It would appear, however, that when disproportionately large quantities of eggs are not shed, the ovary is unable to resorb them. The individual may then become egg-bound, and the mouth of the oviduct plugged by a mass of fibrous tissue. Future passage of eggs may be barred.

SEX DIFFERENCES (FIG. 10.4)

Anyone can tell a male from a female fish when milt or eggs are running. A light press on the belly of most ripe fishes will bring the whitish milt, or the eggs, into view near the anus. Both humans and fish, however, also use features other than these to tell the sexes apart. The characteristics of sexual difference or sexual dimorphism that enable identification of the sexes are classed as primary and secondary. Primary sexual characters are those that are concerned actually with the reproductive process. Testes and their ducts in the male, and ovaries and their ducts in the female constitute primary sexual characters. The primary sexual characters often require dissection for their discernment, which makes the secondary sexual characters often more

SEXUAL DIMORPHISM

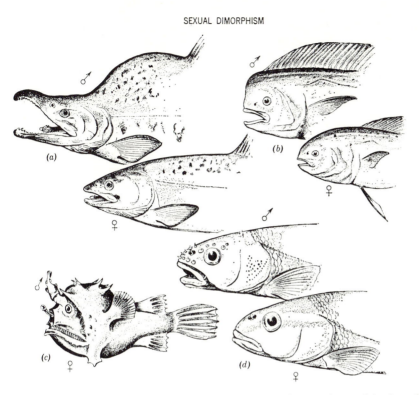

Fig. 10.4 Sexual dimorphism among fishes as shown by: (a) humped back and hooked jaws of the male in the pink salmon (*Oncorhynchus gorbuscha*); (b) domed forehead and anterior position of the dorsal fin in the dolphin (*Coryphaena hippurus*); (c) parasitic male of the deepsea anglerfish (*Photocorynus spiniceps*); (d) nuptial tubercles on the snout and forehead of the creek chub (*Semotilus atromaculatus*).

useful, although sometimes not so positive. Secondary sexual characters themselves are really of two kinds: those which have no primary relationship with the reproductive act at all, and those which are definitely accessory to spawning. Some outstanding points of difference of a dimorphic nature between the sexes follow.

A genital papilla marks the male of such fishes as lampreys (Petromyzonidae), Johnny darters (*Etheostoma nigrum*), and white bass (*Morone chrysops*). The papilla is a veritable penis in one group of sculpins, the Oligocottinae. Body shape is an important secondary sexual character. Ordinarily the female is much more pot-bellied than the male, particularly when she is ripe or near-ripe for spawning, because of the relatively larger quantity of sexual products which she contains. Pearl organs or nuptial tubercles appear on the males of many fishes to mark the sex; examples are the smelt (*Osmerus*), and most species of minnows (Cyprinidae) (Fig. 10.4) and suckers (Catostomidae). These tubercles are little horny excrescences that become evident just before the spawning season and disappear shortly after, under the influence of hormonal secretions. The fins often provide characteristics distinctive of the males; on the average they are larger than in females. Besides, in one North American fish, the fantail darter (*Etheostoma flabellare*), the tips of the spines of the dorsal fin of the breeding male end in fleshy knobs, giving them a clubbed shape, whereas the female normally has ordinary rays. In some other species males have over-all fleshy rays (bluntnose minnow, *Pimephales notatus*), whereas the rays in the fins of the female are just ordinary, fine ones. In some fishes, the caudal fin may show sexual dimorphism; for example, the lower lobe is greatly extended in the males of the swordtail (*Xiphophorus helleri*) of the aquarium fancier, and is somewhat enlarged in the white sucker (*Catostomus commersoni*).

An accessory sexual character marks the males of several species in which the anal fin becomes enlarged into an intromittent, copulatory organ. This organ, designated the gonopodium (Fig. 10.5), occurs in such fishes as the mosquitofish (*Gambusia affinis*), the guppy (*Lebistes*), and in other live-bearing topminnows (Poeciliidae). The anal fin in some viviparous perches (Embiotocidae) bears glandular structures that are of some use in intromission. In the male of the white sucker (*Catostomus commersoni*), the enlarged anal fin serves to convey the watery milt into the bottom of the current-plagued nest in streams, thereby improving the chances for contact between the sperms and the eggs. In the clasping cyprinodont (*Xenodexia*) the modified pectoral fin or heterochir is evidently used in mating for holding the gonopodium in position for insertion into the oviduct of the female. The pelvic fins are variously modified in the sharks and their relatives (Elasmobranchii) as intromittent structures, the myxopterygia (Fig. 10.5), that help to insure internal fertilization which is widespread in this group of fishes.

Obviously, coloration in fishes often serves as a mark of sexual distinction and recognition; it is termed sexual dichromatism. In general the males are brighter or more intense in color than the females. This is also true among the birds but, as among the birds, there are some exceptions to the rule. An outstanding example of sexual dichromatism in the central North American fish fauna is the orangespotted sunfish (*Lepomis humilis*). In this sunfish the male has more numerous and brighter orange spots on the body than does the female. In the bowfin, *Amia calva*, a dark eye-spot develops in the tail region in the young of both males and females but becomes diluted and subdued in the adult females whereas in the males it becomes intensified and develops a brightly colored ocellation (eye-like ring) around it. Wrasses (Labridae) and parrotfishes (Scaridae) in the marine realm are prime examples of sexual dichromatism and in some cases also of dimorphism.

INTROMITTENT ORGANS

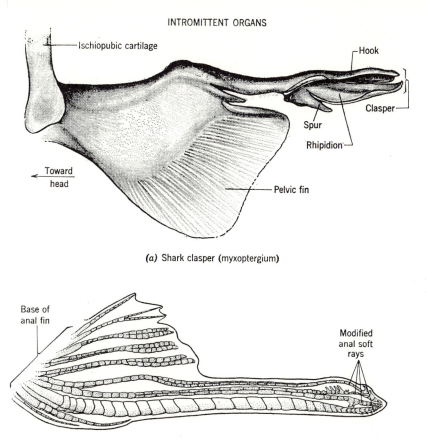

(a) Shark clasper (myxoptergium)

(b) Gambusia gonopodium

Fig. 10.5   Male intromittent organs of two fishes: (a) left pelvic fin of a horn shark (*Heterodontus francisci*) showing its modification into a myxopterygium, ventral view; (b) anal fin modified into a gonopodium in the mosquitofish (*Gambusia affinis*), lateral view. Part (a) based on Daniel, 1934, and (b), on Turner, 1941.

Several different head characteristics also serve to distinguish the sexes among fishes. In the chimaeras (Chimaeridae), the male develops a spiny, stout, retractile knob, the frontal clasper, on the upper part of the head; a similar structure characterizes the forehead of the forehead brooders (Kurtidae; Fig. 10.6). In the salmons and the trouts (Salmoninae), the breeding males typically develop a knobby hook or kype near the tips of both the upper and lower jaw.

A few accessory reproductive structures among females serve as sexual characteristics. An outstanding example is the egg-laying tube or ovipositor in the females of the European bitterling, *Rhodeus amarus*, and an Asiatic lumpsucker, *Careproctus*. In a few species of fishes there have been discovered differences in fin-ray and vertebral counts between males and females. Neoteny, the attainment of sexual maturity by larval individuals, is a sexually dimorphic feature of males of the deepsea stalkeyed fishes (*Idiacanthus*) and among certain European lampreys (Petromyzonidae).

Probably the most extreme form of sexual dimorphism occurs in the deepsea angler fish (Fig. 10.4c) where the male is minute and little more than a parasitic sperm factory attached to the much larger female.

SEXUAL MATURITY

Several factors influence the first attainment of ability to reproduce, or sexual maturity, in fishes. Among these are differences in species, in age and size, and in individual physiology. In general, species of small maximum size and short life span mature at younger ages than do kinds of large maximum size. Some fishes are sexually mature at birth; males of the reef perch (*Micrometrus aurora*) and of the dwarf perch (*M. minimus*) spawn shortly after birth. Although the females of these species receive sperm soon after they are born, they do not bear young until a year later. Other fishes, such as the guppy (*Lebistes*) and the mosquitofish (*Gambusia affinis*), become sexually mature at ages less than a year and lengths of an inch or less, although they are not born (or hatched) in an adult state. Many fishes mature at the age of 1 year. Examples of this in the American freshwater fauna include the inch-long least darter (*Etheostoma microperca*) and other species less than 6 inches long, such as the brook silverside (*Labidesthes sicculus*); a marine example, of which there are many, is the California grunion (*Leuresthes tenuis*). Many species attain sexual maturity for the first time at ages of 2 to 5 years, with lengths from 3 to 12 inches or more. Among these are trouts (*Salmo*), black basses (*Micropterus*), and sunfishes (*Lepomis*). Eels (*Anguilla*) become sexually mature at 10 to 14 years of age and 2 feet or more in length, and sturgeons (*Acipenser*) at 15 years or more, with lengths of about 3 feet (1 meter) and more. Both age and associated size are thus significant factors in the determination of adulthood.

Once a fish is sexually mature, the sex products must still become ripe and the reproductive act must take place. Many factors interplay to bring about these events. The forces at work can be grouped roughly into those that arise within the fish (intrinsic; Chapter 11) and those that are in the surroundings (extrinsic). The individual fish has essentially no control over either set. Among the intrinsic factors determining the attainment of sexual maturity and subsequent breedings and ripenings are the kind of fish and its heredity, foods chosen, and finally the whole physiological makeup of the individual, which of course, would be very difficult to separate from its heredity. Within a species, sex of the individual may determine importantly how it reacts to the factors that set up the drive to spawn. In many kinds of fishes, the males mature and ripen earlier than the females, either in the entire life-cycle or in the course of a single spawning season. For example, the sex glands of some yearling males of certain Pacific salmons (*Oncorhynchus*) may ripen, and the individuals may spawn, although the species normally first spawns 2 or 3 years later. Some of the early-ripening males, the so-called grilse, which often do mate with females, ripen a year earlier than the females of the same age. In many freshwater fishes like trouts (*Salmo*), suckers (*Catostomus*), and creek chubs (*Semotilus*), the males arrive first on the spawning grounds, which for the kinds listed are the riffles of streams for the stream-spawners. Here the male starts building the nest or just awaits the arrival of the females. This early ripening of males has been held to be due to their higher rate of metabolism, perhaps through endocrine influence in response to genetic and environmental forces.

The internal physiological rhythm of gonadal maturation, involving predominantly pituitary-gonadal interactions, is adjusted to insure that breeding will occur at a time when environmental conditions are most favorable for survival of the offspring. Changes in the duration of daylight (photoperiod) and the temperature affect the rhythm of gonadal maturation and spawning of fish in the Temperate Zone and higher latitudes. Near the equator where changing day length could not possibly account for the many examples of strictly seasonal reproduction, reproductive activity in many tropical fishes often coincides with extensive flooding during the rainy season. Other extrinsic factors that may affect reproduction are presence of the opposite sex, current, tide, stage of the moon, and the presence of spawning facilities. In the bitterling, *Rhodeus amarus*, it seems that the presence of a freshwater mussel, into which the eggs are to be laid, stimulates ripening. And the currents from the exhalent syphon of the mussel finally stimulate egg laying by the fish.

REPRODUCTIVE CYCLES

Reproductive behavior in most animals is cyclic—more or less regularly periodic; this is so for nearly all fishes. The reproductive act in some fishes occurs only once in a very short lifetime; an example of this is the brook silverside (*Labidesthes sicculus*). In still other fishes, it occurs once in a moderately long life span: in the Pacific salmons (*Oncorhynchus*), once in 2 to 5 or more years; in the sea lamprey (*Petromyzon marinus*), once in 5 or 6 years; and in the freshwater eels (*Anguilla*), once in 10 to 14 years. Most fishes, however, have a yearly cycle of reproduction, and, once they have begun it, they follow it until they die. Several other species spawn more than once in a year and more or less continually. The guppy (*Lebistes*) may bring out broods at approximately 4-week intervals; also, broods may be several per year in certain tilapias (*Tilapia*). Around Hawaii, the threadfin (*Polydactilus sexfilis*) spawns once a month between May and October. The grunion (*Leuresthes*) of the California coast may spawn every couple of weeks in season during the spawning period until it is spawned out. In all of breeding the objective is to bring about union of the sperm and egg. Other adaptations then lead to the survival of the fertilized egg.

BREEDING

Fish have developed many different ways for gaining nearness of sperms and eggs to each other in order to facilitate and insure their union. Outstanding among these ways are the nicely timed relations of the various reproductive functions. In the prolific open-water spawners, for example, the millions of eggs do not all mature at one time; just enough mature in each batch so that at the final stage of rapid growth of the eggs, the female is not too greatly distended with the enlarged sex products. If all ripened simultaneously, they would occupy a space much larger than the parent fish, so batches are timed in ripening to accommodate the parent. Many fishes have a very short breeding season, occurring only once a year as we saw, yet the males and females are in fully ripe condition at the same instant and both then are capable of exhibiting their complex breeding reactions. The secondary sexual characteristics and accessory structures which are used in courtship, clasping, or intromission develop simultaneously with the readying of the primary sexual products.

Perhaps the most famous case of accurately timed relationships in reproduction is in the palolo worm which swarms to the surface of the open seas once a year at a certain phase of the moon to breed. A few fishes show similar marvelously timed relationships for breeding. The best known of these is the grunion (*Leuresthes tenuis*) of the California coast. This fish spawns just after the turn of high tide at certain times of the year. It spawns almost literally out of water, as far up on the beach as the largest waves will carry it at this time of high tide. It deposits eggs and sperm in pockets in the wet sand at just such a time and in just such a position that the eggs are not likely to be washed out by waves until two weeks or a month later at the time of the next high tide. At this time the waves come up on the sand again and when they hit the places where the nests are deposited and stir up the sand, the young hatch almost instantly and go out with these waves before the tide can recede and make it impossible for them to do so. This is certainly a remarkable instance of timed reproduction as well as development and hatching. Regarding timed relationships, too, we should think that somewhere in a continent such as North America, some fishes are spawning at almost any time, because of the influence of latitude and altitude, as well as individuality of species and races or varieties that compose them.

Most fish species have definite seasons for spawning as a part of their timed reproductive relationships, and are generally grouped as follows. Warmwater fishes are summer spawners and coldwater fishes, fall and winter spawners. Species tolerating intermediate temperatures are generally spring spawners. Some tropical species spawn the year around. Fixed spawning seasons are roughly correlated with the developmental period that the fish require. Warmwater fishes (such as the black basses and sunfishes, Centrarchidae) use only a few days to hatch and emerge into an environment generally favorable to growth and survival. Several months may be required for a char (*Salvelinus*) or a whitefish (*Coregonus*). Some trouts (*Salmo*), at a mean temperature of 50°F (10°C), may average around 50 days for their developmental period. The whitefish (*Coregonus clupeaformis*), at a temperature in the low 30's F (1°C), may take as much as 130 to 150 days for development to hatching. Both seem to be timed to bring the young forth into conditions favorable to them. Timing of reproduction also makes it possible for more than one species to use the same breeding grounds in a calendar year. Thus in American streams that both occupy, spring-spawning suckers (*Catostomus*) may use the same nest (redd) sites as fall-spawning brook trout (*Salvelinus fontinalis*).

Timed relationships, no matter how good, would not be very useful to a fish if it could not recognize a mate of the opposite sex when present. When a scientist is sexing fish he uses such characters as structure, form, and color as described previously. With live fish, he may also use behavior. Fish themselves are also aware of form, color, and behavior when recognizing their mates. An outstanding experimental study of the value of these attributes for sexual recognition to fish is that by Tinbergen, who has demonstrated all of these points by observing sticklebacks (*Gasterosteus*) in aquaria.

In addition to the evolution of timed relationships, and abilities in sexual recognition, there have developed among fishes several useful devices for insuring either external or internal fertilization (Fig. 10.5). For external fertilization, proximity of two individuals of the opposite sex for spawning is the most common means employed. Actual pairing and some form of holding (amplexus) is sometimes used as a special development of proximity. In pairing, some fishes come side by side in actual contact and simultaneously emit eggs and sperms and in other instances the male twists his body around that of the female, in a semicircle, or even in a corkscrew spiral for a fish with a much-elongated body such as a lamprey (Petromyzonidae).

The parasitic dwarf male of deepsea anglerfishes (*Photocorynus*) adheres to the skin of the female by its mouth (Fig. 10.4). Here the male has solely a reproductive function, and has virtually lost the digestive tract as a result of his parasitic adaptation. Concentration of eggs in masses is a feature that insures external fertilization. We see eggs roped together with watery jelly in perches (*Perca*; Fig. 10.6). Eggs are massed underneath stones in such fishes as the Johnny darter (*Etheostoma nigrum*; Fig. 10.6) and the bluntnose and fathead minnows (*Pimephales*). The marine scorpionfishes (*Scorpaena*) make eggs into a two-lobed balloon. There are relationships here with the features of viability and specific gravity of sperms and eggs and also of their abundance. Mass aggregations for spawning are also a means employed for insuring external fertilization.

For internal fertilization, several devices have evolved in fishes. Most common among them is the placement of sperm by the male into the reproductive tract of the female in the process of intromission, for which important adaptations include: special modifications (Fig. 10.5) of the pelvic fins (as in the shark group, Elasmobranchii), and of the anal fin (as in the topminnow-livebearer group, Poeciliidae); and the development of particular genital organs in the region of the genital pore (as in the blind cavefishes, Amblyopsidae) are important. The function of all of these structures is to bring the sperms into the oviduct of the female. The normal composition of the sperm mass in these forms is fluid milt. In at least one fish, *Horaichthys*, of India, a further step has been taken to insure the transfer of a mass of sperm. The spermatozoa are packaged by the male into a tight mass similar to the spermatophore that is employed among salamanders for getting packages of sperm from the male to the female. The sperm ball, which carries an arrow-tipped stalk, is impaled on the skin of the female near the opening of the oviduct, in a position ready to insure fertilization as the female passes one egg at a time. The basking shark, *Cetorhinus maximus*, also has spermatophores.

Still another manner in which the proximity of eggs and sperms is guaranteed is in the stimulus provided by one sex to the other to release the sex products at the right time. This is brought about in many fishes by definite courtship behavior patterns, in which sound emission and pheromones play a role and the male may swim circles around the female or prod her, bunt her, rub her, fan her, herd her, or do most anything to signify that "now is the time." Interestingly, fish with the best-defined and strongest courtships appear to have produced smaller numbers of eggs than those in which there is little or none. Furthermore, in fishes in which the males are very active courters they usually assume the task of caring for the eggs. In addition to the foregoing, reproduction in fishes is also affected by such obvious conditions as water levels, substrate or bottom type, weather, natural and manmade barriers, predation, food supply, and pollution.

EGG FATE

LAID AT RANDOM

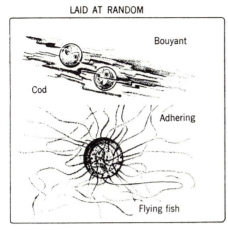

Bouyant

Cod

Adhering

Flying fish

LAID IN MASSES

Perch

PLACED IN ANOTHER ANIMAL

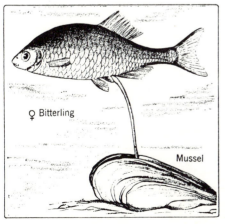

♀ Bitterling

Mussel

Fig. 10.6   Diversity in disposition of eggs and young in fishes.

EGG FATE

PLACED IN NEST

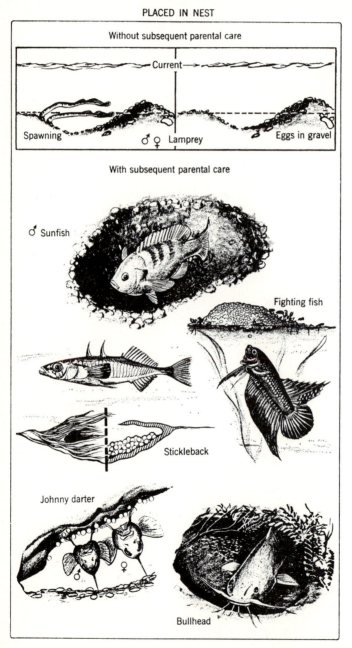

Fig. 10.6 (*Continued*)

EGG FATE

CARRIED BY PARENT

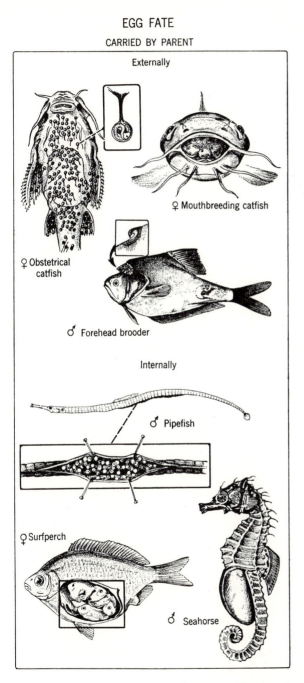

Externally

♀ Mouthbreeding catfish

♀ Obstetrical catfish

♂ Forehead brooder

Internally

♂ Pipefish

♀ Surfperch

♂ Seahorse

Fig. 10.6 (Continued)

Lest the reader assume that breeding habits of fishes are well known, it must be pointed out here that information is lacking or incomplete on most of the 20,000 species.

CARE OF EGGS AND YOUNG

The effective production of eggs and sperms alone would not be enough to bring about survival of species. Whereas many fishes, especially marine ones, rely simply on "safety in numbers," there have evolved various means for affording care to fertilized eggs and young by one or both sexes (Table 10.1). One of the means evolved has been to get the eggs into the right place—the place for which their developmental machinery is suited. Thus, such anadromous fishes as the sea lamprey (*Petromyzon marinus*), sea-run sturgeons (*Acipenser*), and salmons (*Salmo, Oncorhynchus*) ascend freshwater streams to spawn. Many adult salmon home to the very streams in which they hatched and spent their youth, after travels of thousands of miles over periods of 5 years or more. The freshwater eels (*Anguilla*) have a catadromous habit, descending into the ocean to bring the eggs into the salt-water habitat to which they are bound for development.

Once on the spawning grounds, the selection of the spawning site takes place. This process appears to be a particularly careful one in nest-building fishes and is of great importance for successful reproduction. Once the site has been chosen, some fish will fight to the death to defend it. An outstanding example of this is the Siamese fighting fish, *Betta splendens*, well known to aquarium hobbyists as a species that defends its territory to the death. Many fishes, however, do not give special care to the eggs at all. In this instance, the eggs themselves are adapted to survival. They may float, either with or without a little filamentous appendage that will give them some attachment during development. Pelagic eggs which are transparent appear invisible to potential predators. The eggs may be in masses such as the hollow ropes of eggs of the perch (*Perca*), so that they are not scattered, do not fall into the mud, and have some protection by mass-size. The transparent egg balloon of the scorpionfish (*Scorpaena*) mentioned earlier, would be of some value in affording protection, as well as in providing for dispersal. The various attachment devices, as well as adhesiveness, also help the survival of eggs that are deserted by their parents.

Of those fishes that actually build nests, there are broadly two kinds: those that make a nest and then desert it and those that build one and stay with it to guard the site, eggs, and/or young. A few examples of fishes that build excavated nests then protectively fill them with stones interspersed with eggs are the creek chubs (*Semotilus*), the trouts (*Salmo*), and the salmons

Table 10.1. Diversity in Nests and Spawning Places as Shown by Selected Examples from Fresh Waters of North America

I. Nest building species.
  A. Those exhibiting parental care.
    1. Nest a circular depression.
       a. In bed of mud, silt, or sand, often in and among roots of aquatic flowering plants. Sunfishes (Centrarchidae) including the sunfishes proper (*Lepomis*), largemouth bass (*Micropterus salmoides*), crappies (*Pomoxis*), rock basses (*Ambloplites*), and warmouth (*Chaenobryttus*); bowfin (*Amia*); most bullheads (*Ictalurus*).
       b. In gravel bottom. Smallmouth bass (*Micropterus dolomieui*); rarely, largemouth bass, rock basses.
    2. Nest excavated under a stone or other submerged object with the eggs attached to the underside of the roof of the nest. Johnny darter (*Etheostoma nigrum*), fantail darter (*Etheostoma flabellare*), stream-dwelling sculpins (*Cottus*), bluntnose and fathead minnows (*Pimephales*).
    3. Nest made of plant materials and spherical or mound shaped. Sticklebacks (*Culaea; Gasterosteus*).
    4. Nest a tunnel. In bank, channel catfish (*Ictalurus punctatus*); in bottom, yellow bullhead (*Ictalurus natalis*).
  B. Those deserting nests after spawning; nest a well-defined structure in gravelly bottom. Lampreys (Petromyzonidae); stream trouts including the brook trout (*Salvelinus fontinalis*), cutthroat trout (*Salmo clarki*), rainbow trout (*S. gairdneri*), and brown trout (*S. trutta*); salmons (*Oncorhynchus* and *Salmo salar*); creek chub (*Semotilus atromaculatus*); fallfish (*S. corporalis*); river chub (*Hybopsis micropogon*); hornyhead (*H. biguttata*); rainbow darter (*Etheostoma caeruleum*).
II. Species that do not build nests.
  A. Scattering eggs, usually over aquatic plants, or their roots or remains. Northern pike (*Esox lucius*); carp (*Cyprinus carpio*); goldfish (*Carassius auratus*); golden shiner (*Notemigonus crysoleucas*).
  B. Scattering eggs over shoals of sand, gravel, or boulders. Whitefishes and ciscoes (*Coregonus*); lake trout (*Salvelinus namaycush*); suckers (*Catostomus*); walleyes (*Stizostedion*).
  C. Depositing eggs in single masses of definite form. A "rope" of eggs, yellow perch (*Perca flavescens*).
  D. Spawning near surface without reference to vegetation or type of bottom. Brook silverside (*Labidesthes sicculus*); alewife (*Alosa pseudoharengus*).

(*Oncorhynchus*). Some fishes that excavate a nest and then defend it against invaders are the bluntnose and fathead minnows (*Pimephales*), and the sunfishes and black basses (Centrarchidae). The latter family, of North American fresh waters, is characterized by this nest building and defensive habit in which the male is the aggressive partner—all except one genus, the Sacramento perch (*Archoplites*), which builds no nest. Bullheads (*Ictalurus*) dig a nest and stay on guard. The Siamese fighting fish (*Betta*) blows a surface bubble nest and defends it, as sticklebacks (*Culaea* and *Gasterosteus*) defend theirs made of twigs, leaves, and detritus on a substrate. The eggs of the bitterling (*Rhodeus*) are deposited in the mantle cavity of a freshwater mussel and those of the lumpsucker (*Careproctus*), beneath the carapace of the Kamchatka crab.

To carry egg protection to its highest degree, some fishes have evolved various types of internal incubation or gestation. Sometimes protection by the parent body is accorded the young by the male, as in the seahorses (*Hippocampus*) and pipefishes (*Syngnathus*), in which the eggs are placed by the female into the brood pouch of the male. In a Brazilian catfish (*Loricaria typus*) the male parent develops an enlarged lower lip to form a pouch in which labial incubation of the eggs takes place. The marine catfishes (Ariidae), including the gafftopsail catfish (*Bagre marinus*) and the sea catfish (*Galeichthys felis*) employ the mouth as an oral incubator. Each male parent carries from ten to thirty developing eggs in his mouth cavity where the hatchlings may still be found after they have begun to feed independently. Oral incubation has also evolved in the cardinalfishes (Apogonidae), among others, that may thus carry a hundred or more eggs.

Truly internal incubation is that which occurs in livebearing fishes. Livebearing, with its many variations of embryonic and nutritive exchange, has evolved independently in several groups of fishes. Among the sharks, livebearing occurs in more than a dozen families. Most are nonplacental or ovoviviparous, but there are a few species that have well-defined nutritive and respiratory exchange structures among the genera *Triakis*, *Carcharhinus*, and *Mustelus* of the requiem sharks (Carcharhinidae), and *Sphyrna* of the hammerhead sharks. Among the rays, skates, and relatives (Rajiformes), ovoviviparity is the rule, with the oviparous skates (Rajidae) being the exception. Interestingly, the oviparous sharks and relatives are bottom dwellers of relatively shoal waters and are never large, whereas the livebearing ones range from small to gigantic and are widely variable in their habits and distribution.

Livebearing among bony fishes is most highly developed in the top-minnows (Cyprinodontiformes), especially in the families of the guppy (*Lebistes*) and mosquitofish (*Gambusia affinis*), Poeciliidae, Mexican livebearers (Goodeidae), the foureyefishes (Anablepidae), and jenynsiids (Jenynsiidae). It also appears in the halfbeaks (Hemiramphidae) and in several families of the perch-like fishes (Perciformes): surfperches (Embiotocidae), clinids (Clinidae), eelpouts (Zoarcidae), brotulas (Brotulidae), and scorpionfishes and rockfishes (Scorpaenidae). In the latter it is universal among more than fifty species of *Sebastodes* of California coastal waters. The living coelacanth, *Latimeria*, is also ovoviviparous.

In the literature of ichthyology it has often been implied that the development of livebearing fishes takes place in a uterus. Some refinement of this generalization is necessary. If it is allowed that a uterus is fundamentally an enlarged oviduct, the organ is present in livebearing sharks (Squaliformes). Among livebearing bony fishes (Osteichthyes), however, development takes place within the ovaries. In some, the development of the embryo is intrafollicular. Here the egg follicle is vascularized, dilated, and fluid-filled. The embryo, which is without an enveloping chorion, develops within the follicle of its original egg and ovulation and birth are simultaneous. This is the situation in the livebearers (Poeciliidae) and other families including the halfbeaks (Hemiramphidae) and the foureyefishes (Anablepidae). In some fishes, however, the egg may be fertilized while still in its follicle but completes development after leaving the ovigerous follicle within the cavity of the ovary; examples of this include the surfperches (Embiotocidae) and the families Jenynsiidae and Goodeidae. In some cyprinodontiforms, more than one brood may be present simultaneously in an ovary—a condition termed superfetation.

Adding to care of eggs, we should think for a moment of the care of the hatchlings which come from them. Hatchlings are actively defended by many fishes, for example, by members of the catfish family (Ictaluridae) and of the sunfish (Centrarchidae) and stickleback (Gasterosteidae) families. Actual herding of the young into places of shelter and away from enemies, even the kinds of antics that lead enemies away from the young, are sometimes practiced by fishes. The antics of the loon (*Gavia*) and other water birds, in distracting menaces to their young, are well known to naturalists. A largemouth bass (*Micropterus salmoides*) and a bowfin (*Amia*) will behave in much the same way, creating diversionary splashes and so forth in an opposite direction while the little school of black young moves away. Some African tilapias are called "mouthbrooders" because the young when they are hatched escape at time of danger into the oral cavity of either sex.

DEVELOPMENT

A hundred years ago, descriptive aspects of development were exciting the biological sciences. Now, although descriptions of the embryology and postnatal development have still not been made for most species of vertebrates—certainly not for most fishes—excitement of discovery centers in learning about the mechanisms of the developmental processes, including the inter-action of genetic and environmental forces in determining various character-istics of the individual.

In its broadest sense, development is an endless process, continuing from generation to generation. For the individual, the start is fertilization of the egg; attainment of maturity and reproduction afford continuity; and the end of development is death. Spermatogenesis and oogenesis are thus near-terminal stages, preparatory for the next generation. For convenience, how-ever, the successive periods of development may be regarded as: (a) early embryonic; (b) transitional embryonic or larval; and (c) postembryonic. Because of predation, limitations of food supply, and greater sensitivity of embryos and larvae to changes in chemical and physical properties of their environment, these stages of development sustain greater mortalities than later stages in many egg-laying fishes.

The early embryonic period starts when the egg is fertilized by a sperm and ends when the embryo has attained the generalized organ systems as they appear in common among all fish embryos.

The transitional phase of embryonic development involves the transforma-tion of the generalized organ systems and body form of the early embryonic stage into those resembling the adult. Thus the definitive body form is attained at or near the end of this stage. During this period of development, two kinds of larvae, free-living and non-free-living, are represented among fishes. Free-living larvae are characterized by having their existence outside of protective embryonic structures. Non-free-living larvae complete their tran-sition inside the egg membranes or within the body of either female or male parent. In species with indirect development, involving metamorphosis, the completion of metamorphosis ends this stage. In some fishes a late embryonic phase covers those developmental changes that occur after definitive body form has appeared. In live-born fishes, the period may be passed within the reproduc-tive tract of the mother. In species with free-living larvae, adult shape is attained either soon after hatching or after a passage through several stages of development in which the young bear relatively little resemblance to the adult (e.g., convict surgeonfish, *Acanthurus triostegus*).

Postembryonic development has juvenile, sexually mature adult, and sene-scent phases. In a juvenile the organ systems, particularly the reproductive system, complete their development. The body form is essentially that of an adult, although some differential growth of parts may go on. As an adult, the individual has the typical mature form of the species and is capable of reproduction. Adulthood passes into old age, senescence and degeneration. In old age, the reproductive potential declines and/or ends, other activities lessen, and various bodily parts slowly degenerate. The period terminates in death.

Embryonic Development

Impregnation.   For present purposes we may think of embryonic development as beginning at the time of impregnation. In the process of impregnation, an egg is penetrated by a sperm. Activation of the egg results from this entry. Almost immediately there is a cortical reaction in the egg resulting in a block to the entry of additional spermatozoa. In some eggs the micropyle, which is the principal port of sperm entry through the vitelline membrane (often called the chorion) surrounding fish eggs, is sealed by material extruded from large vesicles, the cortical alveoli in the egg cortex. The entire vitelline membrane or chorion may also be made impervious to sperm entry. As a result, the process of polyspermy which is found in some other animals is not common in fishes. After completion of the cortical reaction, the vitelline membrane becomes the fertilization membrane and the process of water hardening goes on in externally laid eggs. Independent of impregnation, the egg absorbs water and its chorion is at least temporarily adhesive because of the osmotic processes involved. When water uptake ends, the egg is turgid and firm to the touch.

At various periods of development fish eggs are especially sensitive to handling. They are quite often tender until the eyed stage is reached. Jolts, jars, and rapid changes of temperature are all particularly harmful to them during this early sensitive period. Once pigment has appeared in the iris of the eyes, the embryos are called eyed eggs. In this condition they are quite hardy and can even be transported for long distances if proper conditions, including those of moisture and temperature, are maintained. They remain in a fairly rugged condition until a few days before hatching. Then, perhaps because of softening of the eggshell in preparation for emergence, the embryos again become tender.

Fertilization.   Actual fertilization of an egg is not complete until the nucleus of the egg cell and that of the sperm have made union in the cytoplasm of the egg. Fusion of the male and female pronuclei from sperm and egg respectively completes the process of fertilization. Now the chromosomes which carry hereditary factors, called genes, are brought together in a process called amphimixis and the genes from both parents begin to exert their effects on the development of the embryo.

Cleavage (Fig. 10.7).   Both the processes and the results of embryonic development are similar among the major fish groups although detailed information about them is spotty. For all groups, the processes include cleavage, which is the division of the egg into successively smaller cells called blastomeres. Cleavage ultimately results in the formation of a transitory blastula stage, which consists of a single layer of cells called blastomeres arranged as

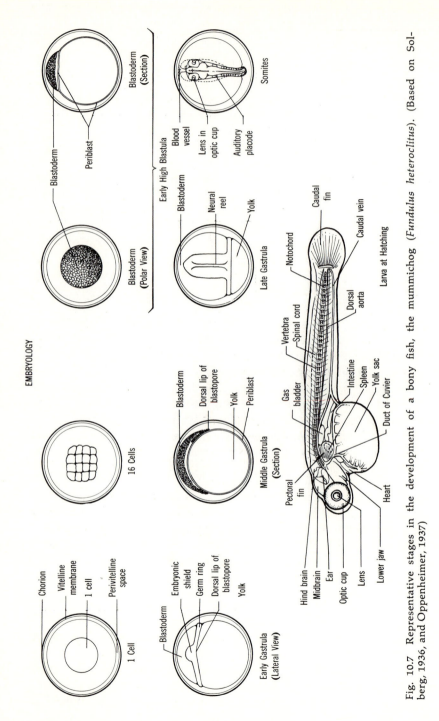

EMBRYOLOGY

Fig. 10.7 Representative stages in the development of a bony fish, the mummichog (*Fundulus heteroclitus*). (Based on Solberg, 1936, and Oppenheimer, 1937)

the rim of a sphere containing a central cavity, the blastocoel, or spread out as a flat plate on the upper surface of a large yolk mass. In the blastula, some areas of cells are already destined to form certain organs. Gastrulation follows blastulation and converts the embryo to a two-layered stage consisting of an outer epiblast and an inner hypoblast. The antero-posterior axis becomes established at this time. Subsequently the gastrula elongates and the major organ-forming areas are extended. The foregoing processes establish the primitive generalized body form. Their rate is under genetic and environmental control, and details differ among major groups of fishes in accordance with the amount and distribution of yolk in relation to cytoplasm within the egg cell.

Typically the cleavage of chordate eggs is either holoblastic or meroblastic, but many intermediate types exist. In the holoblastic type, the entire cell divides into two, then four in meridional planes. The third cleavage plane is latitudinal, resulting in eight cells of nearly equal size. In meroblastic cleavage, only a small disc-like part of the egg divides to produce the blastoderm. The yolk material remains undivided, later to be enveloped by growing tissues. A classical example of holoblastic cleavage is seen in the cephalochordate amphioxus (Branchiostoma). Here the cells divide nearly equally and result in a neat blastula which is a hollow ball of cells; the cavity within the blastula is the segmentation cavity or blastocoel. The blastula in turn gastrulates simply by having one of its sides invaginate to yield a gastrula resembling a rubber ball pushed in at one side. In the lampreys (Petromyzonidae) there is more yolk than in amphioxus. The eggs exhibit a polarity with more yolk at the lower or vegetal pole, an egg-cell condition termed telolecithal. As a result of yolk impedance, the cleavage is not as ideally holoblastic as in the amphioxus egg but is still classified as a holoblastic type. The yolk-free cells of the upper or animal pole (the micromeres) divide faster than those of the lower or vegetal pole of the egg (the macromeres), and overgrow the macromeres. Gastrulation is by convergence of cells toward the blastopore at the posterior end of the embryo, and subsequent turning-in of the micromeres at the border between them and the macromeres. Cleavage is somewhat holoblastic in the South American lungfish (Lepidosiren), the sturgeons (Acipenser), the gars (Lepisosteus), and the bowfin (Amia), but it exhibits various meroblastic features. The early development of these fishes is thus intermediate between holoblastic and meroblastic cleavage types.

Meroblastic cleavage is the prevalent form of egg division in fishes and other vertebrates. In the primitive hagfishes (Myxinidae) and in the sharks and relatives (Chondrichthyes), the egg is strongly telolecithal and cleavage is restricted to a small disc of cytoplasm at the animal pole. The same is true for the bony fishes (Osteichthyes) other than those described above. Cleavages result in the formation of two kinds of cells, blastoderm and periblast.

The blastoderm cells are distinct and produce the embryo. The periblast or trophoblast cells lie between the yolk and cells of the blastoderm and cover the entire yolk mass, having originated from the most marginal and outlying blastomeres. They become syncytial and are involved in mobilization of yolk reserves. The blastulae of the hagfishes, of the sharks and their relatives, and of the teleosts (Actinopterygii) are disc-like and hence called blastodiscs. The blastocoele or segmentation cavity is between the outer blastoderm and the central periblast that covers the yolk beneath the blastoderm.

Through work in experimental embryology, it has become possible to identify, already in the blastula, regions destined to give rise to specific organs. The areas of presumptive gut-forming, notochordal, neural and epidermal cells have been thus established, as has a region of potential mesoderm (Fig. 10.8). The term mesoderm refers to the middle germ layer of the embryo. It is formed during gastrulation when the hypoblast subdivides into mesoderm and endoderm. The outer germ layer of the early gastrula, the epiblast, is termed the ectoderm in the three-layered embryo.

Gastrulation. The process of gastrulation is generally similar among most fishes with meroblastic cleavage. There are, however, some differences among

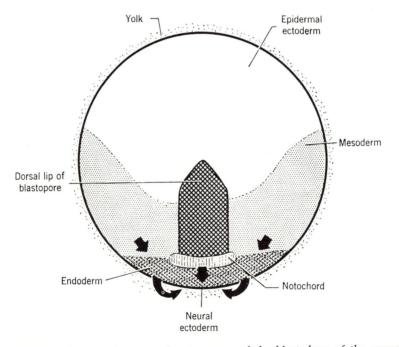

Fig. 10.8 Presumptive organ-forming areas of the blastoderm of the mummichog (*Fundulus heteroclitus*). Arrows indicate direction of cell movement. (Based on Oppenheimer, 1936)

bony fishes, for example, in the bowfin (*Amia*), as well as between bony fishes and the sharks and their relatives (Chondrichthyes). A brief description of the basic process must suffice here.

As the time for gastrulation comes on, the under-rim of the blastodisc thickens to form the "randwulst" or marginal ridge which comes to have an inner layer termed the germ ring. Caudally, the ring is thickest and is recognized as the embryonic shield. At the beginning of gastrulation, the presumptive endodermal cells at the caudal edge of the embryonic shield turn inward beneath the blastoderm and then stream forward beneath it to become the endodermal portion of the hypoblast. For this incursion of cells, an opening, the blastopore, appears at the tail end of the embryonic shield. Also moving inward over the dorsal lip of the blastopore are cells of the prechordal plate and notochord that establish the embryonic axis. Presumptive mesodermal cells also migrate inward over the lip of the blastopore and position themselves on both sides of the embryonic axis beneath the ectoderm. Initially the involuted cell mass is termed the hypoblast. During subsequent proliferation and delamination, the endodermal and mesodermal components of the hypoblast become organized to establish the primordia of the internal organ systems.

While the embolic processes of involution of cells go on, cellular overgrowth of the yolk (epiboly) also proceeds. In epiboly, randwulst and germ ring cells not involved in involution, accompanied by presumptive ectodermal cells, grow to cover the yolk mass outwardly as epiblast; the periblast provides through growth an inner covering of the yolk mass. The initial periblast cells and epiblast in covering the yolk form the yolk sac. As epiboly is proceeding, the presumptive neural plate, forerunner of the central nervous system, transforms into a ridge-like keel along the ectodermal middorsal line. This solid keel, in contact with notochordal cells beneath, gradually becomes overgrown by epidermis and eventually becomes tubulated to form the hollow neural tube.

Gastrulation is complete in bony fishes when the yolk mass has been overgrown (Fig. 10.9). During this process, some of the mesodermal tissue along both sides of the notochord has become organized into segments, the somites. Externally one can distinguish the boundary between embryo and yolk sac.

Organogenesis and Hatching.   Subsequent to gastrulation in fishes, early embryonic development continues until a primitive vertebrate body form is established (Fig. 10.7). The body at this time is more or less cylindrical and basically bilaterally symmetrical. Head and pharyngeal regions project from the yolk mass anteriorly; the trunk lies over the yolk; and the tail projects posteriorly. Five fundamental organ-forming tubes have appeared and have consistent relationships to the axial notochord.

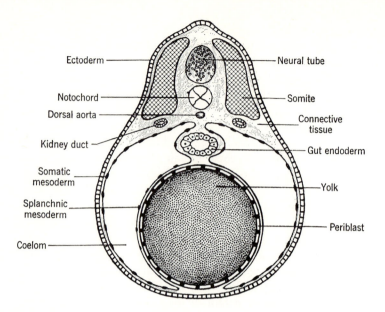

Fig. 10.9 Diagrammatic cross section of a fish embryo removed from egg shell during late embryonic stage to show how yolk sac is covered. (Source: Lagler, 1956).

Tubulation accounts for five tubes of tissue: epidermal (body-covering), neural, endodermal (primitive gut), and two mesodermal (somitic and coelomic cavities). These five tubes of tissue, plus the unspecialized mesenchymal cells of chiefly mesodermal origin, have key roles in organogenesis.

The ectodermal tube becomes the body covering or epidermis. Included in its derivatives among fishes are the outer layer of teeth, the olfactory epithelium and nerve, the lens of the eye, and the inner ear. The neural tube, arising from the sunken neural keel, is also ectodermal in origin. It gives rise to the brain and spinal cord and remaining parts of the eye, including the retina and optic nerve. Associated neural crest cells, among other things, provide assorted ganglia of the nervous system and contribute to the mesenchyme. The pigment cells known as melanophores appear to originate from the neural crest.

The mesodermal tubes of the trunk soon segregate into dorsal, intermediate, and lateral divisions. Dorsal mesoderm is early divided into the block-like somites already mentioned. Each somite is subdivided into three regions called sclerotome, myotome, and dermatome. The sclerotome contributes to the formation of the axial skeleton. The myotome develops into trunk musculature, the appendicular skeleton, and the appendages and their muscles. The dermatome develops into the connective tissue of the dermis of the skin, and its derivatives including scales.

Intermediate mesoderm gives rise to the kidneys, part of the gonads, and their ducts. Lateral mesoderm splits into inner and outer layers enclosing the coelomic cavity. Linings of the pericardial and peritoneal cavities come from lateral mesoderm, as do the heart, blood, trunk blood vessels, and all layers of the gut except its epithelial lining, which comes from endoderm. Mesenchyme in the head contributes to outer layers of the eye, the head skeleton, head musculature, and the dentinal layers of the teeth.

Endoderm supplies the primary sex cells (the germ tract) to the gonads in fishes and makes up the inner epithelium of the entire digestive tract and its associated glands. It also contributes to the development of several endocrine glands that are derived from the primitive gut, including the thyroid and the ultimobranchial bodies. Hatching results from a softening of the chorion by enzymes or other chemical substances that are secreted from ectodermal glands usually on the anterior surface or from endodermal glands in the pharynx. The activity of the embryo assists in breaking through the chorion. Because the chemical composition of the chorion is not well understood, there is doubt about the mode of action of the enzymes (chorionases). A considerable part of the nutrient material in the chorion may be utilized by the embryo via the perivitelline fluid.

Recapitulation by Stages.   When embryos and larvae of different species of either the same size and age are compared, most often the developmental stages are different. Comparisons are most useful when based on equal developmental stage. Even so, problems arise because of difficulties in defining these developmental stages. Nevertheless, in reviewing embryonic development, we may follow a sequence of stages of demonstrated utility in making comparisons among species, at least among oviparous ones, as derived for the mummichog (*Fundulus heteroclitus*), a killifish (Cyprinodontidae; Fig. 10.7). Approximate age in hours in the mummichog is cited for eggs held at 27°C (77°F):

Stage 1.   The unfertilized egg.

Stage 2.   One-celled embryo (Fig. 10.7), the newly fertilized egg (1 hour).

Stage 3.   Two-celled embryo (1½ hours).

Stage 4.   Four-celled embryo (2 hours).

Stage 5.   Eight-celled embryo (2½ hours).

Stage 6.   Sixteen-celled embryo (3 hours, Fig. 10.7).

Stage 7.   Thirty-two-celled embryo (3½ hours).

Stage 8.   Early high blastula. The blastoderm is elevated into a domed, cap-like structure and the periblast is being established (5½ hours, Fig. 10.7).

Stage 9.   Late high blastula. Cells of blastoderm are smaller than in Stage 8.

Stage 10.   Flat blastula. Blastoderm has become a flattened disc, rather than an elevated bulge as formerly (8 hours).

Stage 11.   Expanding blastula. Blastoderm has begun to grow over yolk.

Stage 12.   Early gastrula. Germ ring and embryonic shield formed and blastopore just opening (16 hours, Fig. 10.7).

Stage 13.   Middle gastrula. About half of yolk covered by blastoderm, neural keel just visible in midline of embryonic shield (19 hours, Fig. 10.7).

Stage 14.   Late gastrula. More than half of yolk covered by blastoderm, embryonic shield narrowing, neural keel more clearly visible (22 hours, Fig. 10.7).

Stage 15.   Closure of blastopore. Little embryonic differentiation except formation of rudiments of central nervous system and perhaps of optic vesicles and first somites (26 hours).

Stage 16.   Expansion of forebrain begins (for formation of the optic vesicles). Three primary vesicles visible in brain.

Stage 17.   Formation of cavity in optic vesicles, mesodermal segmentation provides one to four somites.

Stage 18.   Formation of the auditory placode. Somites range from four to fourteen and extra-embryonic coelom appears as a cavity developed by yolk-sac epithelium (36 hours).

Stage 19.   Cavity appears in neural cord behind brain. Ectoderm thickens to form lens of eye and olfactory pit. Somites range from fourteen to twenty (42 hours).

Stage 20.   Expansion of midbrain to form optic lobes. Somites range from twenty to twenty-five. Melanophores appear about neural cord and are present over yolk. Pericardium established. Heart visible, tubular, and pulsing. Location of pectoral fin becoming visible by concentration of cells.

Stage 21.   Motility. Muscular contractions come at about twenty-eight somites.

Stage 22.   Circulation. Circulation starts with about thirty-five somites. Forebrain walls forming cerebral hemispheres.

Stage 23.   Otoliths appear in ear. Melanophores appear in pericardium and blood flows in yolk vessels.

Stage 24.   Pectoral-fin bud pointed.

Stage 25.   Formation of urinary vesicle (as an outgrowth of hindgut).

Stage 26.   Formation of liver and peritoneal cavity.

Stage 27.   Pectoral fin becomes rounded.

Stage 28.   Pigmentation of peritoneal walls.

Stage 29.   Circulation established in pectoral fin. Fin becomes motile.

Stage 30.   Rays appear in caudal fin. Lower jaw is formed.

Stage 31.   Formation of gas bladder (diverticulum of gut). Eyes and jaws become motile.

Stage 32.   Hatching. Some yolk is still present (11 days, Fig. 10.7).

Stage 33.   Pigmentation and growth of gas bladder.

Stage 34.   Yolk absorption completed (12 days or more).

In subsequent larval development, skin and scales have yet to complete their differentiation, the axial skeleton must still be finished, the dorsal and anal fins and adult kidneys are yet to form, and definitive bodily proportions and pigmentation are still to be realized.

Nutritive and Respiratory Relationships.   Embryos of the many different kinds of fishes have different nutritive and respiratory relationships depending on whether the young are born alive or hatch from externally laid eggs.

The embryos of egg layers essentially are totally yolk dependent for their food. Many species have one or more oil droplets in or about the yolk mass, which serve both as a potential source of energy and as a flotation or righting organ. Supplemental dissolved organic material and mineral salts may be absorbed directly from the water. Respiration and excretion of eggs laid in water is through the semipermeable egg membranes and shell.

For the embryos of all livebearers, there is nutritive and respiratory exchange *in utero* or *in ovario*, even when a placental arrangement is lacking, as in the ovoviviparous sharks or bony fishes. In these latter groups the uterine or ovarian juices make a physiological union between mother and embryo. The exchange is more intimate, however, among placental viviparous fishes. Histologically, the uterine lining varies from a mucus-secreting layer of cuboidal epithelial cells in species that depend on egg yolk for nourishment (*Acanthias vulgaris*) through forms with moderately folded serous-secreting linings (*Torpedo*) to those with uterine linings beset with villi (trophonemata) of varying lengths and complexities that secrete an abundance of fat (*Trygon violacea*). Yolk is mainly digested within the intestine of embryos which depend on it for nourishment (*Squalus acanthias*). True viviparity with yolk sac placenta is confined to certain species of two families of sharks (Carcharhinidae and Sphyrnidae). Embryos of *Sphyrna tiburo* develop elaborate circulatory networks in the walls of their yolk sacs, which become greatly folded. The uterine mucosa becomes folded, and the interdigitation of these two series of folds with a thinning of their epithelia bring maternal and fetal circulations into close proximity. The embryo is attached to the placenta by a modified yolk stalk (umbilical cord).

Among viviparous bony fishes, embryonic nutrition and respiration may successively involve several sources and organs, some of them quite special. In the Mexican livebearers (Goodeidae), the embryonal yolk sac, pericardial sac, and trophotaeniae variously take over the tasks. Trophotaeniae are finger-like extensions of rectal tissue. In the jenynsiid topminnows (Jenynsiidae), similar relationships exist between embryo and ovarian cavity and tissues, but trophotaeniae are replaced by trophonemas which are wormlike extensions of the ovarian wall that grow into the gill chambers of the embryo. In embryonic nutrition, the brotulas (Brotulidae) involve the yolk sac, body covering, gills, and finally the gut proper of the embryo through ingestion of the intraovarian fluid and by feeding on ovarian tissues, including egg cells. In the jenynsiids, cannibalism is practiced by some embryos on others within the same intraovarian cavity.

In the eggs of trouts and salmons (Salmoninae), the carbohydrate level increases gradually during development, although the glucose level falls temporarily at hatching. Glycogen is probably synthesized near hatching and is stored in the liver of the embryo near the end of the yolk-sac stage. Fat is used as a fuel, possibly 70 to 80 percent being consumed over the whole period of development. Fat contributes an estimated 68 calories per egg by the time the yolk is absorbed. Protein contributes some 45.9 calories to the process. Among various trouts 60 to 63 percent of the available protein in the yolk supplies energy and 16 to 40 percent of the fat. The energetic requirement of development is some 80 calories, and metabolic combustion uses 114 calories. The efficiency is high; about 40 percent of the energy originally incorporated in the egg is converted into growth of the embryo; the remainder, about 60 percent, goes for such activities as osmoregulation, secretion, circulation, and movements.

Transitional or Larval Development (Figs. 10.10 and 10.11)

Although in common usage a young fish is an embryo until birth or hatching, as indicated previously there are, according to species, varying amounts of subsequent development to be undergone in the earliest free-living stages. Generally, developmental stages prior to the adult stage but following hatching or birth are termed larval, and the young during this period are called larvae. Larval fishes are also called fry. The period of larval life may range from a few moments to several years in length. The ammocete larvae of the sea lamprey (*Petromyzon marinus*), for example, require 5 or more years before metamorphosis.

In fishes, larval development is commonly, though not universally, divisible into prolarval and postlarval stages. Prolarvae are distinguished by the presence of the yolk sac and are commonly called sac fry by fish culturists. In some species, when the yolk sac disappears the little fish is a diminutive adult, commonly called an alevin or an advanced fry by fish culturists. Such a direct development is characteristic of the trouts and salmons (Salmoninae), the North American freshwater catfishes (Ictaluridae), the hagfishes (My-

LAMPREY METAMORPHOSIS

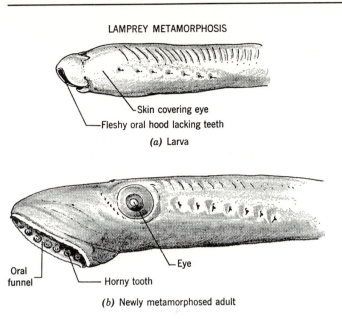

Skin covering eye

Fleshy oral hood lacking teeth

(a) Larva

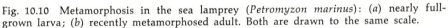

Oral funnel

Eye

Horny tooth

(b) Newly metamorphosed adult

Fig. 10.10 Metamorphosis in the sea lamprey *(Petromyzon marinus)*: *(a)* nearly full-grown larva; *(b)* recently metamorphosed adult. Both are drawn to the same scale.

xinidae), and the sharks and their relatives (Chondrichthyes), as examples. In other fishes, a distinct postlarval stage follows the prolarval, sac-fry one. A postlarva must undergo a metamorphosis to lose larval structures and gain adult features. For many species recognition of the postlarval stage is highly subjective. An indirect development has been recognized in many families including, among others, the South American lungfishes (Lepidosirenidae), lampreys (Petromyzonidae; Fig. 10.10), bowfin (Amiidae), herrings (Clupeidae), suckers (Catostomidae), carps (Cyprinidae), sunfishes (Centrarchidae), goosefishes (Lophiidae), molas (Molidae), deepsea anglerfishes (Ceratiidae, rockfishes (Scorpaenidae), and eels (Anguillidae). Differences between postlarvae and adults in some fishes are so trenchant that they have resulted from time to time in taxonomic confusion. Thus larvae of the eel *(Anguilla)* were once described in the genus *Leptocephalus* and those of the lampreys (Petromyzonidae) in the separate genus *Ammocetes!*

Features of postlarvae are many and diverse. They range from distinctive pigmentation as in many carps (Cyprinidae) and transparency as in the leptocephalus postlarvae of the eel *(Anguilla)*, through profoundly anatomical features. Many marine fishes have a very sharp transition from a pelagic postlarva to a bottom-dwelling juvenile and adult. Such change is nowhere better shown than in the shift from pelagic bilaterally symmetrical, upright young of the flatfishes (Pleuronectiformes) to bottom dwellers after meta-

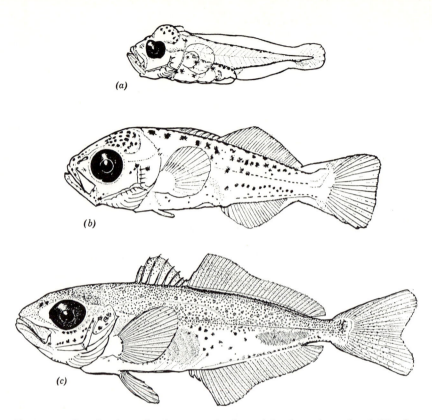

Fig. 10.11   Post-hatching development of a bony fish, the jack mackerel (*Trachurus symmetricus*). (*a*) Larva of 4.9 mm. lacking pelvic fin and development of rays in dorsal and anal fins. (*b*) Larva of 10 mm. with pelvic fin just appearing but with rays established in other fins; note changes in pigmentation over (*a*). (*c*) Juvenile of 28 mm. showing establishment of adult body form and further development of pigmentation over (*a*) and (*b*). (Source: Ahlstrom and Ball, 1954).

morphosis, with both eyes on one side of the head, swimming with the eyeless side down. Other striking postlarvae have elaborate flotational devices. For example, the pelvic fins of larval goosefishes (*Lophius*) are enlarged, and the molas (Molidae) have bodily spines (as well as an evident caudal fin) which are lost at metamorphosis. The skin balloons outward in members of the deepsea anglerfish family (Ceratiidae) and the scorpionfish and rockfish family (Scorpaenidae). The leaf-shaped, leptocephalus larvae of the eels (Anguilliformes), and larvae of the ladyfish (*Elops*), the bonefishes (Albulidae) and their relatives, and many herrings (Clupeidae), all have a watery, mucoid dilution of the muscles which renders them very transparent and makes them float. Enlargement of anterior dorsal-fin rays as flotational structures in larvae is shown by a sea bass (*Diploprion*; Serranidae), flounder

(*Laeops*), the dealfish (*Trachipterus*), the pearlfishes (*Carapus*), and the goose-fishes (*Lophius*). The ammocete larvae of lampreys (Petromyzonidae) are burrowers before metamorphosis, essentially blind and equipped with a bonnet-like oral hood and specialized ciliated cells in the pharynx for wafting food particles toward the esophagus. Both behavior and morphology change when ammocetes assume the adult condition. The adults are free-swimming, have no oral hood or larval feeding apparatus, and develop a sucking-disc mouth, a buccal funnel with teeth, and evident, functional eyes. The teeth are strong in such parasitic species as the sea lamprey (*Petromyzon marinus*) but weak in the nonparasitic ones, such as the American brook lamprey (*Lampetra lamottei*), which do not feed after metamorphosis.

Feather-like external gills are premetamorphic features of the African and South American lungfishes, respectively *Protopterus* and *Lepidosiren*, and also of the bichirs, *Polypterus*, of Africa, and certain elasmobranchs (Fig. 8.7). Postlarvae of the gars (*Lepisosteus*) and the bowfin (*Amia*) have adhesive organs on their snouts. In the dealfish (*Trachipterus*) fin rays that are modestly elongated in the adult differ both in greater number and elongation in the young before metamorphosis. The swordfishes (Xiphiidae) and their close relatives the marlins, sailfishes and spearfishes (billfish family Istiophoridae) all lack, when very young, the great prolongation of the snout into a bill. The elongation of the lower jaw in the needlefishes (Belonidae) takes place well before the upper jaw grows forward to match it.

A few fishes have successfully eliminated either free-living larval stages (by a most direct development) or rarely the adult stage (by neoteny). The males of some viviparous surfperches (Embiotocidae) are sexually mature at birth, thereby doing away with the hazard of a pelagic larval life. Neotenic fishes that mature sexually while retaining larval characteristics further include some pelagic gobies (Gobiidae), eastern Asiatic icefishes (Salangidae), and larvae of the blenny-like genus *Schindleria* which may be mature while still a postlarva only 12 millimeters long. A relative of the deepsea mouth-fishes (stomiatoids), the stalk-eyed fish (*Idiacanthus*) has sexually dimorphic neotony that is very marked; the mature males are less than one-sixth of the size of the females and are larvoid.

Juvenile to Adult Development

Typically metamorphosis or termination of embryonic development is followed by a period of growth in size accompanied by maturation of the gonads. The females usually attain the largest sizes but the males usually mature earliest. An individual fish passes from a young, miniature adult stage through a juvenile condition, to sexual maturity. The period involved may be short, as the 4 to 10 weeks (varying with temperature) required by a guppy (*Lebistes*), or long, as the 6 to 12 years needed by the freshwater eels (*Anguilla*), or the 15 or more years among sturgeons (*Acipenser*). In a few fishes, most of the growth is achieved by larvae prior to metamorphosis. This holds, for example, in the nonparasitic lampreys (*Lampetra* and some species of *Ichthyomyzon*). In these lampreys, there is no feeding after metamorphosis, hence no growth but only maturation of the gonads, one season of spawning, and death.

Any consideration of development after hatching must be concerned with the onset of feeding. As previously stated, it is in the early life stages of fishes that the largest mortalities occur (sometimes exceeding 99 percent). In this period, food is a critical factor in survival. Small fishes at hatching require small foods. Hence, the first foods for most fishes are plankton, bacteria, or tiny bits of organic debris, at the bottom or free in the water depending on where the larvae dwell (Fig. 5.1).

Another feature of early stages in the life history is their pigmentation (Fig. 10.11). Many kinds of fishes at particular developmental stages have chromatophores, especially melanophores, distributed in patterns that permit ready identification. Most fishes with demersal eggs go through a period after hatching when their movements are awkward as a result of the heavy burden of the yolk sac. As the yolk is absorbed, young which hatched on the bottom may soon begin to rise and move about with directed movements and become more and more adept at feeding in the natural environment.

Senescence

Senescencce refers to the deteriorative changes that accompany aging in an organism. In fishes it is characterized by a slowing of growth in length, an acceleration in mortality rate, gradual loss in reproductive capacity, increase in abnormalities among offspring, and sometimes attainment of elevated relative robustness or fatness. Generally males undergo senescence and die at earlier ages than females. The period of old age also typically shows a slowing of activity with accompanying changes in feeding, distributional, and other habits. The period may be long in nature, but it may be prolonged in captivity. For example, the northern pike (*Esox lucius*) has been recorded as approaching the age of 25 years in nature, but, in captivity, it has lived 75 years. Senescence may set in when the individual is very small, as in the tiny Philippine goby (*Mistichthys luzonensis*) which seldom exceeds half an inch in length. It may also come when the individual is relatively young as it must do in another goby (*Latrunculus pellucidus*), the life-span of which is only a year.

In wild populations of fishes, death comes soonest to the fastest growing individuals.

Rates and Factors of Development

In practical fish culture the period of embryonic development, or the developmental period, extends from impregnation to the average hatching date of a batch of eggs or birth of a brood. This interval varies considerably according to species and even somewhat among the offspring of individual parents within species. Its duration is strongly influenced by environmental conditions. By definition, the developmental period ends when half of the eggs in a batch have hatched. The hatching (or birth) period is the interval of time over which a complement of eggs actually hatches (or young emerge from the female parent), from the first to the last.

Many forces affect the whole developmental process. They influence success and determine failure. They also affect rate and determine form and structure. Outstanding among environmental factors is temperature. For example, temperature affects rate of development or the fraction of the developmental process achieved per day. This rate has been expressed simply as the reciprocal of the developmental period in days. The larger the fraction, the faster the development. Thus a fish with a developmental period of 88 days has a rate of $\frac{1}{88}$, whereas one developing in 9 days would have $\frac{1}{9}$. To provide even more ready comparison among species, this rate may also be expressed as a decimal value (for example, 0.011 or 0.111 respectively), for the fractions just considered.

Temperature. Both the developmental period and the hatching period are generally shorter at higher temperatures than at lower ones. Interestingly, many species normally develop in nature under temperature conditions which are not optimal as determined by laboratory experiments. Species differ in their temperature optima and tolerances during development; but for all kinds there are temperatures which are too low and too high for development to proceed. Extremes or sudden changes may be lethal. The range of temperatures over which normal development can proceed may be wide. The mummichog (*Fundulus heteroclitus*), for example, develops at temperatures from 12°C (53.6°F) to 27°C (80.6°F) with 2 percent or fewer abnormalities.

How developmental rate differs among species is shown in the following. At an average temperature of 43°F (6.1°C) during the developmental period, brown trout (*Salmo trutta*) require 88 days; brook trout (*Salvelinus fontinalis*), 80; sockeye salmon (*Oncorhynchus nerka*), 75; rainbow trout (*Salmo gairdneri*), 61; white perch (*Morone americana*), 20; cod (*Gadus morhua*), 14; and pollock (*Pollachius virens*), 9.

There has long been interest in the direct relationship of temperature to development of individuals within a species. Such a relationship was determined more than a hundred years ago for the brook trout (*Salvelinus fontinalis*). For this fish, the developmental period decreases with rise of average water temperature (in degrees Fahrenheit) as follows: 37°, 165 days; 41°, 103 days; 48°, 56 days; 50°, 47 days; and 54°, 32 days. An early generalization for eggs of the brook trout was that the developmental period was 50 days for an average water temperature of 50°F, and that, for each degree warmer or cooler, 5 days, respectively, less or more, would be required. For a long time, this scheme was the rule of thumb for making estimates at trout hatcheries.

Subsequently it has been realized that the time necessary to reach a definite stage of development multiplied by temperature is a constant (K) in the formula $yT=K$ as variously proposed in the past. When the developmental period is plotted as the abscissa and temperature as the ordinate, the resultant figure is a rectangular hyperbola (Fig. 10.12). In the foregoing temperature equation, y is time to reach a certain stage of development following fertilization (for example, hatching), and T is the temperature at which the development is taking place. The corresponding intercept equation for rate of development is $v=K_0+K_1T$ where v is the velocity or rate (that is, 1000 times the reciprocal of y) and K_0 and K_1 are in order the constants for intercept (the biological zero) and slope. A plot of rate (Fig. 10.13) against temperature gives a straight line useful in predicting deviations; however, this plot is curvilinear over wide ranges of temperature—simple linearity applies over a narrow range.

Temperature has an effect on the length of the hatching period, as well as on the total developmental interval. An example of the influence of tempera-

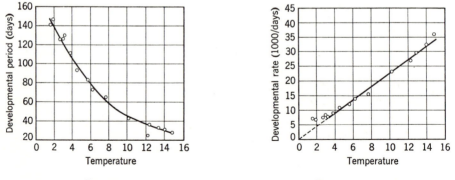

Fig. 10.12

Fig. 10.13

Fig. 10.12 Effect of average water temperature on the developmental period (fertilization to mean date of hatching) in the brook trout (*Salvelinus fontinalis*). (Based on Hayes, 1949)

Fig. 10.13 Rate of development (1000/developmental period in days) in relation to average temperature during developmental period. Data same as Fig. 10.12. (Source: Hayes, 1949)

ture on the hatching period of different samples from a single lot of eggs of the rainbow trout (*Salmo gairdneri*) held at different temperatures follows.

Temperature		Hatching Period to
°F	°C	Nearest Day
42.3	5.7	14
46.8	8.2	8
55.2	12.9	6
63.7	17.6	3

The rate of hatching within the hatching period may be expected to follow a bell-shaped curve of distribution. It is at the peak of this curve under ordinary situations when 50 percent of the eggs will have hatched and the developmental period will have ended.

Temperature affects the efficiency with which food stored in the egg is converted into body weight of the embryo. In an experiment with hatchlings of the Atlantic salmon (*Salmo salar*), gains in weight of the larvae were compared with losses in weight of the yolk sac in fish kept in a series of tanks differing in temperature from 0.2 to 16°C. Efficiency of yolk conversion into fish flesh was low and constant at about 42 percent in the coldest tanks. It started to increase at 5°C and reached a maximum efficiency of about 60 percent in the warmest tank, when calculated for a 10-day interval.

Light.   Light can influence the developing fish embryo, but existing data are equivocal. Although some experiments with trout and salmon (Salmoninae) eggs have indicated that exposure to fluorescent light results in increased mortality, other experiments show no effect on embryos exposed to low levels of fluorescent lights (which are normally used in hatcheries). Juvenile cutthroat trout (*Salmo clarki*) even showed the fastest growth in light.

Dissolved Gases.   Gases dissolved in the water are also factors in development, especially of the eggs of oviparous fishes. It may be expected that both optimum concentrations and intolerable extremes of dissolved oxygen exist and vary according to species. The generally accepted range for development lies between 4 and 12 parts per million (ppm) of dissolved oxygen. Coldwater, stream-spawning species have higher requirements than sluggish or stagnant, warmwater ones. Conditions and requirements of deepsea fishes are quite unknown, primarily because of difficulties in obtaining healthy fish from the deep ocean and holding and experimenting on them under high barometric pressures.

Oxygen pressure can influence the number of meristic elements. In the brown trout (*Salmo trutta*), decreasing oxygen pressure during embryonic development produces an increase in the number of vertebrae.

At least two gases in water are toxic to fishes and their embryos—carbon dioxide and ammonia. Crude information suggests that under ordinary conditions of dissolved oxygen, concentrations of carbon dioxide up to 8 or 9 ppm have little effect on development. In concentrations from 10 to 30 ppm gradual impairment of the process may be expected, whereas concentrations greater than 30 ppm may arrest development and lead to death. Interestingly, raising carbon dioxide pressures during embryonic development decreases the number of vertebrae in the brown trout (*Salmo trutta*). Very high levels, between 55 and 80 ppm, increase the mortality of posteyed embryos and also increase the number of deformed larvae. Ammonia can be toxic at low concentrations. Water quality specialists recognize 1.5 ppm as tolerable for aquatic life but express concern over greater concentrations. Water supersaturated with nitrogen such as occurs below some dams can result in gas-bubble disease in fry.

Salinity.   Salinity can de damaging to the eggs of freshwater fishes and, vice versa, fresh water can harm eggs of marine species. This is clearly shown in the barriers which either fresh or marine waters constitute for fishes primarily restricted to the other environment (Chapter 14). If the salt content of water is intolerably high, the eggs of freshwater fishes immersed therein would lose water and die by shriveling. Similarly, if marine fish eggs are placed in fresh water they may imbibe water and burst. Salinity also has a selective role in development of some structures, as shown experimentally in the effect of its variation on the number of bony plates that develop on the sides of the threespine stickleback (*Gasterosteus aculeatus*).

Endocrines.   The importance of endocrine factors in development is well established. Of particular interest are the roles of the pituitary and thyroid in metamorphosis (Chapter 11).

Amount of Yolk.   The amount of yolk present in the egg has a relationship to the rate of development. Ordinarily in fishes it seems that the larger the amount of yolk, the slower the rate of development. Among animals generally, great parental dependence seems to be associated with relative slowness in development but it is not known if this applies to fishes.

Fixation of Meristic Elements

Rate of development also has something to do with the determination of meristic elements such as vertebrae, fin rays, gill rakers, and numbers of scale rows. For example, the golden shiner (*Notemigonus crysoleucas*), a

North American minnow, in the North has more meristic elements than it does at the southern extent of its range near the Gulf of Mexico. Because less energy is expended in metabolism at the lower temperatures of the northern latitudes, more of the nutritive material of the egg may be available for synthesis of meristic elements. It is likely that environmental factors change the relation between growth and differentiation. If differentiation is late, more tissue is available to be differentiated, leading to a higher count. Analyzed experimentally, the mechanism of such control is found to be more complicated than when judged from field studies. In the brown trout (*Salmo trutta*), subsamples of fertilized eggs from one pair of parents (both with the same number of vertebrae) were treated with low, intermediate, and high constant temperatures during development. The outcome in numbers of vertebrae was lowest at the intermediate temperature and highest at temperatures above and below. In the dorsal, anal, and pectoral fins the highest number of rays appeared at the intermediate temperature but was less at both the higher and lower temperatures. In the same group of experiments it was learned that the numbers of vertebrae and anal-fin elements are fixed in the embryonic stages before the eye is completed, thus well before hatching; the dorsal and pectoral rays are set later, as is the number of scales in the lateral line of the rainbow trout (*Salmo gairdneri*).

Biogenetic Theory

No discussion of the embryology of fishes would be complete without mention of the evidence that embryos afford in support of the biogenetic theory, which holds that ontogeny recapitulates phylogeny. Possible support for biogenesis is well illustrated by development of a heterocercal condition of the tail skeleton, among other features, in fishes. Heterocercal termination of vertebrae appears in embryos of many kinds of fishes, regardless of whether or not the definitive tail is heterocercal or some other such as the predominant homocercal type. This sequence in development has been held as evidence that the heterocercal condition is the more primitive on the ground that ontogeny or embryonic development is recapitulating phylogeny or evolution of the structure.

SPECIAL REFERENCES ON REPRODUCTION AND DEVELOPMENT

Ahlstrom, E. H., and O. P. Ball. 1954. Description of eggs and larvae of jack mackerel (*Trachurus symmetricus*) and distribution and abundance of larvae in 1950 and 1951. *U.S. Fish and Wildlife Serv. Fish Bull.*, 56 (97): 209–245.

Armstrong, C. N., and A. J. Marshall. 1964. Intersexuality in vertebrates including man. Academic Press, New York.

Battle, H. 1940. The embryology and larval development of the goldfish (Carassius auratus L.) from Lake Erie. *Ohio Jour. Sci.,* 40 (2): 82–93.

Blaxter, J. H. S. 1969. Development: Eggs and larvae. *In* Fish physiology, 3: 177–252. Academic Press, New York.

Breder, C. M., Jr., and D. E. Rosen. 1966. Modes of reproduction in fishes. Natural History Press, Garden City, N.Y.

Echelle, A. A. 1972. Developmental rates and tolerances of the Red River pupfish, *Cyprinodon rubrofluviatilis. Southwestern Nat.,* 17 (1): 55–60.

Gerking, S. D. 1957. Evidence of aging in natural populations of fishes. *Gerontologia,* 1 (5): 287–305.

Hayes, F. R. 1949. The growth, general chemistry, and temperature relations of salmonoid eggs. *Quart. Rev. Biol.,* 24 (4): 281–308.

Hoar, W. S. 1969. Reproduction. *In* Fish physiology, 3: 1–72. Academic Press, New York.

Hubbs, Clark. 1967. Analysis of phylogenetic relationships using hybridization techniques. Symposium on newer trends in taxonomy. Bull. Nat. Inst. Sci. India, 34: 48–59.

Hubbs, Clark, and D. F. Burnside. 1972. Developmental sequences of *Zygonectes notatus* at several temperatures. Copeia, 1972 (4): 862–865.

Kerr, J. G. 1907. The development of *Polypterus senegalus* Cuv. *In* Budgett Memorial Volume, pp. 195–284.

———. 1909. Normal plates of the development of *Lepidosiren paradoxa and Protopterus annectens.* Normeltafeln z. Entw. Gesch. der Wirbelt., 10: 1–163.

Mathews, S. B. 1965. Reproductive behavior of the Sacramento perch, *Archoplites interruptus.* Copeia, 1965 (2): 224–228.

Nelson, O. E. 1953. Comparative embryology of the vertebrates. The Blakiston Company (McGraw-Hill Book Company, New York).

Oppenheimer, J. M. 1937. The normal stages of Fundulus heteroclitus. *Anat. Rec.,* 68 (1): 1–16.

Orton, G. L. 1953. The systematics of vertebrate larvae. *Systematic Zool.,* 2 (2): 63–75.

Schultz, R. J. 1971. Special adaptive problems associated with unisexual fishes. Am. Zool., 11: 351–560.

———. 1973. Unisexual fish: laboratory synthesis of a "species." Sci., 179 (69): 180–181.

Solberg, A. N. 1938. The development of a bony fish. *Progr. Fish-Culturist,* 40: 1–19.

Tåning, A. V. 1952. Experimental study of meristic characters in fishes. *Biol. Rev.,* 27: 169–193.

Tinbergen, N. 1953. Social behaviour in animals. Methuen and Co., London, John Wiley and Sons, New York.

van Tienhoven, A. 1968. Reproductive physiology of vertebrates. W. B. Saunders Company, Philadelphia.

Wilson, H. V. 1891. The embryology of the sea bass (*Serranus atrarius*). *Bull. U.S. Fish Comm.,* 9 (1889): 209–278.

11

Integration

THE ROLE OF THE NERVOUS AND ENDOCRINE SYSTEMS

Vertebrates respond to environmental stimuli through sense organs, the brain, and/or the spinal cord which relay impulses to muscles or glands. In fishes the muscular responses frequently result in movements of the entire body rather than of the appendages only. Certain cyclic and slow changes in the environment, such as seasonal temperature rises and changes in day length also affect one or more endocrine glands, by way of the nervous system. The hormonal secretions of endocrine glands (Table 11.3) act on specific target organs, which may be endocrine themselves, or the secretions have a diffuse effect on metabolism in general.

The nervous and endocrine systems are highly interdependent and often act together. Reproductive behavior of many fishes, for example, is influenced by the senses through their perception of the environmental changes that come with approaching spring; several endocrine glands are affected and in turn act on the ovaries and testes. There is further interaction among sense organs, central nervous system, and endocrine glands at the time of physiological readiness for spawning. At spawning time, more or less complex behavioral patterns are released through visual, chemical, or auditory signals. The spawning act itself, through nervous impulses, readjusts the state of endocrine tissues.

The difficulty of describing briefly the complexity of the interaction of the components of the nervous and endocrine systems in fishes is amplified by the extreme numbers of fishes and their diverse adaptations to the great variety of habitats they occupy. Fishes have more varied habitats and therefore greater anatomical differences than any other group of vertebrates. Extensive reviews of the frequently controversial information on fish nervous and endocrine systems exist; they, as well as selected specific articles, are cited at the end of this chapter or are listed among the general references at the end of Chapter 1.

THE NERVOUS SYSTEM

The basic pattern and progress of cephalization, the concentration of stimulus receiving and integrating units in the head, can be followed from the lamprey-hagfish group (Cyclostomata) through the shark-like fishes (Chondrichthyes) to the bony fishes (Osteichthyes). The general organization of the nervous system in each of these three major fish groups has been described in Chapter 3 (Fig. 3.20). The general anatomy and location of the organs of special sense and the endocrine glands have been given in the same chapter. We shall now consider certain details of the anatomy and physiology of the nervous system, sense organs, and the endocrine glands.

Connections of the various parts of the central nervous system are summarized in Figs. 3.20 and 11.1, and Table 11.1.

The Telencephalon (Smell Area of the Brain; Fig. 11.1; Table 11.1)

The most anterior region of the brain, the telencephalon or forebrain, in fishes is prominently devoted to the reception, elaboration, and conduction of smell impulses. Its relative size varies with the degree to which smell plays a role in the life of the kind of fish examined. The olfactory nerve, or cranial nerve I, enters the brain from the sensory epithelium of the olfactory plate or placode that is situated in the nostril or nasal pit. In sharks and their relatives (Elasmobranchii) and bony fishes (Osteichthyes), the right and left olfactory nerves are accompanied by another pair of nerves, the small terminal nerves (0). They are believed to have vasomotor rather than sensory functions. Although the organ of smell of lampreys has secondarily assumed a median position and is unpaired, the underlying brain structures are paired as in other vertebrates. Anteriorly and on each side, the telencephalon consists of an olfactory bulb followed caudally by an olfactory lobe (Fig. 3.20); its two internal cavities are brain ventricles I and II. Elasmobranchs especially and those bony fishes that depend largely on smell for feeding or social interaction have pronouncedly enlarged olfactory lobes. Ventrolaterally these lobes contain a large ganglion, the corpus striatum, which is a center of

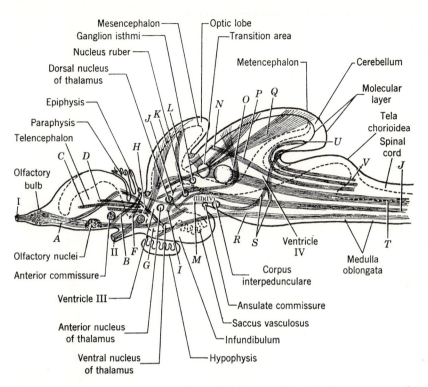

Fig. 11.1 Diagrammatic section of bony-fish brain with main fiber tracts, as shown in a carp (*Cyprinus*). Nerve tracts: *A*, olfactory; *B*, olfactohabenular; *C*, hippocampothalamic; *D*, striothalamic; *E*, striohypothalamic; *F*, preopticohabenular; *G*, olfactohypothalamic; *H*, pallial; *I*, spinohypothalamic; *J*, spinothalamic; *K*, opticotectal; *L*, tectobulbar; *M*, fasciculus retroflexus; *N*, tegmentocerebellar; *O*, diencephalocerebellar; *P*, spinocerebellar; *Q*, mesencephalocerebellar; *R*, lobopedunculospinal; *S*, cerebellotegmental; *T*, facialicerebellar; *U*, lateralicerebellar; *V*, vestibulocerebellar. Cranial nerves are: I, olfactory; II, optic; III, oculomotor; and IV, trochlear. (Based on Neal and Rand, 1939).

correlation mainly to relay smell impulses to posterior sensory or somatic nerve centers. The anteriormost large bundle of nerve fibers that connects the two sides of the brain is the anterior commissure. This commissure is located in the telencephalon and runs, in part, through the so-called terminal lamina, the median anterior wall of the entire neural canal system.

Although olfaction is an obvious function, it is not the sole concern of the rayfin fishes (Actinopterygii) telencephalon which is thought to serve an additional generalized function facilitating activities of lower brain centers and mechanisms. The morphogenetic development of the forebrain of the rayfins is entirely different from all other vertebrates in that it forms by eversion of the lateral walls of the dorsal (pallial) area as opposed to evagination which

Table 11.1. **Important Nerve Fiber Tracts in the Brain and the Spinal Cord**[a,b,c]

Divisions of Central Nervous System	Regions of Brain or Spinal Cord Connected by the Fibers
Telencephalon	Telencephalon ↔[d] Telencephalon
Telencephalon Diencephalon	Medial and lateral walls of olfactory lobe → Habenular nuclei
Telencephalon Diencephalon	Preoptic area → Habenular nuclei
Telencephalon Diencephalon	Somatic area of corpus striatum → Hypothalamus
Telencephalon Diencephalon	Olfacto-somatic area of forebrain → Ventral thalamus
Telencephalon Diencephalon	Olfacto-somatic area of forebrain ↔ Dorsal thalamus
Telencephalon Diencephalon	Hypothalamus ↔ Primordial hippocampus and pre-commissural septal areas of forebrain
Diencephalon Mesencephalon	Lateral geniculate nuclei ↔ Optic tectum
Diencephalon Mesencephalon	Habenular nuclei → Interpeduncular nuclei in base of midbrain
Mesencephalon Diencephalon	Optic tectum ↔ Thalamus
Mesencephalon Diencephalon	Optic tectum → Hypothalamus
Mesencephalon Metencephalon	Optic tectum → Cerebellum
Metencephalon Mesencephalon	Cerebellum → Tegmentum
Mesencephalon Myelencephalon	Nuclei of nerve V → Optic tectum
Mesencephalon Spinal cord	Optic tectum → Spinal cord
Mesencephalon Myelencephalon	Lateral line nuclei → Torus semicircularis and valvular of of medulla nuclei in tegmentum
Mesencephalon Spinal cord	Spinal cord → Tegmentum
Myelencephalon Mesencephalon	Gustatory lobes → Secondary gustatory nucleus of midbrain of medulla and also hypothalamus
Spinal cord Diencephalon	Spinal cord → Thalamus
Metencephalon Myelencephalon	Vestibular nuclei and fibers of Nerve VIII → Cerebellum
Metencephalon Myelencephalon	Lateral line nuclei of medulla ↔ Cerebellum
Myelencephalon Spinal cord	Nuclei of nerves VII to X in medulla → Spinal cord
Di-, Mes-, and Met-encephalon Spinal cord	Several parts of brain → Spinal cord of trunk and tail

Name of Tract	Function of Tract
Anterior commissure	Connects the two sides of the forebrain
Median and lateral olfacto-habenular	Project smell impulses to habenular nuclei as way stations for feeding reflexes
Preoptico-habenular	Tertiary olfactory tract, a feeding reflex path
Strio-hypothalamic	Conveys forebrain impulses to hypothalamic centers
Ventral peduncle of lateral forebrain bundle	Evolutionary beginning of regulation of motor responses by forebrain
Dorsal peduncle of lateral forebrain bundle	Primitive discharge path between thalamus and basal ganglia of forebrain
Medial forebrain bundle	Leads visceral, including taste, sensations to and from olfactory forebrain region
Geniculo-tectal Tectogeniculate (branch of superior colliculus)	Relays optic impulses
Fasciculus retroflexus (habenulo-interpeduncular)	Part of discharge paths of olfacto-somatic areas to motor centers (feeding reflexes)
Thalamo-tectal-Tectothalamic	Relate the highly correlative optic tectum with the developing dorsal thalamus
Tecto-hypothalamic	Connects sight impulses with autonomic centers and pituitary region
Tecto-cerebellar	Relates sight and other sensory impulses with muscle tonus for locomotion
Cerebello-tegmental	Portion of relay system to motor centers regulating muscle tonus
Bulbo-tectal	Coordinates tactile sense organs on head with vision
Tecto-spinal	Coordinates sight and other sensory impulses with locomotion
Acoustico-lateral lemniscus	Projects lateral line impulses onto midbrain
Spino-tegmental (probably multisynaptic)	Regulates locomotion
Secondary and tertiary ascending gustatory tracts	Coordinate gustatory and olfactory feeding stimuli, especially in catfishes (Siluroidei)
Spino-thalamic	Projects general body sensations on thalamus (the presence of this tract in fishes has been doubted)
Vestibulo-cerebellar and direct vestibular root fibers	Part of equilibration system
Latero-cerebellar Cerebello-lateral	Coordinate lateral line and locomotory impulses
Vestibulo-spinal	Relays lateral line and vestibular impulses to lower medullary and spinal centers
Giant cells of Mauthner	Coordinate trunk and tailbeat in swimming

results in lateral ventricles. Consequently, the actinopterygian forebrain retains a single, median, ventral ventricle (subpallium) in contrast to the paired ventral (basal) ventricles in lower fishes and other vertebrates.

The telencephalon of fishes is extensively innervated by secondary olfactory fibers from the olfactory bulb. These fibers project throughout the subpallial portion of each hemisphere, and also project to the posterior basolateral areas of the pallium. The remainder of the pallium of the actinopterygian telencephalon appears not to receive secondary olfactory fibers. There is increasing differentiation and subdivision of the dorsal area through the phylogenetic series from Polypteriformes to Teleostei.

Ablation of the forebrain has shown it to be concerned with activities other than integration of olfactory input. Forebrainless goldfish (*Carassius auratus*) show less initiative and spontaneity than intact animals in exploring their environment. Certain tilapias (*Tilapia*) neglect their young when the forebrain is removed and certain minnows (Cyprinidae) so treated are less wary and distrustful of a new situation than their normal experimental running mates. Forebrain ablations result in suppression of aggressive, sexual, and parental behavior in the threespine stickleback (*Gasterosteus aculeatus*). They also suppress spawning and nest-building behavior in the paradise fish

---

[a] The lamprey central nervous system has a large olfactory forebrain with bare beginnings of non-olfactory centers, a light-sensitive dorsal diencephalon and epiphysis, and, compared to bony fishes, has less developed tactile and gustatory centers. There are prominent connections from lower (medullary) centers to the optic tectum that may have other than direct sight functions.

Corresponding to the cells of Mauthner of bony fishes there are in the lampreys the bundles of Mueller with cell bodies in the medulla and efferent fibers to the body musculature. Both Mauthner and Mueller cells are found in the spinal cord of lampreys. The dendrites come from different higher brain centers and the entire system coordinates sensory input with the sinuous swimming movements of the animals.

[b] Elasmobranch brains and spinal cords, with the sharks better known than rays in this respect, have a more pronounced olfactory area (forebrain) than many bony fishes. Correspondingly there are prominent fiber tracts leading to and from this area of the brain.

Rays swim by undulating their large wing-like pectoral fins; not only is the spinal cord in the corresponding regions well differentiated to allow for multisegmental reflexes but connections with the medulla have also increased in importance, compared to the more elongate sharks whose pectoral fins are stabilizers only.

[c] We wish to thank Dr. Elizabeth C. Crosby who helped in the preparation of this table.

[d] Fibers going one way as indicated by arrow: → . Fibers ascend and descend, as indicated by arrow: ↔ .

(*Macropodus opercularis*). Forebrain extirpation results in a selective suppression of color-vision discrimination in goldfish.

The Diencephalon (Sight Area I; Fig. 11:1; Table 11.1)

A dorsoventral and often also a pronounced lateral constriction marks the transition from fish telencephalon to diencephalon or 'tween brain. Dorsally the parencephalon, or the thin saccus dorsalis, forms the roof of the cavity within the diencephalon, also called the third ventricle. In some fishes such as lampreys (*Lampetra*), gars (*Lepisosteus*), and the bowfin (*Amia*), the saccus dorsalis is enlarged and extends laterally and frontally, even over the telencephalon. There are two landmarks along the diencephalic midline in the lampreys (Petromyzonidae), the parapineal and pineal organs, also known as the epiphysial organs, located slightly left and right of the midline respectively. However, in adult sharks and relatives (Elasmobranchii) and in higher bony fishes (Actinopterygii) only the pineal organ is developed. Embryos of some bony fishes, such as the whitefishes (*Coregonus*), show both organs during early embryonic development, but the parapineal is subsequently lost and only the pineal remains. Wherever both or one of these evaginations persist their structure and function have been linked to the reception of more or less diffuse light stimuli.

In lampreys and hagfishes (Cyclostomata), the pineal organ is connected to the right habenular ganglion (see below), and has a retina, pigment cells, and a lens-like structure. There is an orifice in the cranial roof above it, just behind the nasal opening, and a depigmented spot in the skin so that light stimuli can reach the pineal and parapineal organs. Some sharks (Squaliformes) have an unpigmented spot in the skin of this region of the head and some of them also have a perforation of the cartilaginous cranial roof. However, in the sharks there is generally a less well-developed pineal organ than among cyclostomes and some bony fishes. Particularly well-developed pineal spots are found in vertically migrating, mid-depth fishes such as the hatchetfishes (*Argyropelecus*), in many catfishes (such as *Arius* and *Macrones*), and also in halfbeaks (*Hemiramphus*).

In spite of numerous studies on the pineal body of fishes, no distinct function has been unequivocally demonstrated. The current hypotheses may be divided into two categories—one emphasizing primarily a sensory role and the other a secretory function. The following hypotheses relate to sensory roles: the pineal body is a photosensory structure; it acts as a baro- or chemoreceptor for cerebrospinal fluid; or it functions as a mediator in the olfactory response to exohormones. In secretory roles the pineal gland may be primarily a gland of external secretion related to the chemical composition of the cerebrospinal fluid or the metabolism of brain tissue or, alternatively, the pineal may be a gland of internal secretion with an endocrine function.

The rest of the diencephalon can be divided into an epithalamic region with its habenular ganglia, a thalamus, and a hypothalamus as its largest and most important component. The optic nerves enter the brain and cross anterior to the diencephalon, and posteriorly this region of the brain is joined to the ventral portion of the mesencephalon through the convoluted, thin-walled saccus vasculosus. Below the hypothalamus lies the hypophysis or pituitary gland (Fig. 11.1). This gland is closely attached to the floor of the brain in lampreys but among sharks and higher bony fishes it is carried on a stalk or infundibulum (Fig. 11.8). The thalamus serves as a relay center for the transfer of olfactory and striate-body impulses to thalamo-medullar and thalamo-spinal tracts. In its ventral region, toward the brain floor and the hypothalamus there lie the geniculate nuclei or ganglia (Fig. 11.1) which receive some optic nerve (II) connections before the optic tectum is reached by the axons of this cranial nerve. These nuclei are well-developed in sharks, even to an extent that warrants their being called geniculate lobes; however, nothing certain is known about their function. In the optic chiasma of all fishes the nerve fibers from the retina of one eye cross to the optic tectum on the other side of the mesencephalon, the decussating or crossing fibers from one eye passing over or even through the bundle of fibers from the other eye.

The diencephalon of fishes seems to be an important correlation center for incoming and outgoing messages relating to internal homeostasis. The hypothalamus especially affects the endocrine system through the pituitary gland. Neurosecretory cells have been described in some diencephalic nuclei and special paths have been found that lead secretions from the diencephalon to the hypophysis.

The Mesencephalon (Sight Area II; Fig. 11.1; Table 11.1)

The mesencephalon or midbrain of fishes is relatively large; it consists of the dorsal optic tectum, appearing in dorsal view as the two optic lobes, and of the ventral tegmentum. The tectum is made up of a number of zones of different-sized nerve cells or neurons. Most of the fibers of the optic nerve end in the tectum in such a manner that the frontal part of the retina is projected onto the contralateral caudal portion of the tectum and the converse applies to the anterior retina, whereas dorsal retinal elements of one side are connected to the ventral portion of the tectum on the other side of the fish. Fishes, like other vertebrates, have convex lenses in their eyes that create an inverted image on the retina; through the above tectal pattern the image may become projected essentially as it is in nature. In the lampreys (Petromyzoniformes), the midbrain is covered with a vascular choroid plexus that helps in nourishing the brain and connects with the fluid-filled ventricles and the central canal. Among higher bony fishes (Actinopterygii), tectal gray matter extends into the ventricle as a paired torus longitudinalis and joins the posterior commissure between the two hemispheres which lie at the border of the diencephalon. The torus is presumably connected with vision since it is very small in a blind goby (*Trypauchen*).

Electrical stimulation and removal of portions of the midbrain indicate that the tectum contains functional neural units that correlate visual impressions with muscular responses such as would occur, for instance, in facing prey or adjustment of swimming to objects moving in the visual field, in short, to various complex goal-seeking movements. Crucian carp (*Carassius carassius*) were unable to distinguish the positions of light (spatial orientation) after the tectal hemispheres were separated by a longitudinal midline incision. These carp also appeared to have difficulties in locating the position of sound stimulus. Experiments using a puffer (*Sphoeroides*) in which one eyeball was rotated front to back and the other eye was blinded, led to persistent forced circling even when other parts of the brain, except the medulla, were removed. Thus there is strong evidence for the optic tectum as an eyebody coordinating center. Electric stimulation of the tegmentum, however, leads to generalized but relatively uncoordinated locomotory responses. The optic tectum also bears some relation to learning. Responses to smell stimuli have been associated with visual signals; upon midbrain removal the olfactory responses are extinguished.

In view of the complete crossing over of the fish optic nerves it should be noted that the goldfish (*Carassius auratus*) and a goby (*Bathygobius*) could be trained to sensory optic transfer. Somatic and autonomic responses showed that the untrained eye recognized the stimulus learned with the other eye but that motor responses to the transfer are difficult to obtain.

Because of its functional importance in the central nervous system in many fishes and its widespread connections and because it has several layers of nerve cells, the optic tectum has been loosely compared to the cerebral cortex of mammals.

The Metencephalon (Cerebellum, Equilibrium, Muscle Tonus; Fig. 11.1; Table 11.1)

In the metencephalon, the cerebellum develops from the underlying enlarged medulla as a dorsal outgrowth of the upper rim of the 4th ventricle of the brain. Its main functions are reported to lie in swimming equilibration, maintenance of muscular tonus, and orientation in space; in some fishes it is by far the largest component of the brain. Several distinct layers or differently shaped nerve cells can be distinguished in the cerebellum and include the molecular or chief receptive cell layer and the so-called Purkinje cells that assume prominence in the cerebellum of the higher vertebrates.

In the lampreys (Petromyzonidae), the cerebellum is only a small dorsal bridge between the acoustic-lateral-line nerve areas. In the hagfishes (Myxiniformes), the cerebellum is represented by a small commissure that consists of eighth nerve and lateral line fibers, the octavo-lateralis system. In sharks and rays (Elasmobranchii) the cerebellum roughly increases with the size of the species. Relatively small sharks, such as the dogfishes (Squalidae), have a simple bilobate cerebellum whereas in the large, swift, and predatory mackerel sharks (Lamnidae) the cerebellum has become the largest part of the brain and is convoluted and has an increased surface area (Fig. 11.2). Prominent lateral or auricular outgrowths may form also on the cerebellum and are mainly connected to lateral-line and vestibular sense organs.

In higher bony fishes (Actinopterygii) the cerebellum acquires a characteristic forward projection, the valvuli cerebelli (Fig. 11.3), which extends under the optic tectum. In the elephantfishes (Mormyridae), many of which have extended and highly motile snouts, the valvuli cerebelli arches far forward beyond the forebrain and becomes the largest component of the brain;

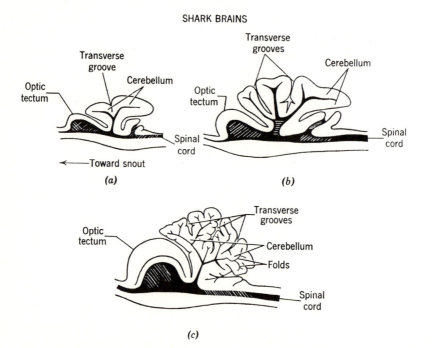

Fig. 11.2   Relative size and development of cerebellum among sharks as seen in diagrammatic sections: (a) a cat shark (*Pristiurus*); (b) a smooth dogfish (*Mustelus*); (c) a sand shark (*Carcharias*). (Source: Voorhoeve in Grassé, 1958).

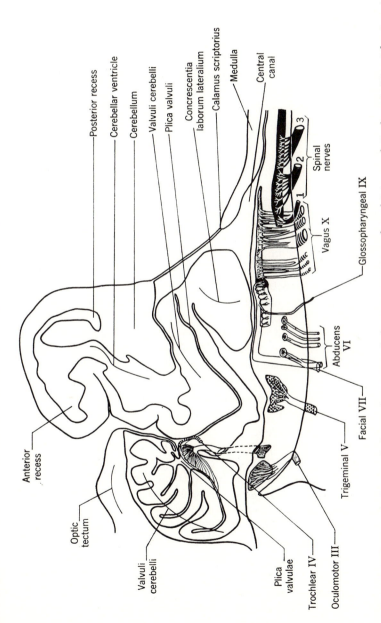

Anterior
recess

Optic
tectum

Valvuli
cerebelli

Plica
valvulae

Trochlear IV

Oculomotor III

Posterior recess

Cerebellar ventricle

Cerebellum

Valvuli cerebelli

Plica valvuli

Concrescentia
laborum lateralium

Calamus scriptorius

Medulla

Central
canal

Spinal
nerves

1

2

3

Vagus X

Glossopharyngeal IX

Abducens
VI

Facial VII

Trigeminal V

Fig. 11.3 The cerebellum and the arrangement of the motor roots and nuclei in a relatively primitive teleost, a tarpon (*Me-galops*). (Source: Ariens, Kappers, Huber, and Crosby, 1936).

among catfishes (Siluroidei), the structure is also well developed. Since the mormyrids emit and respond to weak electrical currents and since there is also at least one catfish with electric discharges (the electrical catfish, *Malapterurus*) it has been suggested that the enlarged cerebellum in these groups deals with electrical impulses.

The ventricle of the cerebellum still prominent in sharks and rays (Elasmobranchii) has almost completely disappeared in the cerebellum of higher bony fishes, as have the sulci or transversal folds so prominent in some large sharks. The lateral auricles and interauricular fibers increase in importance, especially in fishes with well developed lateral-line organs.

A prominent function of the cerebellum apparently is to coordinate muscular tonus and swimming.

The Myelencephalon (Basic Brain Area, Brainstem, Medulla; Fig. 11.1; Table 11.1)

The brain divisions from cerebellum to telencephalon are, in their widest meaning, sensory and sensory-coordinating centers that connect with the brainstem, or myelencephalon, by various neural tracts. These anteriormost four divisions have grown in importance as sensory integrators and have become more and more refined in the course of evolution. The myelencephalon, with the medulla oblongata as its main component, is the center to which lead the sensory nerves except those of smell (I) and sight (II). The boundary between medulla oblongata and spinal cord in fish is indistinct. The medulla can be divided into columns of nerve fibers based on the type of information transmitted. Thus, there are somatic and visceral sensory columns and also somatic and visceral motor columns.

In the brainstem, nuclei of cranial nerves III to X are arranged anteroposteriorly (Fig. 11.1); the tracts of nerves III and IV cross to the contralateral side. The afferent nerve components of the brainstem can be divided into somatic and special sensory nervous inflows with cranial nerve VII (facial), VIII (stato-acoustical), IX (glossopharyngeal), and X (vagus) mostly sensory. There are more somatic sensory fibers in the foregoing nerves in lampreys (Petromyzonidae) than in shark-like fishes (Elasmobranchii). The somatic sensory fibers become further reduced in bony fishes (Osteichthyes). Details of the distribution of the cranial nerves which are, in the main, connected to the brainstem, can be found in Table 11.2.

Depending on the relative importance of the various senses, different portions of the brainstem are enlarged. Such fishes as the Pacific herring (*Clupea pallasi*), the striped mullet (*Mugil cephalus*), and certain carangids such as *Trachurus* have prominent paired swellings designated the cristae cerebelli on the antero-lateral rim of the fourth ventricle. Nerve connections of the cristae are not known but their function may be connected with the strong schooling tendencies of these species. Certain suckers, such as the highfin carpsucker (*Carpiodes velifer*), and minnows, such as the goldfish (*Carassius auratus*), have prominent vagal lobes posterior to the region of the cristae cerebelli from which emerge cranial nerves IX and X. Such fishes also have developed an organ in the roof of the mouth (palatal organ) where food is tested by taste and touch. Another prominent feature, found among minnow-like fishes (Cypriniformes), is the tuberculum impar, or facial lobe, a central hillock posterior to and sometimes partly covered by the cerebellum. The tuberculum impar arises from a fusion of root elements of the facial (VII) and vagal (X) nerves, where gustatory and tactile impulses are correlated with visceral sensory ones as in an European barbel (*Barbus fluviatilis*) and the carp (*Cyprinus carpio*; Fig. 11.4).

The medulla of the higher bony fishes (Actinopterygii) contains a pair of large neurons, the giant cells of Mauthner, at the level of cranial nerve VIII. The lateral dendrites of these giant cells connect with fibers of cranial nerves V, VII, IX, and X and to the cerebellum and the optic tectum. Their axons pass through the spinal cord to the muscles of the tail and are best developed among the good swimmers such as the Atlantic salmon and certain trouts (*Salmo*), whereas they are less prominent in bottom dwellers such as some of the gobies (*Gobius*) and scorpionfishes (*Scorpaena*). In eels (Anguilliformes) and the molas (*Mola*) cells of Mauthner are absent. Apparently the cells are motor coordinators for relaying multiple sensory impulses mainly from lateral-line centers to the caudal and abdominal swimming musculature. In the lampreys (Petromyzonidae), a comparable coordinating system connects the oculomotor levels of the brain with the tail and body musculature through the so-called neurons of Mueller and also cells of Mauthner.

Aside from also being a relaying area between the cord and higher brain areas, the medulla has centers that control certain somatic and visceral functions. Among bony fishes these include respiratory, paling (acting through melanophore aggregation), and osmoregulatory centers. Medullar nuclei are also involved in the maintenance of swimming equilibrium. Tests on medullary control of fin movements in some fishes that use their fins prominently in locomotion, such as wrasses (Labridae), show that the rhythm of fin beats is set by a center in the anterior cord or the brainstem.

THE SPINAL CORD (FIG. 11.1; TABLE 11.1)

The central nervous system of fishes loses much of its structural complexity in the transition from brain to cord. Already in lampreys (Petromyzonidae) but more pronouncedly in sharks and rays (Elasmobranchii), a cross section

Table 11.2. **Cranial Nerves of Fishes and Their Functions**

Nerve Number	Name	Peripheral Connection
0	Terminal	Olfactory bulb
I	Olfactory	Mitral cells; olfactory bulb; nucleus olfactorius anterior
II	Optic	Retina of eye
III	Oculomotor	Superior, inferior, anterior, rectus, and inferior oblique muscles of eyeball
IV	Trochlear	Superior oblique muscle of eyeball
V	Trigeminal divided into deep ophthalmic, maxillary, and mandibular branches	Anterior portion of head, upper and lower jaw regions
VI	Abducens	Posterior rectus muscle
VII	Facial[a] and	1. Dorsal group: postorbital lateral line canal; taste buds on body, where present
VIII	Acoustic[a]	2. Orbital group: supra- and infraorbital lateral line canals; taste buds on snout
		3. Acoustic group: inner ear; temporal lateral line canals
		4. Branchial group: operculo-mandibular oral and jugal line canals; taste inside mouth; taste and touch from opercular hyoid and mandibular regions; head muscles antagonistic to those under control of mandibular branch of V
IX	Glossopharyngeal often fused with	1. Dorsal group: part of temporal lateral line canal; dorsolateral skin sensibility and proprioception, first gill slit region
		2. Branchial group: taste organs in dorsal and ventral pharyngeal mucosa, muscles of first gill slit (third gill arch)
X	Vagus	1. Supratemporal branch: skin and lateral line organs of supratemporal region
		2. Dorsal recurrent branch: joins with body taste bud innervation of VII
		3. Body lateral line branch: from body lateral line organs
		4. Visceral branch: nerves to and from internal organs
		5. Branchial branches: to and from four posterior gill slits or equivalent regions

[a] Cranial nerves VII and VIII, best considered together as the acousticofacialis nerve, divided into dorsal, orbital, acoustic, and branchial nerve groups.

Central Connection	Function
Forebrain; lamina terminalis	Somatic sensory, tip of snout? vasomotor?
Smell centers of forebrain	Special sensory, carries smell impulses
Optic tectum of midbrain	Special sensory, carries visual impulses
Anterior brainstem under optic tectum (Mesencephalon)	Somatic motor, innervates four of the six striated eye muscles and muscles inside eye
Anterior brainstem under cerebellum (Metencephalon)	Innervates one of the six striated muscles of the eye
Lateral column of medulla oblongata region of brainstem, frequently overlaps with centers of VII (Myelencephalon)	Mixed somatic sensory and motor functions. Thermal and tactile sensibility of skin on anterior portion of head; proprioceptors of head musculature
Ventral brainstem behind origin of V	Innervates sixth striated muscle which moves the eyeball
Elongate series of connected nuclei in middle of brainstem (Medulla oblongata; Myelencephalon)	A truly mixed nerve; special and somatic sensory visceral and motor functions. Brings lateral line acoustic and gravity impulses to brain centers, as well as taste and touch impulses from head and mouth; proprioception and innervation of some head muscles
Series of ganglia, or roots, mainly in dorsal and median end region of brainstem (Medulla oblongata; Myelencephalon)	A mixed nerve; taste and lateral line sense in afferent components; innervates anterior gill muscles (first gill slit)
Series of ganglia posterior to but fused with roots of nerve IX; often overlapping with roots of first occipitospinal nerves at posterior end of brainstem (Medulla oblongata, Myelencephalon)	A mixed nerve; motor control of bulk of gill muscles, special fibers from taste and lateral line organs, visceral motor and sensory to and from internal organs; parasympathetic autonomic nervous system

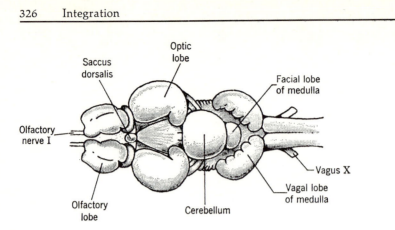

Fig. 11.4   Dorsal view of brain of the carp (*Cyprinus carpio*) showing development of facial and vagal lobes. (Based on Uchihashi, 1953).

(Fig. 11.5) of the cord shows a central region of gray substance consisting primarily of nerve cells and a surrounding area of white matter, the nerve fibers. These fibers are sheathed, or myelinated, and are collected in bundles according to their function and connections. Centered in the gray matter of the cord is the central canal. As in other vertebrates, the gray matter of the cord of bony fishes (Osteichthyes) roughly resembles a letter X with paired dorsal and ventral horns. The dorsal horns receive somatic and visceral sensory fibers and the ventral horns contain motor nuclei, with centers for dorsal or ventral musculature lying in median and lateral positions respectively.

THE SPINAL NERVES

The spinal nerves are segmentally arranged. In the lampreys (Petromyzonidae) the ventral roots appear between the levels of the dorsal roots, thus this group of fishes does not have true mixed spinal nerves; only among the hagfishes (Myxinidae) there begin to occur true mixed spinal nerves like those present, with few exceptions, throughout the jawed fishes—the sharks and rays (Chondrichthyes) and bony fishes (Osteichthyes). The jawed fishes in general, have spinal ganglia for the neurons of the sensory or dorsal root nerves although in many families, such as the minnows (Cyprinidae), cods (Gadidae), perches (Percidae), and drums (Sciaenidae), some afferent fibers come to the cord also from supramedullary and inframedullary ganglia.

From sharks (Squaliformes) to bony fishes (Osteichthyes), there is an increasingly finer differentiation and subdivision of ascending and descending fiber tracts between the brain and the spinal cord. In the searobins (Triglidae), the long and separate anterior rays of the pectoral fins carry special receptors,

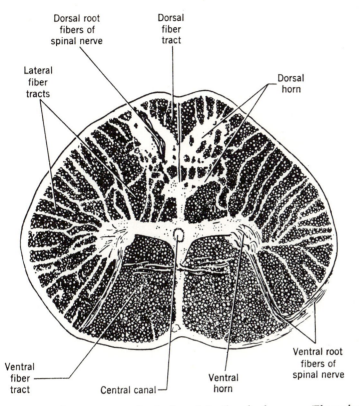

Dorsal root
fibers of
spinal nerve

Dorsal
fiber
tract

Lateral
fiber
tracts

Dorsal
horn

Ventral
fiber
tract

Central canal

Ventral
horn

Ventral root
fibers of
spinal nerve

Fig. 11.5   Cross section of spinal cord in the shark group (Elasmobranchii). (Based on Keenan in Grassé, 1958).

for tactile and chemical clues; sensory nerves from these rays are marked on the spinal cord by separate swellings corresponding to their dorsal root nerves, and their fiber projections also reach the medulla and the midbrain.

Ipsilateral and contralateral reflexes of spinal nerves in one segment of the cord can occur in elasmobranchs; plurisegmental reflexes are common in sharks and teleosts. In view of the important role of the trunk in fish locomotion the question of the nervous control of swimming movements has received considerable attention. There is controversy as to whether a fish requires outside stimulation in order to make coordinated swimming movements or whether swimming is due to the automatic activity of cord and medullar centers, only to be coordinated by peripheral stimuli. Von Holst, after performing many experiments with dogfish sharks (Squalidae), eels (*Anguilla*), loaches (Cobitidae), and minnows (Cyprinidae), suggested that swimming is due to endogenous rhythm from the posterior medulla, in turn passed on to the cord but influenced at all levels by outside "setting" stimuli. Lissmann and Gray, mainly on the basis of experiments with certain cat sharks (Scyliorhinidae), maintain that there is no automatic swimming center but that swimming movements are of reflex nature and depend on the coordination of afferent impulses in spinal cord and medulla. The movements of young ammocoetes of *Lampetra* (a lamprey) are described as entirely reflexogenic.

THE AUTONOMIC NERVOUS SYSTEM

In the autonomic nervous system of lampreys and hagfishes (Cyclostomata), there is no ganglionated sympathetic chain. Nevertheless, in these fishes a diffuse system of nerves and some plexi of sympathetic function have been described. In the lampreys and hagfishes, both the heart and the organs of digestion are innervated by the vagus. Subcutaneous nerve plexi that are innervated by spinal nerve fibers, apparently of the sympathetic system, are found concentrated in the walls and at the orifices of the slime glands of *Myxine* (hagfish). In the sharks and rays (Elasmobranchii), sympathetic ganglia are found along the sides of the vertebral column; they are irregular and do not extend into the head region. The smooth muscles of the gut and arterial walls are innervated by them. However, there is no connection between them and the nerves which supply the skin such as in mixed communicating rami of bony fishes (Osteichthyes) and higher vertebrates. In all shark-like fishes, the hearts and the internal organs of digestion are abundantly supplied by vagal fibers. In the bony fishes, segmentally arranged ganglia form a sympathetic chain as far forward as the trigeminal nerve (cranial nerve V); most preganglionic fibers for sympathetic ganglia of the head arise from the spinal cord of the trunk region and pass forward. Sympathetic connections to the skin nerves are found in the highest bony fishes (Teleostei).

The parasympathetic system is almost exclusively represented in portions of the widely ramifying branches of the vagus (cranial nerve X). A parasympathetic innervation of eye parts (in cranial nerve III) has already been mentioned. A sacral parasympathetic system similar to that of mammals is suggested in skates by the response of the claspers to stimulation of the posterior spinal nerves. The sympathetic cardio-accelerating fibers of higher vertebrates are absent in all fishes and consequently vagal stimulation has a powerful inhibitory action on the sinus venosus and the auricle but no effect on the ventricle. The stimulation of branches of the vagus leading to other internal organs as well as the administration of cholinergic (para-

sympatomimetic) and adrenergic (sympatomimetic) drugs has produced conflicting results, especially in action on the gas bladder and certain portions
of the alimentary canal.

The nerves from the paling center run in the spinal cord and leave it at
various levels in different fishes (in the minnow *Phoxinus*, segments 12 to 18;
in the salmon-trout genus *Salmo*, segment 26; and in the searobins, *Trigla*,
segment 3). The melanophore nerves then go to the subvertebral sympathetic
chain where they divide into an anterior bundle leading to the color cells of
head and nape, and a posterior bundle, to the trunk and the tail. Little is
known about melanophore control in cyclostomes; however, certain elasmobranchs have at least some sympathetically innervated melanophores. The
presence of a darkening center in the midbrain and of parasympathetic
melanin disperser nerves has been postulated.

SUPPORTING TISSUES OF THE CENTRAL NERVOUS SYSTEM

Some embryonic ectodermal derivatives and some mesodermal tissues in the
region of the brain and spinal cord assume supporting, investing, and nourishing functions of the nerve cells proper. Ectodermal ependymal cells line
the cerebrospinal canals of all fishes, permeate the body of the cord and the
brain, and are especially concentrated along the ventricular walls of the
thalamus of bony fishes (Osteichthyes) where they carry cilia. The histology
of these cells suggests that they have a secretory function. Neuroglial cells
of special shape and function have been reported from the spinal cord of the
sea lamprey (*Petromyzon*). In the spinal cords of sharks and rays (Elasmobranchii) occur many branched, astrocyte or spidery neuroglial cells. In the
cerebellum and in the spinal cord of bony fishes various neuroglial elements
are also differentiated.

The choroid plexus, a highly vascularized brain area, especially prominent
in the gars and relatives (Holostei) and elasmobranchs, consists mainly of
mesodermal tissue often in several layers and therefore called leptomeninx.
Choroid plexi or tela choroidea are filled with cerebrospinal fluid, are richly
supplied with blood vessels, and participate in the nutrition of the nervous
tissue they surround and cover.

INTELLIGENCE AND BEHAVIOR

Sensory components of the fish nervous system are highly discriminative,
and the sensitivity of the sense cells is often surprisingly great, as, for example, smell, vision, and temperature perception. Most sensory mechanisms
work by summation, and full efficiency at the normal level of stimulation

may be achieved only with a threshold lower than that which would normally be received. Such low thresholds imply the existence of central processes that allow the animal to disregard random "white noise" in most or in all sensory modalities, but they also imply that the animal can make choices with a high degree of accuracy and confidence of goal-oriented success.

Many acts of discrimination or choice in nature are based on conditioning of the fish to single or multiple stimuli and, even without experiments, there would be no doubt that fishes are easily conditioned. The ability to associate stimulus and response is present even in simple invertebrates; but relative levels of "intelligence" can be distinguished among other criteria, on the basis of the speed of learning simple tasks and on the retention-span of learned responses. Aquatic analogs of the "Skinner box" and the shuttle box are increasingly popular devices used to study instrumental learning in fish.

Electrical shock conditioning of the goldfish (*Carassius auratus*) to visual stimuli and of the fathead minnow (*Pimephales promelas*) among others to olfactory stimuli lead to learning accomplishments. The accomplishments are detected by altered heartbeat (an autonomic response) after thirty-five trials, or by swimming movements (a motor response) after fifty or more trials. Great individual differences of response exist. Some fish in an experimental group are slower than others to learn, and some seem unable to learn even "simple" tasks.

Conditioning by a reward such as food can be established quicker than by punishment; ten to twenty trials are sufficient, depending on the task, to accomplish visual responses in a minnow of the genus *Phoxinus* or to have *Tilapia macrocephala* learn space reversal of food location. Downstream migrating salmons (*Oncorhynchus*) learn quickly to swim a constant course in a circular channel. These responses are probably established speedily because the fish are in physiological readiness for oceanward migration which involves swimming in schools for long distances. One-trial learning has been shown with tagged carp (*Cyprinus carpio*) which became "hook-wise" for months after being caught only once.

Not only higher bony fishes (Actinopterygii) such as the foregoing examples are capable of reflex learning, but also the tiger shark (*Galeocerdo cuvieri*). This shark learns to strike a target and to swim a staked-out path for a food reward. The bull shark (*Carcharhinus leucas*) can be trained to associate sounds of 400 to 600 cycles per second with food.

Second-order conditioned reflexes are also established by fishes. The goldfish is able, after considerable training, to transfer a response learned on the basis of olfactory signals to visual clues. It can also make a similar transfer from auditory to thermal clues.

Experiments of sensory discrimination which first and foremost test sensory acuity rely on generalization, after a conditioned response has been established; that is, they depend on the ability of the animal to respond to a stimulus similar to, but not identical with, that used in the original training procedure. Generalization for different values of brightness and hue as components of vision are readily accomplished by various species in several families of fishes; generalization of olfactory clues by minnows (*Pimephales*) has also been observed.

Retention of learned responses without subsequent reinforcement (that is, long lasting memory) may be another criterion of relative intelligence. Salmons (*Oncorhynchus* and *Salmo*) may retain stream odors, learned during a crucial juvenile period, throughout their entire lives of 2 to 6 years or more. The home-stream odor memory is acquired by an imprinting process. Because the home-stream waters are greatly diluted by the time they reach the river mouth, a multiple imprinting hypothesis has been proposed. A salmon may remember the distinctive odors of a number of points on his downstream migration. The return voyage would then be a series of subvoyages with the salmon reading out the series of imprinted odors in reverse order as "subgoals." While there is strong evidence that imprinted odors are the principal source of guidance in the remarkable homing performance of salmon, the use of additional, perhaps visual, sensory cues is quite possible.

Apprehension and conceptual use of spatial relations are attributes of insight, by laymen credited only to the higher mammals because such abilities are more easily demonstrable in animals with prehensile limbs than they are in other vertebrates. Nevertheless, apprehension of spatial phenomena and insight learning exist in the frillfin goby (*Bathygobius soporator*) and probably in many other fishes. The frillfin goby swims about over submerged tidepools during high tide, but later, when the water has receded, it jumps from pool to pool with an uncanny aim and without seeing the next water area at the take-off point. Transplanted fishes that have not swum over the region and are placed in the pools at low tide either do not jump or, when prodded violently, will eventually jump in a disoriented way. After spending 12 hours, overnight and during high tide in a series of experimental pools, the same individuals jumped directly to the largest pools. Effective orientation was retained up to forty days without additional "high tide" experience. Comparable, though less elaborate performances are reported for the green sunfish (*Lepomis cyanellus*).

Delayed responses, successfully learned, have been taken as indication that the experimental subject in question is endowed with insight. Minnows (*Phoxinus*), after having been allowed to become familiar with a three-doored chamber by swimming freely through it for a time, were later required to reach a previously sighted food morsel by swimming forward and doubling back through one or two possible pathways to reach the reward which was then out of sight. After many hundreds of trials, the performance of the fish did not improve over an average of six out of ten correct solutions. The subjects also developed erratic and vicarious trial-and-error behavior at the point of choice. They also developed displacement activities, such as picking at stray pieces of sand at the entrance into the channel which led to the point of decision. These observations suggest that stimulus satiation, a negative "attitude" to the task or the experimental setup, after repeated experimental runs, may have bedeviled these experiments rather than their being an unequivocal demonstration that fishes are unable to learn to solve delayed response experiments.

Field observations that corroborate the findings of the delayed-reaction experiments, in spite of the earlier caution of stimulus satiation, have been made on the escape of trapped snappers (*Lutjanus*, family Lutjanidae) and groupers (*Epinephelus*, family Serranidae). The fishes that had entered the traps would make a hundred or more trial-and-error attempts to find a way through the wire or netting before chance would help them to locate the exit. An hour of confinement within a trap did not lead to any apparently systematic exploration for an escape route or portal. However, some fishes of the foregoing and other families, such as the sunfish family (Centrarchidae), are known to make such traps their homes, and they will then swim in and out freely.

Another, perhaps oblique, approach to the insight or the conceptualizing ability of fishes comes from a comparison of rats and a fish (*Tilapia macrocephala*) with regard to performance on inconsistent reinforcement of stimuli. On twenty trials per day, demanding visual discrimination between horizontal and vertical bars, responses to one stimulus were reinforced 70 percent and those to the other, 30 percent of the time. After 30 training days the previous 70-percent stimulus was consistently reinforced 100 percent. Records of the responses each day to the 70-percent stimulus showed that the rats "gambled" and that their preference for the more frequently reinforced stimulus rose gradually from a 70:30 level to reach 90 percent at day thirty. The fish, however, after reaching the 70:30 level on about the tenth day, corresponding to the more frequently reinforced stimulus, remained at this matching level of performance for the rest of the experiment.

Position training, demanding a choice between two positions of food placement, was conducted in a fashion similar to the foregoing and gave comparable results. Sensory deficiency was ruled out as a cause for the difference in behavior between the mammal and the fish and it was suggested that the rat, in such test situations, makes an all-or-none response selection, an indication not only of a high level of versatility in cerebral processes, but also of a certain abstractive capacity. The fish, in contrast, matched continuously the probability of reward or punishment with the choice of its response and indicated, as in the case of the delayed reaction, that it tends to perform in a more stereotyped and rigid manner than the mammal.

Another aspect of fish behavior, prominently involving the nervous system, is the occurrence of pronounced, often spectacular, highly specialized and stereotyped reproductive behavior (Chapter 10) that has been the subject of much ethological (animal behavior) research. The basis of behavior components are patterns of neuromuscular coordination that lead to movements in the wake of external or internal stimuli. These patterns are genetically determined and the movements that make up the behavior train during reproductive activities recur with surprising homogeneity in all individuals of a species. It has been possible to study fishes well in this respect, because they are easy to keep and behave quite normally in captivity. Detailed studies of reproductive behavior have been made of the threespine stickleback (*Gasterosteus aculeatus*), various cichlids, a sculpin (*Cottus*), the fightingfish (*Betta*), and others.

Fishes perform species-typical movements of courtship and/or nestbuilding even when raised to maturity in isolation. Therefore the pattern and sequence of these movements is just as much a part of the genetic makeup of the fish as is its body shape or coloration. Early ethological literature neglected the role of learning in inherited behavior but it is now indicated that learning plays an important role here also, and that modification, and perhaps refinement, of behavioral components is possible if not the rule (as known, for example, in a cichlid, *Cichlasoma meeki*).

The movements that make up the courtship behavior of sticklebacks (Gasterosteidae), for instance, can be analyzed down to the activity of single muscles or groups of muscles such as those that move a single fin. Such localized movements are components of larger units and are coordinated into hierarchically arranged groups to build up instincts. The implication is that these movements belong to distinct causal and functional systems that act as units. When in action they reduce the activity of other equivalent but functionally different systems. Instinct also implies a course of patterned rather than haphazard activity. This pattern operates through the common working of sense organs, central nervous system, and effectors in such a manner that we encounter a more or less fixed and obligatory sequence of stimuli and responses for the entire activity to go to completion. The most thoroughly studied such patterned sequence is that of the spawning behavior of the threespine stickleback. For this fish it has been repeatedly shown that one particular movement in the behavior train acts as a stimulus for the next (Fig. 11.6). It has also been shown how one may envisage the entire reproductive instinct of this fish to be subdivided into phases and hierarchically organized into components consisting of individual movements.

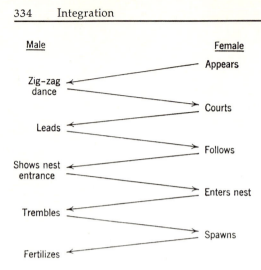

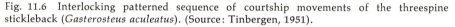

Fig. 11.6 Interlocking patterned sequence of courtship movements of the threespine stickleback (*Gasterosteus aculeatus*). (Source: Tinbergen, 1951).

The interplay between genetic and environmental forces during evolution has so fashioned the pattern and the responses of the nervous system, including the receptors, that simplified, shorthand, or "part-for-the-whole" stimuli often suffice to "release" further movements in a sequence. For instance, the silver color or the swollen outline of a ripe stickleback female elicits the zig-zag dance of the male at the initiation of the courtship. It has been shown that many prominent color patterns of fishes and other animals, especially spots that can be displayed, have such releasing functions as, for example, the black and white markings on tail and flanks of the guppy *Lebistes*. Several releasing signs may be used in one situation, each one adding to the releasing power of the other so that one may fashion a supernormal model with a greater than natural releasing value (principle of heterogeneous summation); even a crudely fashioned model of a female stickleback which is very silvery and has an excessively large abdomen assumes such a function. The principle of stimulus summation also applies to hydrographic and other environmental variables which act as releasers in the spawning behavior of such fishes as the northern pike (*Esox lucius*), whitefishes (*Coregonus*), and the brown trout (*Salmo trutta*).

Early theory on releaser mechanisms probably overemphasized their central nervous system components. Now we must note that the term "releasing mechanism" is operational rather than indicative of a location in the animal. We must also note that the concept stands for the sense organ-nervous system-effector complex rather than for a central pattern or template device.

The course of stereotyped behavior is influenced both by external and internal factors. Among the external factors, the releasing by sign stimuli, such as species-specific colors and shape, has been mentioned above. Among the internal factors in reproductive behavior, it is primarily the hormone levels that determine the intensity and also, to some extent, the course of the behavior. In nonreproductive activities, such as feeding, hunger as an internal factor may determine the extent and sequence of feeding movements. In the analysis of patterns and motivation of movements there appear certain distinct tendencies, such as sex, flight, and aggression, which influence the course of instinctive behavior. The zig-zag dance in the courting male stickleback, for instance, varies in duration of its approaching, the "zig," and its leading of the female, the "zag" phase, according to the level of both the sexual and the aggressive tendencies in an individual male.

In the course of an integrated behavior chain such as courtship, a fish may interpose certain movements such as feeding, chafing, or other comfort movements which belong to entirely different activities or instincts. Thus, for instance, if two male sticklebacks meet at the boundary between their territories they often do not fight but they perform digging movements that belong to the nest-building drive to be activated much later in reproductive behavior and not during the period when they establish their territories. These "displacement movements"—they are displaced out of the normal sequence of events—result from a conflict between different tendencies, in this example, aggression and flight.

Activities not released by their proper stimuli for a considerable time may be performed without any of the adequate releasing stimuli being present—a phenomenon called "vacuum activity." Single male guppies (*Lebistes*), raised in isolation, may, when they become sexually mature, perform sigmoid display movements which they normally show at the sight of a ripe female. Cichlids (Cichlidae) which have not been able to rear their own young can begin to guard a flock of water fleas (*Daphnia*) as an "overflow" reaction.

The nervous system has played a key role in the evolution of species-specific behavioral patterns. At the same time, the study of the behavior of related species reveals considerable plasticity of pattern change and modification so that the behavior derived from one activity can be taken over into another and acquire new significance (ritualization). The "eye spots" that commonly occur on the gill covers, tails, or fins of fishes are prominently displayed in fright and release flight behavior or search for cover by species mates but may also be used in courtship displays. Social releasers, that is, patterns especially adapted to a signal function, play a role in many kinds of behavior. They help the young to find the parent, they serve to maintain a rank order among individuals, and they are instrumental in ritualizing fighting behavior where they allow the beaten animal to show signs of inferiority before it is hurt. They are very important in the synchronization of sexual activities and they may even be operative in interspecific relations. Many territorial fishes such as surgeonfishes (*Acanthurus*) and damselfishes (*Eupomacentrus*) use simple threat displays as well as chasing and ramming behavior to defend their food supply of benthic algae from the invasion of other herbivores. Reef fishes throughout the world perform more or less ritualized solicitation displays to attract the attention of cleaning organisms. These solicitations and displays show a remarkable degree of variability among families but the species within many families show nearly identical symbiotic behavior even though they have been separated from their congeners in different oceans for many thousands of years.

Lack of information on the organization and detailed function of the central nervous system retards the analysis of the seat of coordination of instinctive behavior. Removal of forebrain of the goldfish or a tilapia resulted in massive defects in the acquisition and retention of avoidance performance. Other experiments indicate that forebrain removal depresses many functions but eliminates few. The fish forebrain may serve primarily as a modulator of lower brain centers. Fishes will remain among the best suited research subjects for the ethologists with their threefold study aims: the explanation of behavior in physiological terms, the study of its biological significance in conjunction with ecology, and the understanding of the development of behavior in the course of evolution.

ENDOCRINE ORGANS

The vertebrate endocrine system is present in its essence already in lampreys and hagfishes (Cyclostomata); further similarities with higher vertebrates appear in both the sharks (Elasmobranchii) and in the bony fishes (Osteichthyes). However, the differences are still considerable between the fish and mammal endocrine glands. These glandular differences can probably be correlated with differences in the evolution of other body systems in the two groups and with the exigencies of an aquatic existence. Although mammalian endocrinology is well-advanced, only certain aspects of hormonally controlled functions in fishes have been investigated. Those studied have been the influence of endocrines on color cells, the reciprocal action of sex cells and the pituitary, and thyroid function and adjustment to migration, but the finer details of the action and interrelations of fish endocrines still await elucidation.

The endocrine system, in contrast to the nervous system, is primarily concerned with relatively slow processes of a metabolic nature such as nitrogen metabolism (adrenal cortical tissue, thyroid gland), carbohydrate and water metabolism (adrenal cortical tissue), the maturation of sex cells and reproductive behavior (the pituitary gland and gonadal hormones).

There follows a list (Tables 11.3 and 11.4) of endocrine glands in fishes, with mention of organs of possible endocrine function(s); the individual glands and their action except those of doubtful function will be treated in separate sections; for location of glands in the fish body see Fig. 3.19, and, for interaction, see the diagram in Fig. 11.7.

The Pituitary Gland (Hypophysis)

This endocrine mastergland is found throughout the vertebrates and develops embryonically from a neural element that grows downward from the diencephalon to meet an epithelial element (Rathke's pouch) which grows upward from the buccal cavity. These two elements are thus of ectodermal origin and enclose mesoderm between them, from which develops the future copious blood supply of the pituitary. This blood supply is derived from the internal carotid artery. Anterior to the pituitary gland lies the optic chiasma and posterior, the saccus vasculosus. The stalk or infundibulum that attaches the gland to the floor of the diencephalon increases from the lamprey group (Cyclostomata) to the bony fishes (Osteichthyes); simultaneously, the pit in the bottom of the brain case where the gland is lodged becomes more pronounced (Figs. 3.20, 11.1, and 11.8).

Anatomy and Histology of the Gland.   The pituitary gland or hypophysis in fishes consists of a tripartite adenohypophysis and a neurohypophysis (with variances in terminology of parts). The division of the adenohypophysis into pro-, meso-, and meta-adenohypophysis is based on their relative anteroposterior positions and on different types of their secretory cells. These divisions of the adenohypophysis are also termed the rostral pars distalis, the proximal pars distalis, and the pars intermedia to emphasize their homology with the gland in tetrapods. The posterior, meta-adenohypophyseal portion underlies the thin neurohypophysis; it receives processes of hypothalamic neurosecretory cells. In sharks and rays (Elasmobranchii), the hypophysis projects farther ventrad and has a forward extension, the rostral lobe, which contains pro-adenohypophyseal components of histologic differentiation similar to that in cyclostomes. A down-growth in elasmobranchs is called the ventral lobe. The most anterior portion of the adenohypophysis is its most vesicular division, and cyclic changes in secretory activity with gestation and growth have been demonstrated in it. In the coelacanth lobefin (*Latimeria*), the rostral division of the pituitary (pro-adenohypophysis) is ventrally derived and independently vascularized, an organizational feature strongly reminiscent of elasmobranchs.

The pituitary gland of higher bony fishes (Actinopterygii) has become more compact than in lampreys and sharks, leading to increased difficulty in estab-

Table 11.3. **Parts, Secretions, and Action of Secretions of the Endocrine System**

Gland	Location	Embryonic Origin
Pituitary	Base of Brain	Ectoderm and mesoderm
Posterior lobe (Neurohypophysis)		
Intermediate lobe or zone (Meta-adenohypophysis)		
Transitional lobe or zone (Meso-adenohypophysis)		
Anterior lobe or zone (Pro-adenohypophysis)		
Thyroid	Bony fishes, diffused, or in a few, discrete, along ventral aorta. Cartilaginous fishes, midline between tongue muscles	Endoderm (with mesoderm added later)
Ultimobranchials	Sac-like structures between ventral wall of esophagus and sinus venosus (in characin, *Astyanax,* and many other higher bony fishes)	Endoderm (?)
Suprarenals (chromaffin tissue)	Along kidneys, dorsally	Mesoderm
Adrenal cortical tissue (Giacomini tissue; homologous with mammalian adrenal cortex)	One to three layers of cells lying along cardinal veins in the region of the hemapoietic head kidney	Mesoderm
Pancreatic islets	Gut walls in larval lampreys; hepatopancreas, most bony fishes; discrete pancreas, some few fishes	Mesoderm
Sex glands	Body cavity	Mesoderm
Intestinal mucosa	Small intestine	Endoderm
Corpuscles of Stannius	Attached to or imbedded in kidneys of holosteans and teleosts	Mesoderm

Secretion	Action
Arginine vàsotocin	Osmoregulation and saltwater balance
Oxytocin	Mating and egg laying
Intermedin (MSH)	Melanogenesis (*Fundulus*); melanin dispersion
Growth hormone (GH)	Size increase; metabolic effects (?)
Thyrotropin (TH)	Controls thryroid secretion
Gonadotropin(s?) (LH, FSH)	Controls secretion of gonadal hormones
Prolactin	Melanogenesis together with MSH
Exophthalmos producing substance (??) Melanophore concentrating hormone (?)	Blanching through melanin contraction?
Prolactin	Electrolyte regulation and melanogenesis
Corticotropin (ACTH)	Controls adrenal cortical secretion; melanogenesis (*Carassius*)
Melanophore concentrating hormone (?)	Blanching through melanin contraction?
Thyroid hormone(s)	Hastens maturity; controls metabolism but apparently has no direct control of oxidative metabolism; involved in metamorphic changes and possibly osmoregulation
Calcitonin	Calcium metabolism (parathyroid-like function)
Epinephrine, Norepinephrine	Concentrate pigment granules in melanophores; control blood pressure rise and pupillary dilation; affect pulse rate, etc. (effects comparable to those of sympathetic nervous system)
Adrenal cortical steroids	Promote utilization of stored fat; carbohydrate metabolism; water metabolism; protein catabolism; sodium retention; electrolyte metabolism
Insulin	Carbohydrate metabolism
Androgen(s?) in males	Development of secondary sexual characters; reproductive behavior; maturation of gametes
Estrogen(s?) in females	Development of secondary sexual characters; reproductive behavior; maturation of gametes
Secretin (Pancreozymin)	Stimulates pancreatic secretion
Unknown	Calcium metabolism

Table 11.4. **Fish Endocrine Organs of Doubtful or Unknown Functions**

Gland	Location	Evidence of Endocrine Function
Thymus	Dorsal (and in some ventral) to gill arches	Growth (?)
Pineal body	Diencephalon	Action on melanophores
Pseudobranchs	Gill cavity on hyoid	Chloride balance
Caudal neurosecretory system (Urohypophysis)	End of spinal cord in tail; cells similar to secretory cells of hypothalamo-hypophyseal neurosecretory system	Water and ion metabolism; sodium exchange (?); gas metabolism (?)

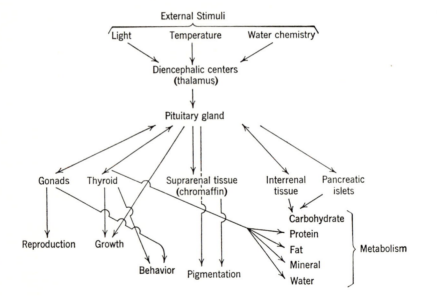

Fig. 11.7   Interaction of some endocrine glands of fishes. (Based on Vivien in Grassé, 1958).

lishing a clear-cut division among its parts. Differential secretory activity can, however, be demonstrated by standard staining techniques. Large ganglion cells occur in the adenohypophysis and neurohypophysis in such fishes as the trouts and salmon (*Salmo*) and the cods (*Gadus*). There is evidence that neurohypohyseal hormones are secreted in the neurosecretory tissues of the hypothalamus and that these accumulate in the neurohypophysis whence they find their way into the bloodstream. The control of the adenohypophysis seems, at least in part, to be based on neurosecretion. The hypophysis of the

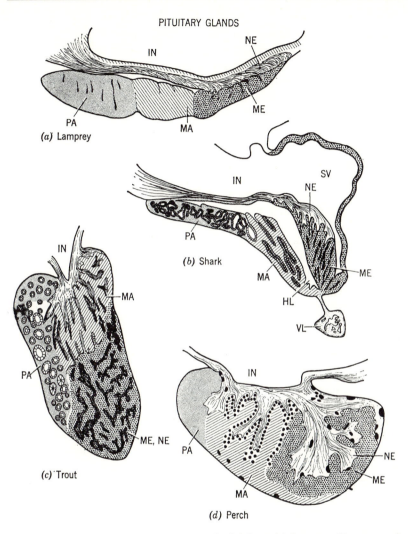

PITUITARY GLANDS

(a) Lamprey

(b) Shark

(c) Trout

(d) Perch

Fig. 11.8   Diagrams of the pituitary gland of fishes: (*a*) lamprey (*Petromyzon*); (*b*) dog-fish shark (*Squalus*); (*c*) trout (*Salmo*); (*d*) perch (*Perca*). HL, lumen of hypophysis; IN, infundibulum; MA, meso-adenohypophysis; ME, meta-adenohypophysis; NE, neurohy-pophysis; PA, pro-adenohypophysis; SV, saccus vasculosus; VL, ventral lobe. (Based on Pickford and Atz, 1957.

lobefin (*Latimeria*) differs markedly from that of other fishes. It is an un-usually elongate, cord-like structure.

The Pituitary Hormones.   Many of the hormones from the fish pituitary have the same or analogous effects as those from the incomparably better-studied mammalian gland. All known pituitary hormones are proteins or polypeptides and in many cases mammalian preparations have been used to establish the role of certain of them in fishes. It is to be noted that minor differences in chemical composition have been established for the same hormone in different families of mammals (pigs and cattle); it is therefore to be expected that similar differences exist between the hormones of different groups of fishes. The specificity of fish pituitary hormones has already been shown in experiments and practical applications with injections and implantation for forcing the spawning of certain fishes of economic importance such as sturgeons (Acipenseridae), trouts (Salmoninae), catfishes (Ictaluridae), and mullets (Mugilidae).

Pituitary hormones of fishes can be divided into two groups, one which reciprocally controls the function of other endocrine glands and the other which influences selected enzymatic processes in one or several types of body cells. In the first group are the adreno-corticotropic, the thyrotropic, and the gonadotropic hormones which are secreted in the pro- or meso-adenohypophysis, depending on the species of fish. In the second group are several hormones, described below, that affect melanophores.

The pro-adenohypophysis is the suspected source of a melanin hormone (MAH) acting on the melanin-bearing cells in the skin to bring about a more permanent background adaptation than would be possible through nervous impulses alone. The melanophore-stimulating hormone (MSH or intermedin) from the meta-adenohypophysis acts antagonistically to (MAH) and expands the pigment in the color cells, thus contributing to long-range background adjustment; melanin synthesis is also furthered by this hormone.

Prolactin, similar to the hormone that influences lactation in mammals, has been demonstrated from the pro-adenohypophysis. In conjunction with intermedin, prolactin promotes the laying down of new melanin in the melanophores of the skin in some fishes such as the mummichog (*Fundulus heteroclitus*). Prolactin is one of several hormones implicated in electrolyte regulation in teleosts, but its importance in maintaining homeostasis appears to vary with species. The adrenocortico-tropic hormone (ACTH, to be discussed later), apart from its target action on the hormone-producing tissue of the adrenal cortex (located diffusely in the kidney of fishes), also has a stimulating effect on melanin production in the xanthic goldfish (*Carassius auratus*). Further work may explain or reconcile these differences in the causation of melanogenesis. A growth hormone from the meso-adenohypophysis promotes increase in body length of fishes. Its mode of action on cell division and protein synthesis has not been explained.

The ventral lobe probably secretes gonadotropins in dogfish shark and skate. A thyrotropic function has also been found in the ventral lobe. The neurohypophysis produces, or stores from hypothalamic neurosecretory cells, endocrine substances that are well known from their effects on mammalian metabolism. Vasopressor and antidiuretic hormones constrict mammalian blood vessels and promote water retention through their action on the kidney. Oxytocin stimulates mammalian uterine muscle, increases the discharge of milk from the lactating mammae, and lowers blood pressure in birds. Fish pituitary material brings about such effects in higher vertebrate test animals or tissues, but it can be presumed that the target organs or the specific sites of action of neurohypophyseal hormones in fishes should differ from those of higher vertebrates where we encounter radically different problems of osmoregulation and therefore of salt and water balance. If fish blood is found to be devoid of neurohypophysial hormones, then it may be that they act as "local hormones" on the adenohypophysis.

The Thyroid Gland

Thyroid hormone contains inorganic iodine, extracted from the blood stream and stored in the follicles of the gland to be released into the circulatory system upon specific metabolic demands. The level of activity of the follicles is under the influence of the thyrotropic hormone of the pituitary which, in turn, is at least in part governed by a combination of genetically determined maturation processes in conjunction with certain environmental variables such as photoperiod, temperature, and salinity. In certain sharks (Elasmobranchii) and most higher bony fishes (Actinopterygii), with exceptions such as the parrotfishes (Scaridae), the thyroid gland consists of diffuse tissue scattered along the base of the ventral aorta (Fig. 3.19); removal or inactivation of the gland, a prerequisite for deficiency studies, is therefore difficult but physiological blocking or radiothyroidectomy using I^{131} has been accomplished. Little is known of the action of thyroid hormones in cyclostomes or in elasmobranchs. Respiratory stimulation by thyroxine is its best known action in mammals, yet most evidence indicates that thyroxine has no such action in teleosts. Physiologically, small doses of thyroxine and tri-iodothyronine cause thickening of the epidermis and paling of the goldfish (*Carassius auratus*). High thyroid titer retards growth of larvae of salmons (*Oncorhynchus*) but transformation into the juvenile smolt stage in the same genus is accelerated by inducing thyroid hyperactivity. In the mudskipper (*Periophthalmus*), a fish that spends much of its time partly outside of the water, induced thyroid hyperactivity leads to the assumption of an even more terrestrial existence with attendant morphologic and metabolic changes. The thyroid gland seems, in part, to influence osmoregulation in salmons (*Oncorhynchus*) and sticklebacks (*Gasterosteus*) and thyroid hyperactivity has been noted in salmons on their spawning migrations. Last but not least there are indications of a thyroid effect on growth efficiency and nitrogen metabolism in the goldfish, expressed through a rise in ammonia excretion.

It appears that thyroid functions in fishes are tied to a wide variety of processes, that some of them are related to growth and maturation phenomena, and that the thyroid, together with other endocrine glands, plays an important role in diadromous migrations of fishes.

The Ultimobranchial Glands

The small, paired ultimobranchial glands originate embryonically near the fifth gill arch but finally come to lie near the thyroid gland. The hormone calcitonin is found in the ultimobranchial glands of fish and in the thyroid and internal parathyroid glands of mammals, in which it produces hypocalcemia. Experiments with mummichogs (*Fundulus heteroclitus*) and with European and Japanese eels (*Anguilla anguilla* and *A. japonica*) have failed to confirm a hypocalcemic effect of calcitonin. Calcitonin may be related to osmoregulation. Eel calcitonin has been shown to decrease serum osmolality, sodium and chloride in Japanese eels. The ultimobranchial glands may be under pituitary control.

Chromaffin Tissue (Suprarenal Bodies; Medullary Tissue)

The chromaffin tissues receive their name from a characteristic staining reaction. In the lamprey group (Cyclostomata), chromaffin cells are arranged as strands along the dorsal aorta and in the ventricle and the portal vein heart, but in sharks and rays (Elasmobranchii) the tissue lies in association with the sympathetic chain of nerve ganglia. In bony fishes (Actinopterygii), location and concentration of chromaffin tissue shows considerable variation ranging from an elasmobranch-like pattern at one extreme, as in the flounders (*Pleuronectes*), to a true adrenal arrangement, as in the sculpins (*Cottus*) where chromaffin and adrenal cortical tissue are joined into one organ, comparable to the mammalian adrenal gland.

Chromaffin tissue extracts of fishes are high in adrenaline and noradrenaline. Exogenous administration of the catecholamines (adrenaline and noradrenaline) results in the following physiological responses: bradycardia, changes in blood pressure, branchial vasodilation, increased blood glucose levels, diuresis in glomerular teleosts, and hyperventilation. These "fight or flight" reactions in fishes are analogous to those of higher vertebrates and establish the suprarenal bodies as having similar function to that of the mammalian adrenal medulla.

Adrenal Cortical Tissue (Interrenal Tissue)

Endocrine cells producing adrenocortical steroid compounds are arranged throughout the body cavity near the postcardinal veins in the lamprey group (Cyclostomata). Among the sharks (Squaliformes) they lie between the kidneys. In rays (Rajiformes) they assume a more or less close association with posterior kidney tissue with some species having the adrenal cortical tissue aggregated near the left and others near the right central border of that organ. In bony fishes (Actinopterygii) there are one to several layers of interrenal cells to be found along the postcardinal veins as they pass through the head kidney (pronephros).

Secretion of adrenal cortical hormones is under control of the adrenocorticotropic hormone (ACTH) of the hypophysis. In the Mexican tetra (*Astyanax mexicanus*), experimental hyper-function of interrenal cells can be produced as well as a condition that resembles the well-described adrenal stress syndrome in mammals. Cortisol, cortisone, corticosterone, and aldosterone known from mammalian adrenal glands, as well as some unknown compounds, have been isolated from blood plasma of the sockeye salmon (*Oncorhynchus nerka*). These findings suggest, as is true in mammals, that two types of secretions are produced by the fish adrenal cortical tissue: (*a*) mineral-corticoids that influence certain aspects of fish osmoregulation; (*b*) gluco-corticoids that influence the carbohydrate metabolism, especially the level of blood sugars. Experimental evidence for the first function comes from the rainbow trout (*Salmo gairdneri*) which excretes higher than normal amounts of sodium ions through its gills but retains more than the normal amount of sodium in the kidneys when subjected to internal salt loads under mineral-corticoid treatment. The second function was verified in the oyster toadfish (*Opsanus tau*) where blood sugar levels are raised upon intramuscular injection of corticosteroid compounds.

The concentration of cortisone in the blood plasma of salmons (*Oncorhynchus*) is high at spawning stages of the life history but low during the more sedentary parr stages. In *Oncorhynchus*, 60 percent of the body protein is catabolized during the migration-spawning phase, and this catabolism is associated with a sixfold increase in plasma corticosteroids and an elevation of liver glycogen. Injection of adrenal cortical hormones promotes lymphocyte release in a tetra (*Astyanax*) and antibody release in the European perch (*Perca fluviatilis*), pointing to the similarity in the function of the adrenal cortical tissues in higher and lower vertebrates. Corticosteroids structurally resemble androgens and have androgenic side effects. These side effects suggest some functional equivalence in fishes. Daily rhythms in concentrations of plasma adrenal corticoids have been reported in several teleost fishes, and these rhythms may play a prominent role in the diel rhythm of teleosts.

Corpuscles of Stannius

The corpuscles of Stannius are endoctrine organs attached to or imbedded in the kidneys of holostean and teleost fishes. Their number varies with the species and histologically they resemble adrenal cortical cells. The corpuscles of Stannius lower serum calcium levels in mummichogs (*Fundulus heteroclitus*) kept in high calcium environments, such as seawater. It appears that the corpuscles work in conjunction with the pituitary gland, which produces a hypercalcemic effect, to maintain a relatively constant level of serum calcium.

The Sex Glands as Endocrine Organs

In fishes the maturation and possibly also the elaboration of gametes requires the presence of sex hormones; in addition secondary sex characteristics such as coloration, breeding tubercles, and the maturation of gonopodia depend on the presence of endocrine substances from the sex glands. Sex hormones are produced by specialized cells of the ovaries and testes at levels determined by the output of gonadotropic hormones from part of the pituitary gland (meso-adenohypophysis).

The male hormone testosterone has been measured in the blood plasma of an elasmobranch (*Raja radiata*) and in salmon, in which there is a correlation between plasma levels and the reproductive cycle. Another gonadal steroid, 11-ketotestosterone, is ten times more physiologically androgenic than testosterone in *Oryzias latipes* (medaka) and sockeye salmon (*Oncorhynchus nerka*). Of the female estrogens, estradiol-17β has been found in many species of fish, often accompanied by variable quantities of estrone and estriol. Progesterone is present in fish, but it is not known if this compound has a hormonal function.

There have been few unequivocal demonstrations for any fish that reproductive behavior is induced or regulated by gonadal steroids. Many researchers have demonstrated effects of exogenous steroids on sex determination and the development of secondary sexual characteristics in fish, but in most cases there is little mention of the effects of these treatments on reproductive behavior.

Upon an injection of mammalian testosterone and estrone, a lamprey (*Lampetra fluviatilis*) develops its cloacal lips and coelomic pores, both structures involved in the reproductive process. In sharks and rays (Elasmobranchii) tests for a similarity between the sex hormones from different vertebrate classes have not shown clear results, but among bony fishes (Osteichthyes) such equivalence seems to exist, although experimenters maintain certain reservations. For instance, ethynil testosterone (pregneninolone), which has mild androgenic and progesterone-like effects in mammals and birds, seems to be highly androgenic in fishes. Testerone injections in immature and female parrotfishes (Scaridae) alike evoke appearance of the male color pattern in the experimental subject. Female sex hormones of fishes are less similar to those of other vertebrates than the testicular hormones which also may strongly affect ovarian development, as shown in a loach, the Japanese weatherfish (*Misgurnus fossilis*). Evidence to date points to less biochemical specificity in both androgenic and estrogenic fish hormones than in mammals and birds and to a greater degree of action on the gonads by sex hormone analogues such as for instance certain corticosteroids.

Prostaglandins, which are potent, hormone-like substances, have recently been discovered in testis and semen of flounder (*Paralichthys olivaceus*), blue-fin tuna (*Thunnus thynnus*), and chum salmon (*Oncorhynchus keta*). In mammals, prostaglandins lower arterial blood pressure and stimulate the contraction of smooth muscle, but their function in fish has not been studied.

The Action of Gonadotropins on the Gonads. Removal of the pituitary (hypophysectomy) performed on several species in all three major groups of living fishes reduces or stops gonadal development, both in the transition from juveniles to sexual maturity, and during the seasonal spawning cycle. Dependence on the pituitary gland for the output of sex hormones by the gonads themselves has thus been clearly established in fishes. Present research favors the existence of only one functional gonadotropin in fishes, for which the term Piscine Pituitary Gonadotropin (PPG) has been proposed. The single teleostean gonadotropin possesses properties that in mammals are distributed between two hormones, LH and FSH. On less complete evidence than is available for mammals rests the assumption that the fish meso-adenohypophysis produces two different gonad-stimulating hormones.

Mammalian luteinizing hormone (LH) promotes release of gametes from nearly ripe gonads in fishes and stimulates appearance of secondary sex characteristics. From this it is inferred that there is a similar hormone in the fishes themselves. Gonadotropins purified from salmon pituitaries resemble LH. Furthermore human chorionic gonadotropins and the urine of pregnant mares, which have LH-like effects, promote the release of eggs in female fishes. Evidence for a follicle-stimulating hormone in fishes (FSH), the second gonatropic hormone known from the mammalian pituitary gland, rests, so far, on standard assay for mammalian FSH performed with fish pituitary extracts. Such assays, however, do not furnish proof of a second pituitary gonadotropic hormone in fishes.

The gonadotropins or whole pituitary preparations from mammals, birds, and reptiles are effective in some fish species in hastening spawning or otherwise influencing sex glands and behavior. When amphibian materials are used success is obtained more often than with mammalian ones. Best results, however, follow the use of glands or gland extracts of the same fish species; however, biochemical interfamilial and interspecific differences of fish gonadotropins seem in many cases to be even greater than those noted between fishes and amphibians.

The Use of Gonadotropins in Fish Culture. In fish-culture operations it may be costly to retain potential spawners over long periods only to have them fail to ripen when desired. Pituitary implants or injections are now widely used to bring about the timely release of eggs and sperm. Among the most extensive users of pituitary-induced forced spawning are Brazil and Russia. In Brazil, the main freshwater-fish crop consists of characins (Characidae). In Russia, the technique is applied to sturgeons (*Acipenser*) for caviar production. Some use is made of the method also for the culture of a variety of fishes, including trouts and salmons (Salmoninae) and carps (Cyprinidae) in Europe, North America, and Asia.

To induce forced spawning, fresh pituitary glands are either injected intraperitoneally after trituration in 0.7 percent solution of salt (NaCl) or they are implanted whole. The glands may be frozen and used within 6 months, or as is frequently done, the glands may be dehydrated in acetone or alcohol, then dried, powdered, and stored in airtight containers for several years. The powder is suspended in acidified saline solution for intramuscular or intraperitoneal injection.

Best results are obtained with glands taken from ripe or nearly ripe donors of the same species although other species can also be used. The dosages depend both on the state of maturity of the donor as well as of the recipient fish. Ripe fish, in general, need only a single dose whereas nearly ripe individuals may require injection of material from a whole gland every second or third day for a month. The use in practical fish culture of the principle of stimulation of gonads by gonadotropins of the pituitary gland is promising. We may expect considerable refinement of procedures as information accumulates on the structural formulas of fish gonadotropins and sex hormones, as already indicated by the effectiveness in inducing spawning in several species of fishes by highly purified salmon gonadotropin.

The Intestinal Mucosa

The intestinal lining (mucosa) produces hormones that join with nervous controls to regulate the pancreatic secretions. The intestinal hormones are secretin and pancreozymin (formerly both subsumed under secretin). Secretin produces flow of enzyme-carrying liquids from the pancreas, and pancreozymin enhances flow of the zymogens. Both hormones are formed mostly in the anterior portion of the small intestine. Their production can be brought about by introducing into the pyloric region of the stomach of carnivorous fishes of an acidified homogenate of fish flesh or through the injection of secretin into the gastric vein which stimulates the secretion of the pancreas. Thus the classic tests for endocrine activity are satisfied for the fish small intestine. In higher vertebrates there exists another intestinal hormone, cholecystokynin, which promotes flow from the gall bladder. A polypeptide with cholecystokinin-pancreozymine-like activity has been found in cells of lamprey intestine.

THE SENSES (SENSE ORGANS) OF FISHES

The sense organs receive physical or chemical stimuli from the environment. Physical changes in heat flow or touch are felt through skin receptors; visual stimuli involve changes in light intensity and quality; and acoustical ones are received through the inner ear or lateral line. Chemical stimuli are those experienced through either smell or taste organs. Pain is probably not experienced as a strong sensation by fishes, though forceful or noxious physical or chemical stimuli evoke violent reactions.

The Cutaneous Senses (Temperature, Touch, and Taste; Figs. 11.9–11.12)

The temperature sense can be used by the fish to remain in a temperature preferendum and may contribute to orientation on long- or short-range movements. However, temperature perception has been thoroughly investigated only among higher bony fishes (Actinopterygii). By conditioned reflex training, several species of both marine and freshwater bony fishes have shown capability of detecting temperature changes of about 0.03°C, provided the rate of heat flow is fairly rapid (in excess of 0.1°C per minute), and the detection of small temperature differences relies on areal summation of impulses from finely branching epidermal nerve endings (Fig. 11.9). Fishes can distinguish a rise in temperature from a fall, but the physiological mechanism for such discrimination is not known. There is evidence that some fishes may be most sensitive to temperature changes on their anterior portions and that temperature sensation is intimately tied to the perception of light (weak) tactile stimuli.

The ampullae of Lorenzini of sharks and rays (Elasmobranchii), a component of the lateral-line system, respond to cooling by a rise and to warming by a lowering in discharge rate along their nerves (Fig. 11.10). It is not known whether they serve as temperature receptors in the intact animal, or whether or not a diffuse cutaneous temperature sense is also present.

Some fishes are sensitive to touch, as in some catfishes (Siluridae). However, as yet there is little evidence of specialized dermal touch receptors like those in mammals and birds (Vater-Paccinian, Krause's and Herbst's corpuscles). In these higher vertebrates, each touch corpuscle has a skein-like nerve net, enclosed in a connective tissue capsule. Comparable structures have so far been described only in the fins of sharks (Squaliformes), where

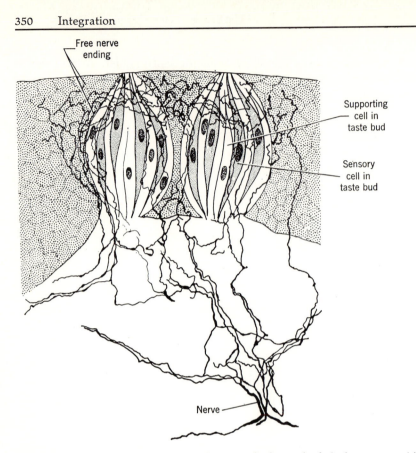

Fig. 11.9   Free nerve endings surrounding taste buds in a barbel of a sturgeon (*Acipenser*). (Based on Dogiel in Bolk, Göppert, Kallius, and Lubosch, 1938).

they act as proprioceptors when the fin is adducted or abducted, and in the snout region of some moray eels (*Gymnothorax*), where they serve for fine touch discrimination in the choice of food (Fig. 11.11). Similar structures also occur on the separated anterior pectoral fin rays of searobins (Triglidae).

Observations of behavior and electric recordings of nerve discharges have shown bullheads (*Ictalurus*) to be highly sensitive to touch on the head and barbels. These regions carry a net of thin, profusely branching fine nerves with free endings, derived from cranial nerves V and VII, but no specialized touch structures have been found in minnows (Cyprinidae) where they appear to react jointly to tactile and thermal stimuli. On the body surface free endings of the sensory roots of the spinal nerves receive the touch stimuli.

Fishes in general have specialized organs of taste in the epidermis of the mouth and lips, in the pharynx, and on the snout (Fig. 11.9). Certain groups which often live in muddy water or which do not feed mainly by sight, such

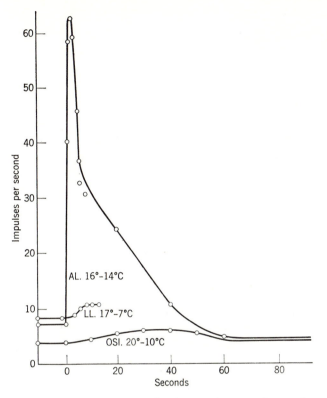

Fig. 11.10 Changes in impulse frequencies upon temperature changes in an ampulla of Lorenzini (AL), lateral-line organ (LL), and an intramuscular nervous end organ (OSI). (Source: Sand in Grassé, 1958).

as the catfishes (Siluroidei), the cods (Gadidae), and the loaches (Cobitidae), have additional taste buds at other places on the body; over 100,000 such units have been estimated on the body of a bullhead (*Ictalurus*). The buds are mainly innervated by a branch (ramus recurrens) or cranial nerve VII, but elements of cranial nerves IX and X also share in the innervation of taste receptors that occur on outlying parts of the body (Fig. 11.12).

The vertebrate taste bud is typically composed of elongated sensory cells arranged like segments of an orange (Fig. 11.9). The three different cell types in fish taste buds are receptor cells, supporting cells, and basal cells. In addition, there are transitional or intermediate forms of cells that may become either sensory or supporting cells. The sensory cells have short, hair-like extensions in contact with the water and extend from the tip of the taste bud to the underlying dermis where they may have a bottle-shaped expansion and where the nucleus is located. In especially sensitive regions, such as the

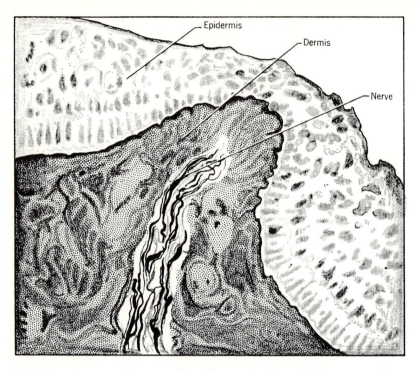

Fig. 11.11 Diagram of presumed touch sensor from the lower lip of a moray eel (*Gymnothorax vicinus*). (Based on Bardach and Loewenthal, 1961).

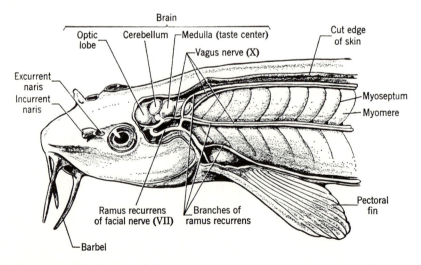

Fig. 11.12 Distribution of the ramus recurrens innervating taste buds on body of a loach (*Noemacheilus*). (Source: Dijkgraaf in Grassé, 1958).

barbels of bullheads (*Ictalurus*) and sturgeons (*Acipenser*) and the lips of conger eels (Congridae) (and amphibians), nerves that reach the taste buds also give off a net of free nerve endings to the surrounding regions. It appears that taste and touch may thus work in close cooperation in the choice of food before ingestion.

In the lamprey group (Cyclostomata), taste buds have been found in the pharynx, the gill cavity, and on the head. In sharks and rays (Elasmobranchii), taste buds occur in the mouth and the pharynx and taste-bud-like structures, probably also for chemical perception, are present in pits on the body surface.

The carp (*Cyprinus carpio*) has taste sensations similar to mammals; it can sense salty, sweet, bitter, and acid stimuli. It also tastes extracts of fish skin and other, so-called sapid, substances. Amino acids singly or in mixtures are particularly effective taste stimulants in fishes. Marine fishes (searobins, *Prionotus*; hake, *Urophycis*) with taste- or chemo-receptors on their modified paired fins also sense sapid substances, for example, the extracts of clams or worms. They appear to taste compounds that normally occur in their food but their receptors do not respond to salty, sweet, or bitter substances though they still react to acids. Histologic examination of the skin of a minnow (*Phoxinus*) reveals a subepidermal nerve plexus that sends branches into the epidermis where the nerve fibers divide and taper into fine endings. Some specialized single cells, presumably chemosensory, have also been found in a minnow (*Phoxinus*). Thus, tactile, temperature and chemical sensations are closely integrated in fish skin. The coarser among these divisions are probably tactile whereas the finer ones are believed to serve temperature and chemical sensations.

Vision

Natural waters are rarely pure; therefore light rays in them undergo multiple scattering in all directions. Furthermore, waters are often turbid, almost opaque, and therefore a sharply defined image is difficult to obtain. Nevertheless, it is of great biological significance for fishes to detect slight movements and color changes that signal prey, predator, companion in a school, or mate. Thus it appears as if the retinal mechanisms of fishes have developed towards contrast enhancement and motion perception besides the formation of a clearly focused image. Basic anatomical features of the eyes of fishes have been described already (Fig. 3.22); here will be added certain details of the functional anatomy and vision of color and form.

The Cornea of the Eye. In the lamprey and shark groups (respectively, Cyclostomata and Elasmobranchii) the cornea is transparent, without pigment, as it is in most bony fishes (Osteichthyes). Some bony fishes, such as the bowfin (*Amia*), the perches (*Perca*), and the wrasses (Labridae), have yellow or green corneal membranes. The sea lamprey (*Petromyzon*) has a cornea composed of two tissue layers separated by a gelatinous substance. This primitive condition is also found in the embryos of higher fishes and in the adults of certain kinds with poor vision such as the conger eels (Congridae). The inner layer of the cornea continues into a corneal muscle that compresses the eye when contracted and brings about modest distance accommodation. No fine intraocular accommodation mechanism exists in cyclostomes. A distinctly two-layered cornea is found in certain flounders (*Solea*), perhaps to protect the eye of such a bottom-dwelling fish from sand and detritus. The mudskipper (*Periophthalmus*) leaves the water for considerable periods of time; it also has a two-layered cornea and folds of skin beneath the eye which retain water and help to prevent drying of the organ. In the lungfishes (Dipnoi) the cornea is rugged but there are no wetting structures.

The Sclerotic Coat.   The sclerotic coat, sclera, or firm layer of the eyeball is variously reinforced in all groups of fishes. In lampreys (Petromyzonidae) the sclera is fibrous and firm but sharks and rays (Elasmobranchii) have cartilaginous elements and also calcified plates in this layer for additional reinforcement. The sclerotic coat in an elasmobranch is highly vascular and has a base of fibrous connective tissue. The sturgeon and its relatives (Chondrostei) carry a cartilaginous support at the corneal border of the sclerotic coat as well as some bony structures. In the sturgeons (*Acipenser*), a half-moonshaped bone, the conjunctival bone of Mueller, partly surrounds the cornea. The highest bony fishes (Teleostei) have fibrous, flexible sclerotic coats within the orbit and around the optic nerve, but there may be cartilaginous or even bony supporting elements in the outer portion of the eye which can surround the entire cornea. The living coelacanth lobefin (*Latimeria*) carries cartilaginous reinforcements in the sclera which thicken towards the optic nerve; there are also calcified scleral plates that form a pericorneal ring.

The Choroid Coat (Fig. 11.13).   Cyclostomes carry much melanin in the choroid coat, especially toward the posterior pole of the eye. The shiny nature of most shark eyes is due to the reflecting material in a well-developed choroid coat or tapetum lucidum. The tapetum functions to collect a maximum amount of light for the retina by scattering reflection from guanine crystals.

In sturgeons and relatives (Chondrostei) vascular elements become more prominent in the choroid layer of the eye than they were in shark-like fishes (Elasmobranchii) and layers of guanophores are also present.

Most holostean and teleostean bony fishes, with the exception of the eels (Anguilliformes) and related groups, have the blood vessels of the choroid coat collected into a horseshoe-shaped choroid gland lying between choroid and sclerotic coats of the eye (Fig. 11.13), in a position around the exit of the optic nerve. The gland has its blood vessels arranged like the rete mirabile in the gas bladder (Chapter 8), for the secretion of oxygen at higher tensions than those found in the blood in order to provide oxygen in abundance to tissues in the vicinity of the gland. It should be noted that the retina lies close by, that it has the highest oxygen demand of any tissue in the body, and that the retinal circulation proper is not especially well developed in those bony fishes which have a large choroid gland. Measurements with polarographic electrodes in the vitreous humor have shown that fishes with a well developed choroid rete such as the bluefish (*Pomatomus saltatrix*) and remoras (*Remora*) have intraocular oxygen tensions (450 to 780 mm Hg) which are several times that of their arterial blood. Oxygen tensions (210 mm Hg) in the eye lower than arterial blood are found in fishes with a small choroid rete such as the tautog (*Tautoga onitis*) and also eels (*Anguilla, Conger*), as well as sharks and rays (Elasmobranchii), which have no choroid rete and have intraocular oxygen tensions (10 to 20 mm Hg) below that of

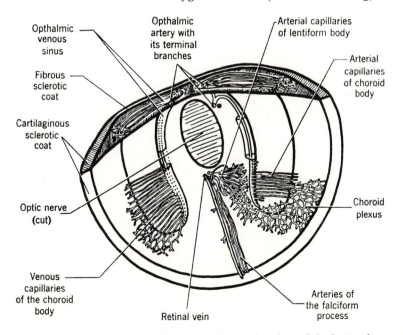

Fig. 11.13 Arrangement of blood vessels in the choroid body in the eye of a trout (*Salmo*). (Based on Barnett in Grassé, 1958).

the surrounding seawater. Thus the choroid gland can provide the retina with a good oxygen supply apart from any other function it may have; it has also been suspected to act as a cushion against compression of the eyeball.

Brilliant guanophores occur in the choroid coat of many bony fishes (Actinopterygii), to maximize the amount of light that reaches the retina. The choroid in the living lobefin (*Latimeria*) also has a brilliant guanine cover.

The Iris.   The iris forms the pupil and controls the amount of light that reaches the retina. In the cyclostome *Mordacia mordax* the iris lacks muscles, but sharks and rays (Elasmobranchii) have muscular elements in their guanine-studded iris and thereby can adjust the shape of the pupil. The remaining groups of fishes, with some exceptions among the flounders (Pleuronectiformes) and eels (Anguillidae) have a fixed pupil, either circular or oval. Their iris contains no muscles, but some guanine and melanin are present. The pupil of the living lobefin (*Latimeria*) is large and noncontractile. Some deep-sea teleosts lack an iris.

The Lens and Accommodation.   The lens of the fish eye is a firm, transparent ball made up of noncollagenous protein. Since the index of refraction of the cornea and the ocular humors are about the same as water (1.33), refraction and image formation depend almost entirely on the lens, which has a very high *effective* index of refraction (1.67). The fish lens is not homogeneous but has an *actual* refractive index from 1.53 at the center to 1.33 near the outside. An almost constant feature of fish eyes is the distance from the center of the lens to the retina divided by the radius of the lens. This ratio is about 2.55. The lens is roughly spherical in most bony fishes (Osteichthyes) and in lampreys (Petromyzonidae). However, sharks and rays (Elasmobranchii) have a horizontally compressed lens and in fishes that leave the water at times the lens face shows modifications; in the four-eyed fishes (Anablepidae), for instance, the lens is pyriform to provide for aquatic and aerial vision. The lens is attached from above to the suspensory ligament and from below to the falciform process with its retractor lentis muscle.

In most fishes, accommodation, which is the adjustment of vision from near to far, results from changing the distance between the lens and the retina rather than altering lenticular shape. Some sharks and rays (Elasmobranchii) accommodate by small changes in lens convexity. Others are capable of moving the lens by means of a muscular papilla on the ciliary body and thus achieve a little adjustment. Rayfin fishes (Actinopterygii) show varying degrees of lens accommodation, depending on the development of the retractor lentis muscle. According to Pumphrey, sight feeders such as the brown trout (*Salmo trutta*) have this muscle well developed and possess other remarkable features of lens and retina which suggest that the vision of this fish is acute over a wide range. Trouts (Salmoninae), in fact, have the ability to focus simultaneously on different portions of the retina objects which are distant and other objects which are nearby. The lens in these fishes is slightly ellipsoid with the longest lens diameter parallel to the long axis of the fish. The lens thus has two focal lengths. One is for light rays reflected from distant objects that are lateral to the fish and come to focus on the central retina. The other, shorter one, is for close objects ahead that are projected on the posterior (temporal) retina. Furthermore, in accommodation, the lens is moved in a plane parallel to the long axis of the fish rather than inwards. Consequently, the distance between lens and central retina remains unchanged, and the focus remains adjusted for distance vision. At the same time the contraction of the retractor lentis muscle brings the lens closer to the posterior retina that represents the frontal field of vision of the fish. In this portion of the visual field the fish is thus capable of obtaining sharp images within distances ranging from a few inches to infinity, or the limit of transparency of the water.

Retinoscopy has shown that many species of bony fishes (Osteichthyes) are farsighted. Refractive errors change by several diopters during such measurements on schooling fishes, such as the silversides (*Menidia*), certain jacks (Carangidae) and others, suggesting that fishes can indeed move their lenses toward and away from the retina or deform their ocular globe for accommodation. Some groundfishes such as flounders (*Paralichthys*) and reef fishes such as the squirrelfish (*Holocentrus*) and butterflyfishes (Chaetodontidae) are emmetropic, that is the images focus on the retinal surface. No retinoscopic tests have been made on small stream and lake fishes such as certain sunfishes (Centrarchidae) but analogies with reef fishes in behavior and habitat suggest that they also might be emmetropic. In the alewife (*Alosa pseudoharengus*) the focal plane of the eye lies an average distance of 0.2 millimeter behind the retina, suggesting that static images which this fish sees may not be as sharp as those of animals with aerial vision.

In deepsea fishes especially large lenses and tubular, upward-viewing eyes are found, as in the pearleyes (*Scopelarchus*), possibly to aid in gathering on the lower retina signals from the ventral photophores of species mates, of prey, or of predators.

In the front of the lens there is a small space with clear saline aqueous humor. The main cavity of the eyeball is filled by the transparent vitreous humor which is secreted by the ciliary body and contains transparent tendinous reinforcements.

The Retina. Fishes have typical vertebrate retinas made up of several partially interlocking cell layers (Fig. 11.14). As light hits the retina after travel-

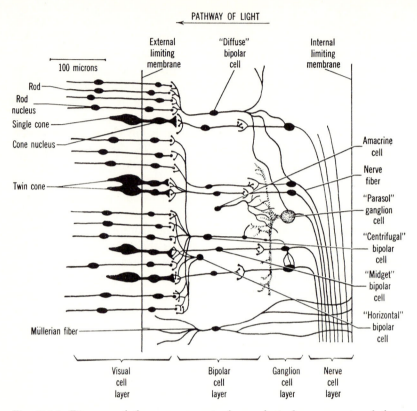

Fig. 11.14 Diagram of the arrangement of neurological components of the retina of a salmon (*Oncorhynchus*). (Source: Ali, 1959).

ing through the lens and the humors, the retinal layers encountered in order are: (*a*) fibers leading to the optic nerve, (*b*) their ganglion cells, (*c*) bipolar cells which, in turn, are in contact with (*d*) the photoreceptor cells, proper, namely the rod and cone cells of the retina. In the retina of lampreys (Petromyzoniḍae) no distinction can be made between the rods and cones but in some sharks and rays (Elasmobranchii) and in all the highest bony fishes (Teleostei) the rods and cones are well differentiated. In certain bony fishes twin cones occur and triple or even quadruple cones have been reported; it is not known whether these cone aggregations, peculiar to the fishes, have any special function. In teleost, but not in elasmobranch retinas there is a fifth layer (*e*) of retinal cells, the epithelium, which contains melanin and borders on the choroid coat.

In bright light the melanin pigment of the epithelial layer of the retina shades the sensitive rods; in dim light, however, the melanin becomes con-

centrated close to the choroid border so that photosensitive cells are fully
exposed to whatever light is available (Fig. 11.15). At the same time the contractile myoid elements in the base of the rods and cones move the sensitive
cell tips in such a fashion that rods are extended away from the lens in
bright light to be engulfed by the epithelial melanin, whereas in darkness the
rod myoids contract and approach the cells towards the lumen of the eyeball. Cones move in the opposite direction toward the lens in bright and
toward the outer epithelium in dim light (Fig. 11.15). Fish eyes show these
photomechanical or retinomotor reactions more pronouncedly than the eyes
of other vertebrate groups. The level of light reaching the retina can also be
controlled by contraction of the pupil. With the exception of a few groups
such as eels and flatfishes, this process is relatively poorly developed in fish.
In sharks and relatives (Elasmobranchii) the photoreceptor layer may be
shielded by melanin which moves in and out in the choroid layer inasmuch
as the retina lacks a pigmented epithelium.

The number of rods and cones connected to one ganglion and thus to one
fiber of the optic nerve varies greatly in different kinds of fishes. Crepuscular
sight feeders and some deepsea fishes have many rods converging on one
optic fiber, as an adaptation for vision in subdued light (summation of impulses); in other groups, such as the pikes (Esocidae), that seem to rely on
good light for feeding, far fewer retinal receptors are connected to one optic
neuron. Relative numbers of rods and cones also vary considerably in different species. The highest cone-to-rod ratios are in sight feeders that are

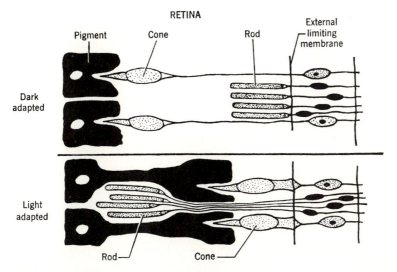

Fig. 11.15 Diagrams of the adaptations of the visual elements of the bony-fish retina
to light and dark. (Source: Wunder in Herter, 1953).

active during the day whereas many crepuscular species have apparently increased the number of rods as compared to the cones. Certain midwater fishes (for example, *Winteria* and *Opisthoproctus*), besides having large eyes with luminous organs adjacent to them, also have very large numbers of rods in the retinas.

Few bony fishes (Osteichthyes) have a fovea (retinal area of clear vision) where cones and ganglion cells are more numerous than in other regions of the retina. The fovea appears to be restricted to fishes that live in clear waters and especially to those which can move their eyes independently or converge them to obtain a wide angle within which they may have binocular vision; examples include the seahorses (*Hippocampus*) and many members of the sea bass family (Serranidae).

The retinas of fishes yield two kinds of light-sensitive pigments, rhodopsin and porphyropsin. Purple-colored rhodopsin has been isolated from the eyes of many marine rayfin fishes (Actinopterygii), whereas the retinas of freshwater rayfins and the marine wrasses (Labridae) contain a rose-colored, light-sensitive substance, porphyropsin. Diadromous fishes, such as the sea lamprey (*Petromyzon marinus*), eels (*Anguilla*), and the Atlantic salmon (*Salmo salar*), possess both rhodopsin and porphyropsin; lamprey (*Petromyzon*) retinas from seaward migrants yield more rhodopsin, whereas porphyropsin predominates in the eyes of spawners in fresh water; converse changes occur in the eel as it travels from lakes and rivers to the Sargasso Sea. Rhodopsin and porphyropsin are synthesized in the dark from Vitamin A. On exposure to light there appears in the rods the yellow pigment retinine which can, however, be used to reconstitute the light-sensitive purple or rose pigments. The light-induced chemical change from purple to yellow pigment in the rods becomes translated in the retina to nerve impulses which travel along the optic nerve. Biochemical and physiological processes occurring in cones and thus connected with color vision are not well known. Some pigments with different sensitivity maxima have been isolated from vertebrate retinas and electrical recordings have been made from the bipolar cell layer of several families of marine rayfin fishes. These findings suggest that the cones in the retinas of shallow-water species have paired responses to stimuli of different wave lengths of light. There is a family of elements sensitive to the combination of yellow and blue and another to red and green stimuli. When the rods are removed from such a retina, the cones respond to white light thus suggesting that color vision in fishes is based on a three-pigment system. Snappers (Lutjanidae) only respond to white light. They may be color blind, but it is to be noted that they live in water depths of 30 to 70 meters (about 100 to 200 feet), where the light consists largely of the blue-green portion of the spectrum and where even to man, with his well-developed color vision, things appear to be of different shades rather than of different hues.

The initial step in vision is photochemical and involves reactions in the light-sensitive pigments of the rods and cones, as already suggested. It is not known how these chemical changes become transformed into the electrical impulses that can be picked up in the retina, nor how these finally come to be the signals which leave the eye in the optic nerve and travel to the brain. Signals in the optic, as in other afferent nerves are of three kinds: "on," "off," and "on-off." They are characterized by changes in discharge frequency along the nerve. The retina of the goldfish (*Carassius*) has its lowest threshold for "on" responses in the blue, and for "off" responses in the red portion of the spectrum, indicating that this type of coding of receptor function permits acute contrast vision. Good discrimination of movement and contrast is also implied for this fish by the following observations. When a pointed mono-chromatic light of 153 microns in diameter is used to test the receptive field of one optic nerve ganglion that covers a field of about 4 square millimeters, the above three types of responses convey highly detailed messages about direction and progress of the moving light stimulus (Fig. 11.16). Progressing diagonally from the lower left to the upper right "on" responses as an increase

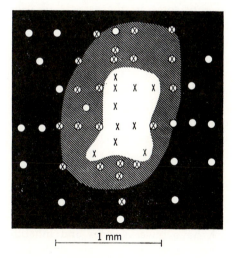

1 mm

Fig. 11.16 Receptive field of a single ganglion cell. Central clear area indicates region where only "off" responses are found. Crosshatched area indicates region where "on-off" responses are found. "On" responses are found only in peripheral area indicated by solid black. Test stimulus was 153 microns in diameter, wavelength 600 millimicrons, intensity 18 microwatts/cm^2. (Source: Wagner, MacNichol, and Wolbarsht, 1960).

in volleys of nerve impulses are recorded as the light appears. Then as the point-light proceeds its progress is announced in an "on-off" fashion by volleys both on appearance as well as on disappearance on a certain area of the field. In the center the retinal receptors cause an increase in discharge rate ("off" response) along the nerve only as the light disappears, that is as it moves upwards. The upper half of the diagonal path shows the reverse sequence of coded messages from those observed in the lower half.

The eyes of lungfishes (Dipnoi) resemble those of amphibians, and in their retinas they have no (or poor) distance accommodation, few rods, and many cones, as well as some twin cones. The living lobefin (*Latimeria*) has large eyes, retinas without melanin, and receptor layers which consist almost exclusively of rods. Furthermore, relatively few rods are connected to each optic nerve fiber, and the eyes altogether resemble those of other mid-depth fishes that live in an oceanic twilight-zone.

The coded message carried by the optic nerve is further elaborated in the brain. These processes, of which little is known from any vertebrate, including man, must finally make the visual stimulus meaningful to the organism.

Sight in the Lives of Fishes.   Many fishes rely on sight for capturing food, for receiving signals that bring on or complete mating behavior, and for finding and relocating shelter, which are only some of the aspects of fish behavior that are sight dependent or sight influenced.

Experimental training has shown that shallow-water fishes have color vision, very much like man; they can distinguish twenty-four different narrow-band spectral hues. Some fishes even exceed man's spectral range into the violet. White is to fishes a sensation basically different from that of hues and they see complementary colors. More importantly still, these trained fishes, mostly minnows (Cyprinidae, including the elritze, *Phoxinus phoxinus*), are capable of experiencing simultaneously brightness and color contrasts. There is also good discrimination for different shades in a series of greys. The difference in electrical responses in the retinas of snappers (Lutjanidae) from deep waters compared to those of shallow-water fishes has been mentioned, and it may well be that not all but only shallow-water fishes have developed color vision. Such a difference among fishes would not be surprising, considering that the mammals, which have fewer species and inhabit an equally varied but different medium, also have some groups with and others without color vision.

Sight-feeding fishes, such as the elritze (*Phoxinus phoxinus*) and the black basses (*Micropterus*), when given conditioned reflex training on different shapes without inherent biological meaning but coupled in training with either punishment or reward, show fair to good form discrimination. The fishes learn quickest and make fewest mistakes if the objects are moved rather than held stationary.

In this connection we should regard closely the feeding behavior of certain sight-feeding fishes such as the Pacific salmons (*Oncorhynchus*). In these fishes as well as in those of the training situation above, good use is made of the peculiar retinal mechanism that is geared to detect contrast and movement. Salmon, like trout (*Salmo; Salvelinus*) hunt with their eyes and switch from cone to rod vision as the light fades (Fig. 11.17). Between 0.1 and 100 or more footcandles of light, the fish relies on the cones to give it maximal information on the hues and shades of its moving prey; when prey are located, they are stalked and taken head first into the mouth. Below 0.1 and close to 0.0001 footcandles (the light equivalent to that of a clear new moon at night), the rods take over, feeding stops, and schools break up. Reaction to light stimuli at these low intensities further suggests that fishes may use vision still at considerable depth—about 700 meters or more in clear ocean waters. In experiments where water fleas (*Daphnia*) are fed to young salmon in low light intensities, the fishes assume an oblique position below their prey so as to see them in maximal contrast against the screen afforded by the water surface, watch them move for a few moments, and then snap them up one by one.

Binocular vision that exists over varying angles in different species is important in the judgment of distance and depth. Some fishes such as the pikes (*Esox*) have veritable sighting grooves on their snouts in front of their eyes. Two imaginary lines projected from the grooves to cross the head in front of the eyes will include the angle within which a particular species has overlapping visual field, and thus presumably the field in which the animal has

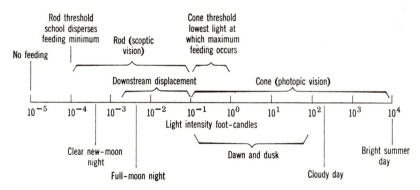

Fig. 11.17 Diagram summarizing rod and cone vision under various light intensities together with some other responses of a salmon (*Oncorhynchus*) to different light intensities. Light intensities under natural conditions are also indicated for comparison. (Source: Ali, 1959).

three dimensional vision. Many fishes can move their eyes independently and adjust this portion of their visual area. Stream fishes, such as most trout (Salmoninae) and many minnows (Cyprinidae), fix their position in the current by visually marking a stone, log, or other object.

Incident and background-reflected light falling on the upper and lower portions of the retina initiate chromatophoric adjustment of body color to background. Light falling on the eye from above in conjunction with the gravity receptors of the inner ear keeps the fish in the proper position for swimming or for resting. In addition to the foregoing treatment of vision, mention must be made of the light-sensitive skin areas of lampreys and hagfishes (Cyclostomata). In the skin of the dorsal tail region of the ammocete larvae of lampreys (Petromyzonidae) and of the cloacal region of hagfishes (*Myxine*) there are round sensory cells with photosensitive pigments of maximal reactivity at light wave lengths of 500 to 530 millimicrons. In the lampreys these cells are innervated by branches of the lateral-line nerves and they function to enhance the negatively phototropic reactions of the animal. The epiphysis and the dorsal region of the brain are also light sensitive in various fishes.

The Sense of Smell

Most odoriferous substances that play a meaningful biological role in the lives of fishes are of organic nature. They are detected through the olfactory epithelium connected to the olfactory nerve in the nasal sac (Figs. 3.21, 11.18, and 11.19), although some of them are perceived through the taste buds. However, the sensitivity spectra of taste and smell are different for the same substances and some organic molecules are only smelled while others are only tasted. The sensitivity of the sense of smell also serves for orientation when smell gradients are followed.

The nasal orifices as well as the entire nasal structure show wide varieties of adaptations among fishes. The single median nostril of lampreys and hagfishes (Cyclostomata) leads into a short nasal tube that bends posteriorly and ends in a nasopharyngeal pouch at the level of the second internal gill slit. The pouch is subject to the rhythmic respiratory movements that cause an exchange of water in it and in its tributary nasal sac. In the hagfishes (Myxiniformes), the nasopharyngeal pouch connects with the gut, but, in the lampreys (Petromyzonidae), it ends blindly. The nasal sac, branching from the nasal tube, is protected by a valve that directs a stream of water inward at inspiration but allows a reversal of flow at the emptying of the nasal pouch in response to the expiratory movement of the gills. The nasal chamber proper is thrown into twenty-five folds in the sea lamprey (*Petromyzon marinus*) and covered with olfactory epithelium where one fold faces another; the

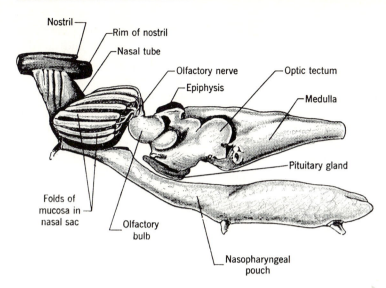

Fig. 11.18 Sketch of a wax restoration of the olfactory organ of a sea lamprey (*Petromyzon marinus*). (Based on Kleerekoper and van Erkel, 1960).

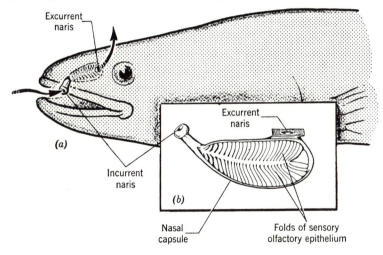

Fig. 11.19 (*a*) Position and extent, and (*b*) section of olfactory organ of an eel (*Anguilla*). Arrows indicate direction of water flow. The stalk of the incurrent naris is erectile. (Sources: (*a*) von Frisch in Grassé, 1958; (*b*) Wunder, 1936).

remaining part of the surface of each fold is covered with indifferent non-sensory epithelium. As in all other vertebrates, sensory cells of the olfactory epithelium give rise to the fibers of the olfactory nerve.

In sharks and rays (Elasmobranchii), the paired pits that lead into the internally folded nasal sac are usually on the ventral side of the snout and divided by a fold of skin into an inlet and an outlet; the latter sometimes leads into the mouth by means of a skin-covered nasal furrow and, as the fish swims and breathes, water reaches the olfactory epithelium. Certain sharks depend largely on their sense of smell for food-finding and follow odor gradients by swimming in circles or following a figure-eight course.

The paired olfactory organs of rayfin fishes (Actinopterygii) are relatively smaller than those of elasmobranchs but have a wider variety of adaptations for intake and exchange of water. The anterior nasal apertures are usually ahead of and somewhat medial to the eyes. The apertures lead directly or through a short tube into the nasal pit with its rosette of folded olfactory epithelium. In most rayfins, the posterior nasal apertures are normally close to the anterior ones; however, in the eels (*Anguilla*; Fig. 11.19) and morays (*Gymnothorax*), the apertures are relatively far apart. Some fishes, such as the sculpins (*Cottus*) and the sticklebacks (*Gasterosteus*), have only one nasal aperture and the nasal pouch is alternately filled and emptied by the breathing movements. An eel (*Anguilla*) has a projecting ciliated tube that leads the water to the anterior nasal opening, over the folds in the egg-shaped nasal sac, and out of the posterior nasal openings near the eyes (Fig. 11.19). In the herrings (Clupeidae) and the catfishes (Siluroidei), there are lymphatic sinuses near the nasal sac which act as a hydraulic system and fill and empty the nasal sac as pressure is transferred onto the accessory pouches during breathing, through the movement of several of the bones of the head. In many fishes, of which the pike (*Esox*) and the cod (*Gadus*) are examples, water is deflected into the nasal pit by the swimming movement of the fishes themselves. The division between anterior and posterior nares is thrown into an erectile fold that conducts in-current water to the center of the olfactory rosette.

The physiological mechanism of odor reception is incompletely understood. Electrode recordings from single units of the fish olfactory tract of *Abramis brama* (European bream) and *Carassius auratus* (goldfish) show the following: a fiber exhibits different patterns of activity when different odors are tested, different fibers respond to the same odor in different ways (spatial coding), and most fibers exhibit adaptation following continuous stimulation with the same odor.

Many experiments have shown the sensitivity and great importance of the sense of smell to fishes in getting food and in orientation. Bluntnose and fathead minnows (*Pimephales*) can distinguish by smell between the rinses of related species of invertebrates and of aquatic plants. Fishes that feed largely by odor, such as bullheads (*Ictalurus*), some sharks (e.g., *Squalus*), and morays (*Gymnothorax*), cannot find their food when their nares are plugged. Pheromones, "special substances," from prey that are tasted by fish include glycerophospholipids and also nucleotide mono- and diphosphates. The bluntnose and fathead minnows can be trained to detect chlorophenols in the water much below a concentration that can be tested chemically. An eel (*Anguilla*) is reported to detect a pure organic substance at a dilution of 2×10^{-18} when only 3 or 4 molecules of the material could have found their way into the nasal sac of the fish. Accordingly, it is no surprise that sunfishes (*Lepomis*) displaced from their home range in a stream, can find it again through their sense of smell, and that stream odors appear to play a decisive role in guiding returning salmons (*Oncorhynchus*) to their home streams. Minnows (Cyprinidae) give off a pheromone alarm substance (Schreckstoff) when frightened or injured. When this substance is smelled by the other members of a school, it causes them to disperse. Fishes with acute olfaction such as the catfishes (Ictaluridae) use their sense of smell in social interactions, differentiating between individuals in a group or social hierarchy by means of the specific odor of the slime of any one fish.

The Inner Ear (Equilibrium and Hearing; Figs. 3.24, 3.25, and 8.12).   The organs of equilibrium and hearing are divided into a pars superior and a pars inferior. The pars superior is composed of the semicircular canals and their ampullae, and a sac-like vesicle, the utriculus. The pars inferior, the structure of sound reception proper, consists of two more vesicles, the sacculus and the lagena. The ampullae contain patches of receptor tissue, the cristae staticae, with sensory cells that resemble those of the lateral line and to some extent also those of taste buds. However, the cristae have longer sensory hairs at the apices of their sensory cells than taste buds and they are covered by a gelatinous cupula (Fig. 11.20). Displacement of the body in any direction sets up shear forces to which the sensory cells of the ampullae react. The cells have spontaneous, rhythmic, nervous discharge. The bending of a cupula in one direction brings about a decrease in the rhythmic activity of the receptor, whereas bending of the cupula in the other direction accelerates the discharge rate. This is the mechanism for the detection of angular displacement.

In bony fishes (Osteichthyes), the utriculus contains an otolith, or earstone, called the lapillus (Fig. 3.24). This and the otoliths of the sacculus and lagena are limy and grow discontinuously, at least in the Temperate zones, and can therefore serve for the assessment of age of the fish. The lapillus rests horizontally on the hairs of the sensory cells of the crista utriculi. The lapillus responds to the force of gravity and activates the sensory cells of the cristae. For the maintainance of balance, they work in conjunction with the lower portion of the retina. Thus with both the eyes and utriculi intact, light from above and gravity from below push and pull, as it were, to keep the fish in an upright position. Upon extirpation of one utriculus, or upon shining a strong beam of light at the eye of a fish at right angles instead of from above, the fish can be made to lean to one side or the other. When a fish has its pars superior removed from both inner ears, it can be made to swim at a $90°$ angle or even upside down if it is illuminated from the side

SENSORY EPITHELIA

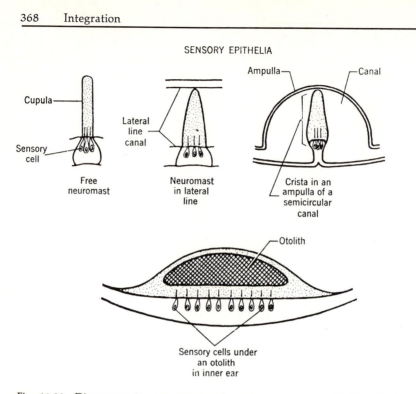

Fig. 11.20 Diagrammatic comparison of various sensory epithelia of the acoustico-lateralis system. (Source: Dijkgraaf in Grassé, 1958).

or from below (Fig. 11.21). The adequate utricular stimulus for swimming upright is the full weight of the lapillus resting on the sensory hair cells with no shear. A displacement of the fish by 90° will render the lapillus weightless but will produce maximum shear on the hairs of the sensory cells under the otolith. The lagenar otolith, and the underlying sensory tissue are, to a small degree, also involved in gravistatic or righting reflexes, mainly through their influence on muscle tonus.

In sharks and rays (Elasmobranchii) an endolymphatic duct leads from the inner ear to the outside. Through this duct sand reaches the gelatinous cupulae over the sensory cells of the inner ear. However, it is not known to what extent such particles from outside assist in the stimulation of the sense cells inasmuch as there are in the inner ear of sharks also otolith-like calcareous concretions of endogenous origin.

The size and arrangement of the semicircular canals and the vesicles are considerably different among various fishes. The canals are relatively large in the elephantfishes (Mormyridae) and in certain blind cavefishes (Amblyopsidae). The sacculus and the lagena are more broadly joined to one another in the sharks (Squaliformes) than in bony fishes (Osteichthyes).

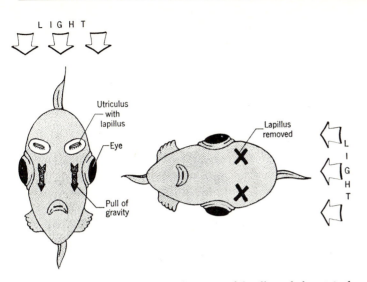

Fig. 11.21 Effects of retina of the eye and lapillus of the utriculus of the inner ear in equilibration of fishes. When the lapilli are removed, the fish no longer orients to gravity and light but orients to light only. (Based on von Holst, 1950).

The relative development of superior and inferior portions of the inner ear also differs in rayfin fishes (Actinopterygii) of different habits. Flying fishes and pelagic fast swimmers that require superior three-dimensional orientation have a better developed pars superior than demersal species where in turn the inferior portion reaches the highest degree of development.

The pars inferior of the inner ear, lagena and sacculus with their otoliths—named asteriscus and sagitta, respectively—is the seat of sound reception. These two structures, as well as the posterior semicircular canal, are innervated by the posterior branch of the auditory nerve (cranial nerve VIII) whereas the superior part of the inner ear and the two remaining semicircular canals are connected to the anterior branch of the same nerve. The otoliths of lagena and sacculus are held vertically over the acoustic sensory maculae by tendinous attachments to the inner walls of the vesicles.

The physiological mechanism of discrimination of different sound frequencies is not known but it is seated mainly in the utriculus and one must distinguish between two types of hearing structures among bony fishes, the common and the cypriniform types. Among the minnows, catfishes, and suckers (Cypriniformes), the inner ear is connected uniquely to the gas bladder through a chain of bones, the Weberian ossicles (Fig. 3.25; Chapter 8). The Cypriniformes have a finer and wider range of hearing than fishes of the normal type, their lagena and sacculus are divided by a narrow constriction, and their sagitta carries a large wing into the endolymph, apparently better to pick up vibrations that are transferred to it from the resonating gas bladder.

Nothing is known of the range and acuity of sound reception in lampreys and hagfishes (Cyclostomata). Sharks (Squaliformes) respond to various noises and sounds, some of pure frequencies ranging from 400 to 600 cycles per second. Rayfin fishes (Actinopterygii) can perceive both intensity differences and a range of vibrations from 16 to 13,000 cycles per second (cps) as in the bullheads (*Ictalurus*). Low frequencies of 16 to 100 cycles are probably sensed through skin and/or lateral-line receptors. The most acute level of sound perception in the creek chub (*Semotilus atromaculatus*), a minnow, lies around 280 cps; this fish is capable also of distinguishing quarter-tone intervals between 400 and 800 cps. Still, true directional hearing seems to be limited to a few dozen meters. The upper range of sound perception of fishes without accessory devices lies near 5000 cps, as in the labyrinthfishes (Anabantidae), and they have less capacity to distinguish between tone intervals.

The Lateral Line (Figs. 11.22–11.24)

The lateral-line system is a set of sense organs restricted to the fishes and aquatic stages of the amphibians. Although it has connections in the brain with the afferent nerve pathways from the auditory system of the inner ear,

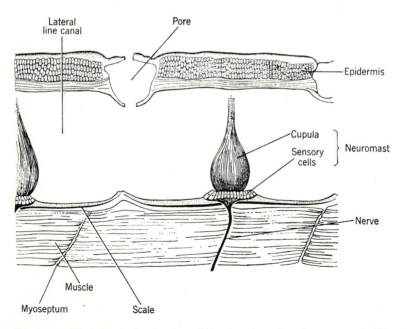

Fig. 11.22 Neuromasts in the lateral-line system of a Japanese eel (*Rhynchocymba*). (Based on Katsuki, Mizuhira, and Yoshino, in Grassé, 1958).

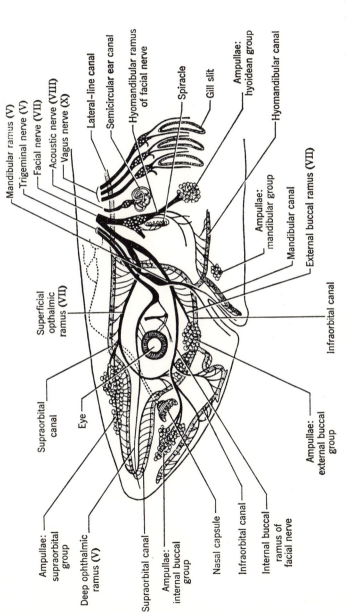

Mandibular ramus (V)
Trigeminal nerve (V)
Facial nerve (VII)
Acoustic nerve (VIII)
Vagus nerve (X)
Lateral-line canal
Semicircular ear canal
Hyomandibular ramus of facial nerve
Spiracle
Gill slit

Ampullae: hyoidean group
Hyomandibular canal

Ampullae: mandibular group
Mandibular canal
External buccal ramus (VII)

Superficial opthalmic ramus (VII)

Infraorbital canal

Supraorbital canal
Eye

Ampullae: supraorbital group
Deep ophthalmic ramus (V)

Supraorbital canal

Ampullae: internal buccal group

Nasal capsule

Infraorbital canal

Internal buccal ramus of facial nerve

Ampullae: external buccal group

Fig. 11.23 Distribution of the ampullae of Lorenzini and sensory canals and their innervation on the head of a shark. (Based on Ewart in Grassé, 1958).

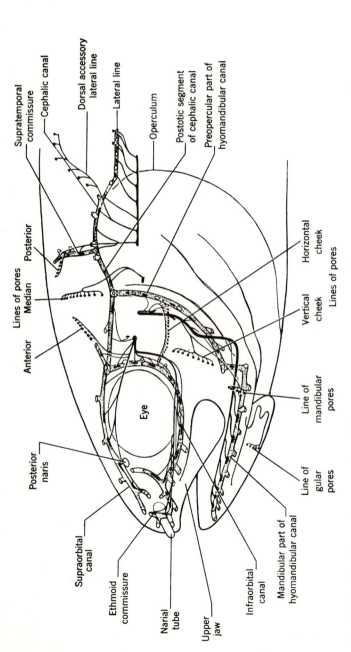

Fig. 11.24 Arrangements of head lateral-line canals and their pores, and innervation on a bony fish, the bowfin (*Amia*). (Based on Allis in Grassé, 1958).

Supratemporal commissure

Cephalic canal

Dorsal accessory lateral line

Lateral line

Operculum

Postotic segment of cephalic canal

Preopercular part of hyomandibular canal

Lines of pores
Posterior
Median
Anterior

Horizontal cheek

Vertical cheek

Lines of pores

Line of mandibular pores

Line of gular pores

Eye

Posterior naris

Supraorbital canal

Ethmoid commissure

Narial tube

Upper jaw

Infraorbital canal

Mandibular part of hyomandibular canal

it is a different system. It was already present in fossil ostracoderms, and is thought to have originally evolved from primitive cutaneous sense cells. Fossil evidence from *Pteraspis* (Agnatha: Pteraspidomorphi) onwards permits some tracing of lateral-line evolution, but embryological evidence is less clear because in some fishes the lateral line arises from the otic or auditory placode, whereas in others, as for example, an European minnow (*Leuciscus*), several separate lateral-line placodes form on head and body.

Apart from the typical lateral-line canals (Fig. 11.22) with their pores in skin or scales on the head and body of bony fishes (Osteichthyes), the following widely differentiated sensory structures are, at present, tentatively ranged within the lateral-line complex:

a. Sensory crypts or pits and papillae between or under the placoid scales of sharks and rays (Elasmobranchii), arranged in groups or lines, usually in the anterior and dorsal body regions; certain histologic peculiarities of their sensory cells suggest that they are taste receptors (Fig. 11.9) rather than a part of the lateral-line system proper.

b. The ampullae of Lorenzini (Fig. 11.23) are more or less sac-like structures in the head region of sharks and rays (Elasmobranchii) and also of *Plotosus*, a southeast Asian catfish, to which various groups of pores (supraorbital, buccal, hyoidean and mandibular) give access through a canal. The structures are jelly-filled and may have few to several diverticula lined with sensory epithelium. The sensory cells are innervated by branches of the facial nerve (VII). Isolated preparations of such ampullae give rhythmic nervous discharges which are accelerated on cooling and slowed on warming. They are also sensitive to mechanical and weak electrical stimuli as well as to small changes in salinity. Their highest sensitivity, however, is to differences in electric potential such as would be generated by muscular movements, for instance, the opening and closing of the gill covers. The ampullae of Lorenzini are now believed to be mainly electro-receptors assisting in the location of prey as well as serving proprioceptive functions in swimming. Their sensitivity is extremely high, reacting to a change of 0.01 microvolts/cm of seawater as shown in the European thornback ray (*Raja clavata*) in which the rate of heartbeat changes when stimulated with such small changes in electric potential. The cat sharks (*Scyliorhinus*) hold similar sensitivity indicating that they and probably other elasmobranchs are capable of detecting the presence of potential prey animals in their vicinity by changes in muscle potential of the prey.

c. The vesicles of Savi, found in several genera of rays (Rajiformes) including certain electric rays (*Torpedo*), mainly on the underside of the snout as linear series of gelatin-filled follicles which enclose gray, granular, amorphous material. These vesicles are innervated by a branch of the trigeminal cranial nerve (V) but their function is unknown.

d. The organs of Fahrenholz are pits in the head region of larval lungfishes (Dipnoi). These pits have hair-bearing sensory cells that resemble those of the lateral-line organs; their function, however, is unknown.

e. The spiracular organs, sensory structures of a basic lateral-line type are of unknown function in the spiracles of the Chondrichthyes, the Holostei, and the Chondrostei.

f. The cutaneous sense cells of the elephantfishes and gymnotid eels (Mormyridae and Gymnotidae, respectively), which are presumably electrosensitive

Only some fishes among all animals are now known to have organs which generate an electric discharge. The voltage of this discharge varies considerably in different groups of fishes as does the nature and origin of the current-generating structure. Two groups can be distinguished: (1) those which produce a strong stunning current such as the electric eel (*Electrophorus*), the electric rays (*Torpedo*), and the electric catfish (*Malapterurus*); and (2) those which produce currents of low voltage (Mormyridae and Gymnotidae with the exception of *Electrophorus*). The second group emits a continuous series of pulses to set up an electric field around the animal. The field becomes distorted when prey enters it or when inanimate obstacles, such as may be encountered in swimming, obstruct it. The device makes possible the location of objects in the turbid habitat of these fishes. The number of discharges per second varies with the species, but in both the South American gymnotids (Gymnotidae) and the African elephantfishes (Mormyridae) certain species discharge with relatively low frequencies of up to 200 cycles per second, whereas others emit up to 1600 low-voltage impulses per second. Species-specific discharge rates may play a role in the recognition of mates.

Generation of an electric field for the sensing of surrounding objects implies the existence of electric receptors. In the elephantfishes there are pores, concentrated in the head region, with narrow canals leading into single pits. Each pit has a patch of tissue, resembling lateral-line receptors. These structures are renewed throughout life, like other skin cells. They may respond to chemical and pressure stimuli aside from being sensitive to electrical impulses. The sensitivity of the mormyrid *Gymnarchus niloticus* to static electric charges is 0.15 microvolt per square centimeter. Gymnotids have comparable skin organs composed of several sense cells that lie in a vesicle which opens to the outside through a jelly-filled canal. Nonsensory cells may also occur in these structures which belong to the lateral-line system. The structures are presumed to act as electrical receptors. Catfishes (*Ictalurus*) have also been shown to respond to weak electrical stimulation with this sensitivity being located in the small pit organs of the lateral-line system. In the remaining groups of fishes the lateral-line organs proper may be arranged structurally as follows:

1. Free, arranged in groups on the head and in lines along the body (the elritze, *Phoxinus*; the loach, *Cobitis*).
2. Sunk in pits, in groups, or in several lines on the head or along the body (many elasmobranchs, cyclostomes) or in bands crossing the ventral surface (elasmobranchs, cyclostomes).
3. In partially open or closed, variously developed canals (Fig. 11.24) located
    a. on the head where there are typically three main canals, the supraorbital, the infraorbital, and the hyomandibular, with branches, extensions, or connections such as the supratemporal commissural canal that connects the canal system of the two sides of the body;
    b. on the body where there is typically one canal extending from the cleithrum of the shoulder girdle to the base of the caudal fin.

Both the body lateral-line canal and that of the head exhibit several modifications. For example, the body canal may be extremely short as in the European bitterling (*Rhodeus amarus*), intermediate in length as in the fathead minnow (*Pimephales promelas*), interrupted as in a parrotfish (*Scarus*), strongly arched towards the back or decurved towards the belly. There may also be several roughly parallel lateral lines as in the greenling (*Hexagammus*) and mullet (*Mugil*) or ones scattered in short segments over most of the side of the body, as in the pikes (*Esox*).

The different parts of the lateral line are innervated by cranial nerves VII, IX, and X and have close central associations with the auditory centers served by nerve VIII. Centers in the medulla oblongata of the brain are divided into a dorsal group for the head lateral lines and into ventral ganglia for the body lateral-line system. Secondary connections in the brain relay impulses from the lateral-line system to the thalamus, the hypothalamus, the optic tectum (especially in sharks, Squaliformes), and the cerebellum (particularly in the Mormyridae). The giant cells of Mauthner, with their final common path as wholesale movers of trunk and tail musculature, have axon tracts that bring them in direct contact with the lateral-line branches of nerves VII, IX, and X as well as with the optic tectum, the sacculus, and the cerebellomotor centers.

The receptor unit of the lateral-line system is the neuromast, an area of sensory tissue made up of pyriform receptor cells each with a hair-like extension at its apex that reaches into a gelatinous cupula (Fig. 11.22). In contrast to the sensory cells of taste buds, which they resemble, they do not reach the base of the neuromast but are cushioned from below by nonsensory, supporting cells.

Like the end organs of equilibrium in the semicircular canals, the neuromasts have a continuous rhythmic discharge (Fig. 11.25). In the skates (*Raja*) two types of receptors can be isolated upon perfusion of a few functional units with a stream of water. One group responds with a strong increase in discharge frequency to a headward slow jet of water whereas tailward perfusion inhibits the discharge activity. The end of headward perfusion is marked by a silent period but the cessation of water flow toward the tail brings about a vigorous resumption of nervous activity; the second group of receptors behaves in the opposite manner. The antagonistic, rhythmically discharging receptors are mixed randomly in each group of sensory cells, but there have also been found, again in the skates, "spontaneously silent" units in the lateral-line system that can be made to emit impulses during perfusion in one direction and at the end of perfusion in the other. In some Japanese eels (*Anguilla japonica* and *Rhynchocymba nystrani*), the sense cells in the center of neuromasts are without resting discharge and adapt rapidly in contrast to the peripheral cells which have a slow adaptation rate and a spontaneous discharge. The central cells are connected to substantial nerve fibers of 15 microns in diameter whereas the peripheral cells are connected to small fibers of 4 microns in diameter.

The anatomical and physiological mechanism underlying detection of hydrodynamic displacements by prey, predators or social partners is as follows. The sensory cells of the neuromasts have hair-like cilia consisting of one elastic kinocilium and numerous softer stereocilia (40-50 in *Lota,* the burbot). Adjacent sensory or hair cells have their kinocilia oriented in opposite directions, alternately towards the head or the tail of the fish, but always along the axis

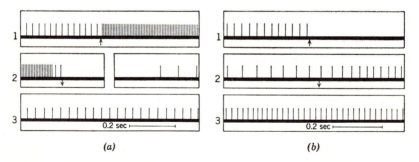

(a)   (b)

Fig. 11.25   Electric discharges in a lateral-line nerve of a skate (*Raja clavata*) with arrows indicating the beginning of stimulation by a jet of water: (*a*) stream of water toward the head; (*b*) stream of water toward the tail. (Based on Sand and Dijkgraaf in Grassé, 1958).

of the lateral-line canal. As head- or tail-ward bending of the cupula would affect the kinocilia of different hair cells differentially one can envisage the pattern of lateral-line nerve discharges to convey to the brain exact information on direction and/or provenience of water displacement in the vicinity of the fish (Fig. 11.25).

Training experiments, operations to eliminate lateral-line reception, and physiological investigations show that the system responds to a variety of stimuli. Irregular pressure waves and low-frequency vibrations of 50 to 100 cycles per second (cps) excite the receptors of the larger lateral-line fibers of the bullheads (*Ictalurus*), whereas a weak but regular water flow or vibrations of 20 to 50 cps act on the receptors connected to the thinner nerves. Several fishes have been trained to locate small vibrating rods or discs close to their heads and sides as well as to react to small localized water currents. After the difficult operation of cutting only nerves of the lateral line on the head and trunk, and not nerves of the touch sense, the capacity of fine discrimination of small vibrations or current disturbances is lost.

It can be concluded that the lateral-line system informs the fish of localized disturbances, caused by small currents, by mechanical vibrations below 100 cps, or by irregular nonviolent displacement of the surrounding water. The system is also involved in "distant touch" location of moving objects such as predators or prey or in the sensing of fixed objects that reflect the water movements brought about by the swimming fish itself. Lateral-line organs also react to monovalent cations (e.g., Na^+) but the functional significance of this sensitivity has not been evaluated.

SPECIAL REFERENCES ON INTEGRATION

A. The Nervous System

Ariens-Kappers, C. U., C. Huber, and E. Crosby. 1936. The comparative anatomy of the nervous system of vertebrates, including man. The Macmillan Company, New York.

Bernstein, J. J. 1970. Anatomy and physiology of the central nervous system. *In* Fish physiology, 4: 1–90. Academic Press, New York.

Eddy, J. M. P. 1972. The pineal complex. *In* The biology of lampreys, 2: 91–103. Academic Press, New York.

Evans, H. M. 1940. Brain and body of fish. The Blakiston Company (McGraw-Hill Book Company, New York).

Fenwick, J. C. 1970. The pineal organ. *In* Fish physiology, 4: 91–108. Academic Press, New York.

Ingle, D. (Ed.) 1968. The central nervous system and fish behavior. University of Chicago Press, Chicago, Ill.

Lissmann, H. W. 1946. The neurological basis of the locomotory rhythm in the spiny dogfish (*Scillium canicula, Acanthias vulgaris*). I Reflex behavior, II The effect of de-afferentation. *Jour. Exptl. Biol.* 23: 143–176.

Nicol, J. A. C. 1952. The autonomic nervous system in lower chordates. *Biol. Rev. Cambr. Philos. Soc.*, 27 (1): 1–49.

Uchihashi, K. 1953. Ecological study of Japanese teleosts in relation to the brain morphology. *Bull. Japanese Sea Reg. Fish Res. Lab.*, 2 (2): 1–180.

von Holst, E. 1934. Weitere Reflexstudien an spinalen Fischen. *Zeitschr. vergl. Physiol.* 21: 658–665.

B. The Endocrine System

Ball, J. N., and B. I. Baker. 1969. The pituitary gland: anatomy and histophysiology. *In* Fish physiology, 2: 1–110. Academic Press, New York.

Bardach, J. E., R. Ryther, and W. McLarney. 1972. Aquaculture: the farming and husbandry of freshwater and marine organisms. Wiley Interscience. John Wiley and Sons, New York.

Bretschneider, L. H., and J. J. Duyvene de Wit. 1947. Sexual endocrinology of non-mammalian vertebrates. J. Elsevier Publishing Company, Amsterdam.

Gorbman, A. (Ed.). 1959. Comparative endocrinology. John Wiley and Sons, New York.

Gorbman, A., and H. A. Bern. 1962. A textbook of comparative endocrinology. John Wiley and Sons, New York.

Hoar, W. S., and D. J. Randall (Eds.) 1969. The endocrine system. Vol. 2, Fish Physiology, Academic Press, New York.

Lagios, M. D. 1975. The pituitary gland of the coelacanth *Latimeria chalumnae* Smith. *Gen. Comp. Endocrinol.*, 25: 126–146.

Nomura, T., H. Ogata, and M. Ito. 1973. Occurrence of prostaglandins in fish testis. *Tohoku Jour. Agric. Res.*, 24: 138–144.

Pickford, G., and J. W. Atz. 1957. The physiology of the pituitary gland of fishes. New York Zoological Society, New York.

C. Sense Organs

I. The Cutaneous Senses

Bardach, J. E., and J. Atema. 1971. The sense of taste in fishes. pp. 293–336 *In* Handbook of Sensory Physiology, Vol. IV (2), L. Beidler, Ed. Springer Verlag, Berlin, Heidelberg, New York.

Bull, H. O. 1936. Studies on conditioned responses in fishes. Part VII, Temperature discrimination in teleosts. *Jour. Marine Biol. Assoc. United Kingdom*, 21 (1): 1–27.

Dijkgraaf, S. 1940. Untersuchungen ueber den Temperatursinn der Fische. *Zeitschr. vergl. Physiol.*, 27: 587–605

Sand, A. 1938. The function of the ampullae of Lorenzini with some observations on the effect of temperature sensory rhythms. *Proc. Roy. Soc. London Ser. B. Biol. Sci.*, 125: 524–553.

Späth, M. 1974. On the processing of the mechanical and thermal information of the receptors in fish skin. *In* Symposium on Mechanoreception, Rheinisch-Westfaelische Akad. d. Wissenshaften. 53: 251–262. Bochum.

Sullivan, C. M. 1954. Temperature reception and responses in fish. *Jour. Fish. Res. Bd. Canada*, 11 (1): 153–170.

Whitear, M. 1952. The innervation of the skin of teleost fishes. *Quart. Jour. Microsc. Sci.*, 93: 289–305.

II. Vision

Ali, M. A. 1959. The ocular structure, retinomotor and photobehavioral responses of juvenile Pacific salmon. *Canadian Jour. Zool.*, 37: 965–996.

Feuerwerker, E., and M. A. Ali. 1975. La vision chez les poissons historique et bibliographie analytique. *Rev. Can. Biol.*, 34 (4): 221–285.

Granit, R. 1947. Sensory mechanisms of the retina. Oxford University Press, London.

Kleerekoper, H. 1972. The sense organs. *In* The biology of lampreys, 2: 373–404. Academic Press, New York.

MacNichol, E. J., Jr., and J. Svaetichin. 1958. Electric responses of fish retinas. *Amer. Jour. Opthalmol.*, 46 (3 II): 26–46.

Munz, F. W. 1971. Vision: visual pigments. *In* Fish physiology, 5: 1–32. Academic Press, New York.

Pumphrey, R. J. 1961. Concerning vision. *In* The cell and the organism, pp. 193–208. Cambridge University Press, Cambridge.

Wagner, H. G., E. J. MacNichol, Jr., and M. L. Wolbarsht. 1960. Opponent color responses in retinal ganglion cells. *Science*, 131: 1314.

Wald, G. 1945. The chemical evolution of vision. The Harvey Lectures, 41: 117–160.

Walls, G. 1942. The vertebrate eye and its adaptive radiation. *Cranbrook Inst. Sci. Bull. No. 19.* Cranbrook Institute of Science, Bloomfield Hills, Mich.

III. Smell

Bardach, J. E., and T. Villars. 1974. The chemical senses of fishes. Pp. 49–104. *In* Perspectives in chemoreception by marine organisms. Academic Press, New York.

Hasler, A. D. 1954. Odour perception and orientation in fishes. *Jour. Fish. Res. Bd. Canada,* 11 (2): 107–129.

Klerekoper, H. 1969. Olfaction in fishes. Indiana University Press, Bloomington, Ind.

Kleerekoper, H., and G. A. van Erkel. 1960. The olfactory apparatus of *Petromyzon marinus. Canadian Jour. Zool.*, 38: 209–223.

IV. The Inner Ear

Kleerekoper, H., and E. C. Chagnon. 1954. Hearing in fish with special reference to *Semotilus atromaculatus atromaculatus* (Mitchill). *Jour. Fish. Res. Bd. Canada,* 11 (2): 130–152.

Lowenstein, D. 1950. Labyrinth and equilibrium. *Symp. Soc. Exp. Biol.*, IV: 60–82.

von Holst, E. 1950. Die Arbeitsweise des Statolithenapparates bei Fischen. *Zeitschr. vergl. Physiol.*, 32: 60–120.

V. The Lateral Line

Dijkgraaf, S. 1933. Untersuchungen ueber die Funktion der Seitenorgane bei Fischen. *vergl. Physiol.*, 20: 162–214.

Dijkgraaf, S. and A. J. Kalmijn. 1966. Versuche zur biologischen Bedeutung der Lorenzinischen Ampuller bei den Elasmobranchiern. *Zeitschr. vergl. Physiol.* 53: 187–194.

Flock, Å. 1965. Electromicroscopic and electrophysiological studies on the lateral line canal organ. *Acta Oto-laryngologica*, Suppl. 199:1–90.

Freihofer, W. C. 1972. Trunk lateral line nerves, hyoid arch gill rakers, and olfactory bulb location in atheriniform, mugilid, and percoid fishes. *Occas. Papers Cal. Acad. Sci.*, 95, 31 p.

Hoagland, H. 1933. Electrical responses from the lateral line nerves of catfishes. *Jour. Gen. Physiol.*, 16: 695–731.

Katsuki, Y., T. Hashimoto, and T. T. Kendall. 1971. The chemoreception in the lateral line organs of teleosts. *Japanese Jour. Physiol.* 21: 99–118.

Katsuki, Y., S. Yoshino, and J. Chen. 1950. Action currents of the single lateral line nerve fiber of fish. *Japanese Jour. Physiol.*, 1: 87–99; 179–194; 264–268.

Murray, R. W. 1962. The response of the ampullae of Lorenzini of elasmobranchs to electrical stimulation. *Jour. Exptl. Biol.*, 39 (1): 119–128.

12

Genetics and Evolution

What we know of the mechanisms of inheritance in fishes is based primarily on intensive studies of very few members of a single order, the Cyprinodontiformes—particularly the guppy (*Lebistes reticulatus*), platyfishes and swordtails (*Xiphophorus*), the medaka or Asiatic ricefish (*Oryzias latipes*), the Amazon molly (*Poecilia formosa*), and Mexican species of the genus *Poeciliopsis*. In recent years, knowledge of the field has been enlarged through researches on carp and goldfish (Cyprinidae), Anatolian cyprinodontids, experiments with trouts and chars (Salmoninae), and on scattered representatives from a few other groups. Much is known of the genetics of the goldfish (*Carassius auratus*), chiefly as recorded in the Japanese literature. There remains the vast majority of fishes including the primitive, jawless vertebrates (Cyclostomata), for which almost no gentic information is available. In other words, we have hardly scratched the surface of potential hereditary data for this large and diverse group of animals.

INHERITANCE IN FISHES

Simple Mendelian inheritance is well demonstrated by the occurrence of *golden*, a recessive mutant trait of the guppy (*Lebistes*) that results from a reduction of about 50 percent of the melanophores characteristic of the wild populations. The loss of black pigment cells uncovers and makes visible the underlying yellow pigment cells present in the skin of all guppies.

Mating a wild type with a golden one produces an F_1 generation that is colored like the wild parent. But crossing two of these gray first-generation offspring (brother–sister or sibling mating) results in a ratio of three wild to one golden offspring. This typical Mendelian ratio is that expected when a simple Mendelian recessive trait is involved.

Significant deviations from a 3:1 ratio may result from secondary complications. For example, the albino guppy (*Lebistes*), characterized by lack of melanophores and pink eyes, when mated with the wild type produces an F_1 generation like that of the wild parent (Fig. 12.1). But in the F_2 generation, resulting as before from a brother–sister mating from the F_1 stock, the ratio

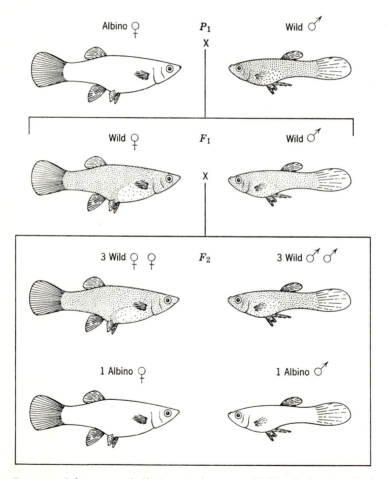

Fig. 12.1   Inheritance of albinism in the guppy (*Lebistes*) showing ideal result (3:1 ratio of wild to albino) that is rarely achieved for reasons given in the text. (Source: Gordon, 1953).

of wild (ordinary) to albino is 53:1. This wide discrepancy from an expected 3:1 ratio is attributed to an inherent weakness of the albino guppy—albinism being a semilethal (recessive) mutation that results in the mortality of a high percentage of the albinistic offspring prior to birth.

Another guppy trait that shows simple Mendelian inheritance is the "zebra" pattern of two to five bars on the posterior part of the body of males (Fig. 12.2). Though unable to express the pattern, females are nevertheless able to transmit the hereditary factor for it to their sons. Traits such as this one, which are carried by both sexes but are expressed in only the male (or the female), are called *sex-limited*. Mating a zebra male to a non-zebra female results in offspring heterozygous for the character—that is, each has one

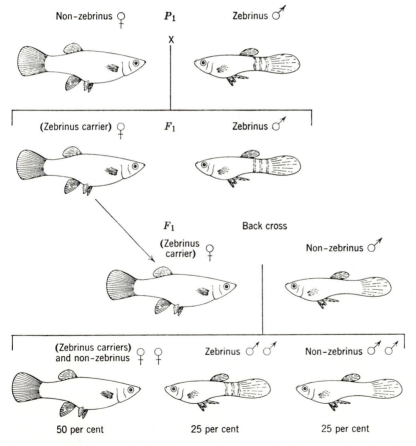

Fig. 12.2 Inheritance of zebra pattern in the guppy (*Lebistes*) showing simple Mendelian sex-limited heredity. (Source: Gordon, 1953).

dominant and one recessive "zebra" gene—but since the character is limited to males, only the sons show it in the phenotype (the daughters are carriers). When a sibling cross of the F_1 generation is made, the resultant offspring, counting sons only, show a 3:1 ratio in favor of the zebra pattern. This indicates that the trait represents a simple, dominant Mendelian factor.

Reversion to wild type of domesticated strains of plants and animals may be commonly observed. In the goldfish (*Carassius auratus*), first believed to have been domesticated in China nearly a thousand years ago, reversion to the ancestral olive-green color persistently occurs. These so-called throwbacks are readily explained in terms of ordinary Mendelian inheritance. The reappearance of ancestral traits is usually the result of the reunion of distinct hereditary characters that had become separated during the process of domestication.

Non-Mendelian inheritance or the appearance of unpredictable results in mating experiments may be due to such phenomena as blending of traits or matroclinous inheritance (from the mother only, see below). For example, when the domesticated gold platy (a mutant recessive of *Xiphophorus maculatus*) is mated to the wild comet platy (which is gray and has a jet-black margin on the upper and lower lobes of the caudal fin), the F_1 are not gray and comet (as would be expected, since these traits are dominant over golden and noncomet) but are gray and "wagtail," in which not only is the entire caudal fin black but also the other fins (Fig. 12.3). In other words, the gray color behaved in inheritance as expected, but the comet pattern did not. The reason for this unexpected F_1 is that the gene that determines the comet condition is affected by another gene. This additional gene, called a specific modifier, radically altered the usual effect of the comet gene by causing a much more extensive pigmentary response.

A type of inheritance unknown to Mendel is that called *multiple allelism*. An allele represents one of several alternate phases of a gene. In a cross involving two pairs of independent contrasting characters, such as comet (*CoCooo*) and one-spot (*cocoOO*) in platies (*Xiphophorus*), all of the F_1 are comet and one-spotted because both comet and one-spot are dominant color patterns. When an F_1 female is mated to her brother, the results expected on the basis of two-factor inheritance would be in the ratio of 9:3:3:1 (9 comet, one-spot: 3 comet only: 3 one-spot only: 1 lacking a pattern). The observed sixteen phenotypes, however, occurred in the ratio of 1:2:1 (4 comet only: 8 comet, one-spot: 4 one-spot only). These deviations from expectancy, although not great, persist with repeated matings and with additional second-generation offspring. The unexpected ratios are significant and may be explained as follows. Instead of one gene having just two phases, one dominant and one recessive, the gene may have two or more expressions in its dom-

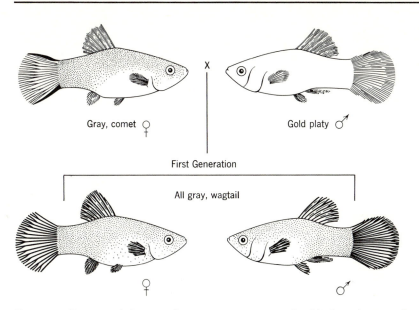

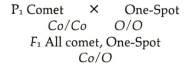

Fig. 12.3 Genetics of the cross between gray comet and gold platy showing the effect of a specific modifier gene on the usual effect of the comet gene. (Source: Gordon, 1953).

inant phase—that is, O and Co may both be dominant to a common recessive. The term used by geneticists for this interesting type of inheritance involving a series of genes is *"dominant multiple alleles."* Hence, in the example just given of the contrasting platyfish color patterns, we are not dealing with two simple dominant and two simple recessive genes but with a series of *multiple alleles.* Consequently, the F_1 generation is not properly expressed by the formula $CocoCo$, since Co and O are alleles; the cross must be expressed by the following formulas:

$$P_1 \text{ Comet} \quad \times \quad \text{One-Spot}$$
$$Co/Co \qquad O/O$$
$$F_1 \text{ All comet, One-Spot}$$
$$Co/O$$

In breeding the F_1 generation, $Co/O \times Co/O$ gives offspring in the ratio of 1:2:1 (1 comet: 2 comet, one-spot: 1 one-spot). There are three comet-patterned platies to one without the comet pattern. Similarly, there are three one-spot patterned fish to one lacking the one-spot pattern.

Although these results do not conform to Mendel's predictions (on the basis of two-factor interpretation), they are nevertheless explainable from his basic principles. The 1:2:1 ratio is merely a slight variation of the fundamental 3:1 ratio.

SEX DETERMINATION

Almost every type of sexual reproduction known in animals may be found in fishes (Chapter 10), including certain specializations without parallel elsewhere among the vertebrates. However, on the whole, the mechanism of sex determination in the majority of bony fishes is still in a primitive labile condition. The sex chromosomes are so little differentiated from the autosomes that they are only rarely distinguishable from them. Sex determination may be controlled, in part if not completely, by genes dispersed among the autosomes. Sex-determining potential may differ between various species such that, in hybrid crosses, the stronger male-producing genes of one species combine with the weaker female-producing genes of another species to result in unbalanced sex ratios heavily, or even wholly, favoring the male.

The available data on sex determination in both wild and laboratory populations of fishes are more extensive than they are in other vertebrates. Whereas in mammals the female is homogametic (XX) and the male heterogametic (XY), and in birds the opposite type prevails (female, WY; male, YY), in fishes both of these systems are represented.

It must be stressed that sex chromosomes as such are exceedingly difficult to confirm cytologically in fishes, for fish chromosomes are tiny, usually numerous, and hard to observe. In many of those examples in which investigators have claimed to distinguish sex chromosomes morphologically, the seemingly distinct structures are probably only artifacts—the result of inadequate fixation. However, a very clear pair of sex chromosomes has been found in the male of the deepsea fish *Bathylagus wesethi*; the longest in the karotype is presumably X, the smallest Y. Two other species of this genus and some species of *Fundulus* have a similar sex chromosome pair as do *Gambusia affinis* and others. In *Megupsilon aporus*, a Mexican cyprinodontid, there is an X_1X_2Y multiple sex chromosome mechanism with a very large metacentric Y. The XY type of sex chromosome was recently demonstrated for a Japanese lancelet, *Branchiostoma belcheri*. The discussion that follows is therefore based on genetic experiments which provide strong theoretical evidence for sex chromosomes but do not prove their existence.

Sex-linked inheritance is that in which the trait in question is determined by a gene (or genes) located on the sex chromosomes. *Sex-limited* or "one-sided" inheritance (Fig. 12.2) results when a trait is handed down only from male to male or from female to female; the gene involved lies on the odd chromosome. An example of a sex-linked, sex-limited character is the inheritance of a dominant color trait, the black spot on the dorsal fin of the male guppy (*Lebistes*). This also illustrates that *Lebistes* has a type of genetic sex determination which conforms to the XX female, XY male pattern known in mammals. In the following formulas, the symbols for the sex chromosomes are capitalized and are within parentheses, whereas the genes that they carry are italicized (the black spot is called *maculatus*, represented by *Ma*):

	Females	Males
P_1	(X)–(X)–	(X)–(Y)*Ma*
F_1	(X)–(X)–	(X)–(Y)*Ma*
F_2, etc.	(X)–(X)–	(X)–(Y)*Ma*

This type of sex-limited inheritance has also been demonstrated for the medaka (*Oryzias*) and the platyfish (*Xiphophorus*).

Cytological Investigations

Chromosome number, along with conventional morphological criteria, data from paleontology, behavioral patterns, ecology, and genetic experiments, provides a further tool for deciphering phylogeny in fishes. The latest summaries show that the diploid number varies from 18 to 104, but workers may disagree on the precise number of a given species. Chromosome numbers determined from mitotic chromosomal figures that show so well during stages of the egg cleavage may differ radically, within a single species, from those determined by studying male germ cells during spermatogenesis. The latter are said to be more accurate. It is to be expected that a close correlation exists between chromosome morphology and structural organization, since evolutionary changes are initiated by changes in chromosomes.

Surveys of chromosome number are less complete for fishes than for other groups of animals. It must be remembered, however, that fishes constitute a greater number and diversity of species than all other vertebrates combined. Serious work on fish chromosomes dates from the 1930's and this field of inquiry has received much of its impetus from papers appearing within the past decade. Approximately four hundred of an estimated twenty thousand species of living fishes have been studied by chromosome cytologists. Growing interest in chromosome number and morphology in these animals promises to yield important contributions to evolutionary theory as well as to fish phylogeny.

There is a general tendency for the chromosomes of related species and genera to show similarity in number and shape, but our knowledge of the subject is still so incomplete that definite conclusions linking chromosomes and classification are only just beginning to crystallize.

HYBRIDIZATION

Natural interspecific hybridization is commoner among fishes, especially those inhabiting continental (Holarctic) fresh waters, than in any other class of vertebrates. It is a particularly recurrent phenomenon in salmonoids, pikes (Esocidae), minnows (Cyprinidae), suckers (Catostomidae), sunfishes (Centrarchidae), and darters (Etheostomatinae, family Percidae). In the world fauna, hybrids are known (1972) in 56 families of fishes. More than two hundred are known among North American freshwater fishes. For our purposes, a hybrid is defined as the offspring from parents of two distinct species (or genera). Altered environment is the most important, but by no means the only, factor influencing the production of hybrids. Because of man's continual modification of aquatic environments and of fish distribution, hybridization is on the increase in freshwater fishes.

Since fertilization is chiefly external, fishes might be expected to hybridize more frequently than higher groups like birds and mammals. Families in which a large percentage of species hybridize are characterized by numerous, closely related forms with similar habits and habitat requirements.

Generally speaking, F_1 hybrids possess a spectrum of characters that lies approximately intermediate between those by which the parental species differ significantly. Uniformly intermediate characters may be taken as a strong indication of sterility in hybrids. Conversely, wide variation in hybrid offspring is an indication of at least partial fertility. Interspecific crosses further commonly show disturbed sex ratios, often exhibit *heterosis* or "hybrid vigor," and in those examples that have been studied cytologically or experimentally, partial or complete sterility occurs in at least one sex.

Evidence of natural hybridization must usually be largely circumstantial. Such evidence, when carefully documented and appraised by one thoroughly familiar with the fauna, may nevertheless be taken as a reliable indication for the hybrid interpretation. In the several groups that have been tested experimentally, the presumed natural hybrids have been convincingly confirmed.

Experimental work with sunfishes (*Lepomis*) has shown that interspecific hybrids grow faster than do either parental species and that they dominate the social hierarchy that corresponds to the peck order in birds. For example, when a mixed group is fed, the hybrids—larger and more active than the pure species—take the food first. Increased vigor of hybrids has also been shown to occur in minnows (Cyprinidae) and trouts (Salmoninae), wherein hook-and-line fishing methods capture relatively more hybrid individuals than those of either of the putative parents.

The great majority of sunfish (*Lepomis*) hybrids have sex ratios preponderantly in favor of the male—often 4 to 1 or more. These males, when mature, construct, fan, and guard their gravel nests with great vigor for long periods of time, but the energy in many such hybrids is wasted, since sterility may be complete, or nearly so, due to grossly abnormal spermatogenesis, among other possibilities.

One of the most widely known examples of aberrant reproduction in vertebrates is illustrated by the so-called Amazon molly, *Poecilia formosa*, which is known to exist in nature in the female sex only (occasionally seen phenotypic males appear to represent sex-reversed females). This fish is believed to have arisen by hybridization between two closely related species of *Poecilia* and to be a permanently-fixed diploid. It persists in nature by mating with the male of whichever species of *Poecilia* it is sympatric with; since 100 percent, rather than about 50 percent, of its population are female breeders, this increased fecundity would seem to give it a definite selective advantage over the bisexual species with which it lives. Embryonic development is initiated only by fertilization (internal in this example); but, although sperms are essential, gametic (nuclear) fusion does not occur. Hence the phenotype of the offspring is wholly dependent upon the genotype of the mother. The young are invariably daughters and are genetically identical with their mothers. The consistent production of female offspring from *P. formosa* occurs not only in first-generation crosses but in all subsequent backcrosses and out-crossings. This reproductive process is called *gynogenesis* and the hereditary mechanism is referred to as *matroclinous inheritance*, since only the chromatin of the mother is passed on to her offspring, which are all daughters.

Recently, other unisexual female types in the same family, the livebearers (Poeciliidae), but belonging to the genus *Poeciliopsis*, have been experimentally demonstrated. However, unlike the situation in the Amazon molly (*Poecilia formosa*), male chromatin is transmitted to the all-female offspring so that the first-generation young express characters of both parents. Hence a different genetic mechanism must be invoked to explain unisexuality in *Poeciliopsis*. It is postulated that a process of selective maturation eliminates the paternal chromosomes during oogenesis. All of the unisexual forms described above may be looked on as perpetual hybrids.

All-female offspring of other viviparous fishes are known from laboratory cultures only. With the aid of marker genes, sex-reversed males presumably bearing XX chromosomes have been noted in the guppy (*Lebistes*); their genetic sex was female, their real sex, male. Normal males in *Lebistes* conform to the XY type. When one of the exceptional XX males was mated to a normal XX female, all of the resultant offspring were females, as would be expected. The same sort of phenomenon has been found in domesticated platyfish (*Xiphophorus*) and in the Asiatic ricefish (*Oryzias latipes*).

A positive correlation exists between success in hybridization and closeness of relationship, such that greater survival of hybrids occurs when the parental species are more intimately related. This is to be expected on the basis of the hypothesis that closely related species possess chromosomes of similar morphology, number, and biochemical traits.

MECHANISMS OF EVOLUTION

Evolution—the continuous and gradual transformation of lines of descent from a common ancestor—is an accomplished fact. The long procession of evolution, extending over two billion years, has resulted from fundamental causes that are still in operation. Whereas Darwin had to deduce his theory mostly from indirect evidence, we can today study and even produce evolutionary changes at will in the laboratory. How evolution has taken and is taking place is the subject of this section.

Because hereditary modifications provide the basic materials for evolution, the way in which inherited variations originate is basic to our comprehension of evolutionary mechanics.

Mutation, along with recombination and selection, is essential for the continued evolution of most, if not all, types of organisms. The genetic raw materials supplied by mutation provide the building blocks from which evolutionary changes may be constructed by natural selection. The origin of variation is a problem entirely separate from that of the action of selection.

Gene Mutation

The gene is the unit of Mendelian heredity. Genes occur chiefly in chromosomes of the cell nucleus and, since they are parts of chromosomes, they typically exist in pairs just as the chromosomes do. A permanent, spontaneous change in a gene is called a gene mutation. Such heritable changes provide an important source of organic variation, and heritable variability is essential to evolution. Contrary to the ideas expressed by the critics of genetics, when that science was in its infancy, it is well established that mutation is a normal phenomenon in nature. Mutation rates probably vary greatly with different genes mutating at different rates even within the same population. Mutation rates may be greatly increased in wild populations by so-called mutator genes.

In experimental as well as natural environments, the majority of mutations has been found to be disadvantageous, such that the mutants are eliminated by natural selection. But there remain many mutations that are "neutral" or advantageous to the organism possessing them, and some seemingly deleterious ones could be of value under changed environments. Since most mutations alter only a single gene at a time, it is the accumulation of small changes (not a sudden, major change) over a long period of time that contributes to changing genetic constitutions. Radiation, for example by X rays, is known greatly to accelerate mutation rates, and it is probable that mutation is partly caused chemically in nature (as by hydrogen peroxide). In general, spontaneous mutation occurs at random or in an undirected manner; this is true of all types of mutations.

Chromosome Mutation

Intensive studies of the giant salivary gland chromosomes of the fruit flies (*Drosophila*) have demonstrated that chromosomes are longitudinally differentiated and that their genes occur at definite loci or positions. Thus chromosome "maps" have been made of the salivary gland chromosomes, and even small segments are identifiable by reference to a standard map. Rearrangement of chromosomes may occur by four processes: (*a*) *deletion*—the loss of a segment of a chromosome; (*b*) *duplication*—the duplication of a chromosomal segment; (*c*) *inversion*—the reversal of a segment of a chromosome due to breakage at two points; and (*d*) *translocation*—the transference of a segment of one chromosome to a nonhomologous chromosome. These architectural changes in chromosomes may be produced experimentally by high-energy radiation, and it is safe to infer that naturally occurring radiation, as well as chemical mutagens, may play a role in producing the changes that have been observed experimentally. Each of the above types of rearrangement has been observed to behave as though it were a gene mutation, thus producing a different, heritable phenotype. Many are lethal when homozygous, but some are not. Translocation of chromosomal segments may lead to a change in gene effect, as from recessive to dominant. Translocations also may affect chromosome number, as in fishes of the family Cobitidae: the Japanese loaches (*Misgurnus* and *Barbatula*) have 26 and 24 pairs of chromosomes, respectively, most of which are rod-shaped. However, two pairs of chromosomes of *Barbatula* are V-shaped, and the reduction (from 26) is attributed to translocations that resulted in uniting originally distinct chromosome pairs.

Natural Selection

The theory of natural selection, Darwin's greatest discovery, forms a central theme in all biological thinking. It is the main evolutionary process and is essentially an attempt to account for adaptation of organisms to their environments. The core of selection is that the carriers of different genotypes in a population contribute differently to the gene pool of the succeeding generations. Some genotypes make relatively greater contributions on the average than do others in the same environment. The relative capacity of carriers to transmit their genes constitutes the adaptive value, or Darwinian fitness, of that genotype. Mendelian populations (not individuals) are the units of natural selection and adaptation.

As the result of natural selection, the gradually changing organism becomes adapted to the conditions under which it must exist because some variants are likely to produce more surviving progeny than others. Adaptation—effected through natural selection—is thus a process of prime importance to the evolving plant or animal. Paleontology especially emphasizes this fact, since a record of progressive adaptations taking place through definite periods of geologic time leaves no doubt as to the importance of adaptation and selection in evolution.

Important changes through natural selection have occurred within historic time in certain species. Thus the melanistic (black) color phases of moths, generally of sporadic occurrence only, have become prevalent in areas affected by the industrial revolution (hence the phenomenon has been termed "industrial melanism"). The distribution centers of the melanistic forms are the large industrial cities. The explanation for the success of the dark- over the light-color phase in smoky areas is that the melanistic moths are less conspicuous prey. However, the darker coloration may be an adaptation secondary to physiological changes, such as resistance to poisonous chemicals in the industrial smoke. Both traits proved to be preadapted to a changing environmental condition and were of little value to the animal until the industrial revolution prospered.

Polyploidy

The occurrence in closely related organisms (a single genus, for example) of chromosome numbers in some species that are multiples of those in others is called polyploidy. For example, various species of wheat have 14, 28, and 42 chromosomes; the two higher sets are multiples of the smallest set. Polyploids may result from the duplication of the genome within a species (autopolyploidy), or an increase in chromosome complement may arise through hybridization and produce unlike genomes (allopolyploidy).

Although of major importance in the plant kingdom, polyploidy is a rare or, at best, an uncommon phenomenon in animals. Bisexual reproduction is generally credited to be the basis for this difference between the two kingdoms. This suggestion receives support from the fact that polyploidy is rare in those plants that have come to have separate sexes (dioecious). Whereas hybridization is important in plant evolution and accounts for the frequency of allopolyploids in plants, the production and maintenance of successful hybrids appear to have played relatively minor roles in animal evolution. It has been demonstrated that sexually reproducing polyploid species do occur in cyprinids and salmonids, in which chromosome sets of some species seem to be multiples of those of others. The majority of animals reported to be polyploids, however, are parthenogenetic, and curiously, all of these are arthropods. The brine shrimp, *Artemia salina*, occurs in tetraploid and octoploid parthenogenetic forms but there is also a tetraploid race that reproduces sexually.

ISOLATING MECHANISMS

Fully distinct, closely related species often occur together without losing their identity. They fail to interbreed because they are kept apart more or less completely by a variety of isolating mechanisms. Indeed, species formation without some previous form of isolation is held to be impossible. In general, these barriers to crossing fall into two major groups: geographic and genetic, or extrinsic and intrinsic.

Geographic isolation involves the spatial separation of populations in different geographic regions, either within a continuously inhabited area or separated by distributional gaps. It is independent of any genetic differences between the populations. Direct gene exchange between populations whose ranges do not overlap is prevented simply because they are allopatric. Allopatric populations may or may not exhibit reproductive isolation if artificially crossed or, as a result of secondary overlap, when their range barriers are removed. Conversely, some sympatric species—for example, *Xiphophorus helleri* and *X. maculatus*, the swordtail and the platyfish—that are not known to hybridize in nature are readily induced to cross-mate in aquaria.

Intrinsic isolating mechanisms involve a great host of devices. Moreover, the interbreeding of a given pair of species is usually prevented not by one mechanism but by several cooperating ones.

Behavioral traits, involving sexual, psychological, or ethological barriers to cross-mating, are probably operative in promoting species segregation in most fishes that are sexually dimorphic and/or dichromatic. Darters (Etheostomatinae), live bearers (for example, Poeciliidae), sticklebacks (Gasterosteidae), and coral-reef fishes include many examples in which such mechanisms play an important role in maintaining species as distinct populations.

Differences in mating time provide temporal barriers to crossing. The Dolly Varden (*Salvelinus malma*) spawns in the fall, the rainbow trout (*Salmo gairdneri*) in the late winter or spring. Thus, even when these two fishes occur together they cannot cross.

Adaptations to different ecological niches prevent the breakdown of genetic barriers between related species. For example, the Johnny darter (*Etheostoma nigrum*) spawns on the undersides of rocks in an upside-down position, whereas the greenside darter (*Etheostoma blennioides*) lays its eggs, while in an upright position, on algae and moss.

Physiological adaptations may work with ecological ones in preventing genetic breakdown. In Belgium, the threespine stickleback (*Gasterosteus aculeatus*) occurs in two ecologically and physiologically distinct populations. One inhabits only fresh water, usually small creeks, whereas the other lives in the sea in winter but migrates into river estuaries in spring and summer to spawn. Although the two forms differ on the average in a number of morphological traits, the variations overlap broadly. The freshwater form has the osmoregulatory capacity to maintain blood chloride content at a rather constant level in waters of low salinity. The migratory form, however, loses chloride ions in waters having a salt concentration less than one-quarter that of the ocean. This quickly leads to death. The morphological intergrades that occur in natural habitats are not physiologically intermediate but belong to either one or the other physiological type. Although the two forms are readily crossed by artificial insemination of the eggs, hybrids rarely occur in nature. Their failure to survive there is due to an unsuitable environment, since waters of proper salinity are scarce.

Physical incompatibility of the genitalia because of architectural differences in these organs affords a mechanical isolating mechanism. This, for example, virtually prevents the cross-fertilization of many species of live-bearers (Poeciliidae).

Biochemical factors may prevent the union of spermatozoa and egg or, in live-bearing fishes, the sperm may have a greatly lowered viability in the sexual ducts of other species. In addition, sperm competition may account for the failure of foreign male gametes to penetrate the egg.

Should hybridization occur, and it repeatedly has in freshwater fishes of the Northern Hemisphere, there are several mechanisms that militate against the successful establishment of the hybrids. The hybrid zygotes may be inviable or adaptively inferior to the zygotes of either parental species—as in most sunfish hybrids. The hybrids may fail to produce functional sex cells; this hybrid sterility is a common and effective isolating mechanism and is presumed to be a major factor in the failure of most natural interspecific fish hybrids to survive beyond the first generation. But, should the F_1 hybrids produce normal gametes, all or part of the F_2 generation or of backcrosses may be either inviable or adaptively inferior to that of the F_1 hybrids.

PARALLEL EVOLUTION OR CONVERGENCE

When animals of different origin and history develop under closely similar environments in well-isolated regions, functionally similar but only distantly related forms may evolve. The superficial resemblance results from parallel evolution, a phenomenon not clearly recognized by Darwin but known for a long time and very common. It is also called convergent evolution or simply convergence.

The distribution of the large, flightless birds, the Ratites (kiwis, ostriches, cassowaries, emus, and rheas) of Africa, Madagascar, South America, Australia, New Guinea, and New Zealand is a classic example. Once thought to be intimately related and hence all placed in a single group, these birds are now known to have sprung from divergent ancestors. A striking example among fishes may be observed in the extraordinarily abundant and diverse characins (family Characidae), a dominant group of South American freshwater fishes. In adapting themselves to varied ecological niches, species of this single family have come to resemble remarkably closely such unrelated North American families as gars (Lepisosteidae), bowfins (Amiidae), pikes (Esocidae), minnows (Cyprinidae), and darters (Etheostomatinae). The loss of eyes and pigment in different families of cave-inhabiting fishes affords another good example of convergence.

TIME RATE OF EVOLUTION

How fast do organisms evolve? This basic evolutionary question cannot be answered simply, for evolutionary rates not only vary greatly in different groups of animals but there is also great fluctuation within classes, families, or genera. Geographic accident, such as that enabling an ancestral form to penetrate an unsaturated environment (an isolated lake, for example), may result in relatively rapid, "explosive" evolution. This type of differentiation has occurred repeatedly in lakes, especially old Tropical ones.

One of the great difficulties in attempting to estimate speciation rates in fishes is determining whether the observed differences between populations are phenotypic or have, in part or in whole, a genetic basis. Another problem is measuring geologic time, but this handicap is becoming increasingly lessened by the innovation of more precise techniques.

In general, for fishes, the time span of some 10,000 to 12,000 years that has elapsed since the close of the last or Wisconsin glaciation is usually associated with differentiation at no higher than the subspecies level. Most species are probably at least a million years old, many much older. However, under special circumstances attainment of full species rank may have taken place within the past 30,000 or 60,000 years, or even in only 4000 years. Speciation may be especially rapid in small, isolated populations. Many of the genera now living had probably evolved by Miocene or Pliocene time, others perhaps shortly after the onset of the Pleistocene. The fossil record of the Cenozoic Era is as yet too inadequate to be of much help in estimating evolutionary rates. Many late Pleistocene fossils, originally interpreted as distinct species, have been found to differ little if at all from their living descendants. In general, periods of major geological change and periods of rapid evolution are correlated, whereas the much longer times of geological uniformity have seen only slow evolutionary change. There was an explosive evolution in the Silurian and Devonian eras, when the land was being colonized by plants and animals. Such evolutionary bursts, though on a much smaller scale, have recurred many times. Species flocks—the evolution of many species of a single genus through adaptive radiation in a restricted area—provide a relatively recent example in African lakes.

SPECIAL REFERENCES ON GENETICS AND EVOLUTION

Blair, W. Frank (Ed.). 1961. Vertebrate speciation. University of Texas Press, Austin.

Brooks, J. L. 1950. Speciation in ancient lakes. *Quart. Rev. Biol.*, 25: 30–60 and 131–176.

Chen, T. R., and A. W. Ebeling. 1966. Probable male heterogamety in the deep-sea fish *Bathylagus wesethi* (Teleostei: Bathylagidae). *Chromosoma* (Berl.), 18: 88–96.

Chen, T. R., and F. H. Ruddle. 1970. A chromosome study of four species and a hybrid of the killifish genus *Fundulus* (Cyprinodontidae). *Chromosoma* (Berl.), 29: 255–267.

Dobzhansky, Theodosius. 1951. Genetics and the origin of species, 3rd ed. Columbia University Press, New York.

Ebeling, A. W., and T. R. Chen. 1970. Heterogamety in teleostean fishes. *Trans. Amer. Fish. Soc.*, 99 (1): 131–138.

Fryer, Geoffrey. 1959. Some aspects of evolution in Lake Nyassa. *Evolution*, 13: 440–451.

Gordon, Myron. 1957. Physiological genetics of fishes. *In* The physiology of fishes. Academic Press, New York, 2: 431–501, figs. 1–16.

Greenwood, P. H. 1965. The cichlid fishes of Lake Nabugabo, Uganda. *Bull. Brit. Mus. Nat. Hist.*, Zool. 12 (9): 315–357.

———. 1973. The cichlid fishes of Lake Victoria, East Africa: The biology and evolution of a species flock. *Bull. Brit. Mus. Nat. Hist.*, Zool., suppl. 6: 1–134.

Myers, George S. 1960. The endemic fish fauna of Lake Lanao, and the evolution of higher taxonomic categories. *Evolution*, 14: 323–333.

Nogusa, Shyunsaku. 1960. A comparative study of the chromosomes in fishes with particular considerations on taxonomy and evolution. *Mem. Hyogo Univ. Agr.*, 3 (1): 1–62, pls. 1–6.

Sadoglu, Perihan. 1957. Mendelian inheritance in the hybrids between the Mexican blind cave fishes and their overground ancester. *Verh. Deutsch. Zool. Gesell.* Graz, 1957: 431–439.

Schultz, R. Jack. 1969. Hybridization, unisexuality, and polyploidy in the teleost *Poeciliopsis* (Poeciliidae) and other vertebrates. *Amer. Nat.*, 103 (934): 605–619.

———. 1971. Special adaptive problems associated with unisexual fishes. *Amer. Zool.*, 11, 351–360.

Schwartz, Frank J. 1972. World literature to fish hybrids with an analysis by family, species, and hybrid. *Gulf Coast Res. Lab.*, Publ. 3: 1–328.

Tax, Sol (Ed) 1960. Evolution after Darwin. Vols. I–III. University of Chicago Press, Chicago.

Uyeno, Teruya, and Robert Rush Miller. 1971. Multiple sex chromosomes in a Mexican cyprinodontid fish. *Nature*, 231 (5303): 452–453.

White, M. J. D. 1973. Animal cytology and evolution, 3rd ed. Cambridge University Press, London.

13

Systematics and Nomenclature

Systematics is the science that deals with the kinds and diversity of living things and with their arrangement into a natural classification. Basically a study of the evolutionary relationships of organisms, systematics lies at the foundations of biology and it is elementary, in insuring a sound structure, to look to the foundations. A system of classification provides the means for attacking the problem of the origin and evolution of life. It is erroneous to suppose, as many have, that classification is an outmoded phase of natural history. Everyone is, at heart, a classifier, whether by virtue of necessity or because of mere curiosity. Taxonomy, the study of the theoretical bases, principles, and procedures necessary to an understanding of relationships, is the source of information for classification. Systematics involves both taxonomy and classification.

All relationships among living things are the direct results of what has happened during their evolutionary history; that history is a real phenomenon, whether or not we may ever be able to reconstruct it precisely. All approaches to understanding the systematics of organisms are limited to the study of similarities and differences and are basically comparative in principle. Renewed interest in the methods and theory of classification has been enhanced by the development of the computer and the coming of whole new areas of inquiry designated numerical taxonomy, including cladistics. Taxonomy is the theory and practice of classifying organisms. It is important at the outset to distinguish between classifying things and naming or identifying them, for these two activities are totally different. Classification involves scientific philosophy that uses inductive procedures, whereas identification involves deductive procedures that allow us to place individuals into previously established taxa. Zoological nomenclature involves the application of distinctive names on the basis of established rules.

The primary function of classification is to create order out of chaos by leading to accurate identification of individuals, and to their ranking or arrangement into various taxa, since it is impossible to discuss or think about organisms without first assigning them names. To do this also requires the application of biological nomenclature to the groups (taxa) that are recognized. One aim of classification is that of convenience. The categories of the systematist are based on degree of similarity so that the more closely two organisms are related the more characters they will usually share. Latin names were applied to animals and plants long before 1735, when Linnaeus first attempted to catalogue all the known kinds, and it no doubt became apparent even in the most primitive societies that it was useless to make observations on a plant or animal unless one knew its name. The Aztecs had considerable knowledge of the species inhabiting their region prior to the Spanish conquest; moreover, they employed names which, by appropriate prefixes, served to indicate relationships.

This leads to a second function of classification—that of serving as a guide to relationships. Modern classification is based upon phylogeny, but classification and phylogeny are distinct and should not be confused. Phylogeny is the actual evolutionary history of organisms and is natural, continuous, and dynamic. Classification is the result of human efforts to interpret or reconstruct phylogeny, and it is arbitrary in that many classifications may be constructed from the same phylogeny. The evolutionary history of a group entails one, and only one, phylogeny; but, in developing a sound interpretation of the origin and evolution of such a group, scientists propose various schemes of classification that undergo change as new light is shed on the course of evolution.

The fundamental problem of systematics is, as Huxley so aptly put it, "that of detecting evolution at work." The modern systematist utilizes information from a great variety of disciplines—genetics, ecology, paleontology, comparative physiology, animal behavior, cytology, zoogeography, biochemistry, biometrics, and related fields. Background information from such disciplines is essential for the solution of many taxonomic problems. Systematics has become the focal point for many fields of biology since it deals with organic diversity, a major integral branch of biology.

SUITABILITY OF FISHES

Fishes are particularly well-suited to systematic studies because: (*a*) they are abundant in species and numbers, widespread, and readily available; (*b*) they possess many characters that are especially well-suited for taxonomic analysis and statistical treatment; (*c*) environmental correlations are well-marked; (*d*) many have proven suitable for experimental analyses under laboratory conditions; and (*e*) intergradation and hybridization are relatively common and provide tests for genetic relationships.

HISTORICAL BACKGROUND

The beginnings of taxonomy no doubt antedate recorded history, for it is one of man's characteristics that he likes to name and arrange things. Aristotle (383 to 322 B.C.) synthesized the knowledge of his time and formulated it into the beginnings of a science. Although he did not propose a formal classification, Aristotle provided a basis for such when he stated that "animals may be characterized according to their way of living, their actions, their habits, and their bodily parts." His philosophy prevailed for nearly 2000 years. Of the zoologists who lived shortly before Linnaeus, John Ray (1627 to 1705), the son of a British blacksmith, proposed a more natural system of classification than his predecessors, and his work had a marked influence on Linnaeus (1707 to 1778). This great Swedish naturalist first consistently applied what is known as the binomial system of nomenclature (see below) in the tenth edition of his monumental *Systema Naturae* (1758), which marks the starting point for zoological nomenclature. His contributions so influenced later students that he is generally called the father of taxonomy. Linnaeus also provided a hierarchy of categories: variety, species, genus, order, class, and kingdom. Subsequently, the principal changes were adding family and phylum. His system was so practical that it was quickly adopted, expanded, and elaborated, and it dominated the field for the next century, during which time species were regarded as immutable.

A century later, Charles Darwin's revolutionary theory of evolution produced a tremendous stimulation of biological thought and work. Workers in the decades immediately following his *On the Origin of Species* (1859) were concerned principally with discovering whether living organisms actually are descendants of common ancestors. Phylogeny was, therefore, the chief preoccupation of this period. The phylogenetic tree was introduced by Haeckel (1866), and proved to be a useful and stimulating method providing taxonomists with a graphic means of expressing presumed relationships. This was a productive and exciting period in the history of taxonomy, and some of the keenest minds were attracted to the field by the rewards of almost daily discoveries of new species and genera and, not infrequently, of new families or orders. However, by or before the end of the nineteenth century, the period of such major discoveries among the higher animals was over. Those who were anxious to describe new orders, families and genera now resorted to refining the classification and splitting the existing categories. As might be expected, some of this splitting was necessary and beneficial, but in other cases it led to a disintegration of natural categories by concealing true affinities. Part of the disrepute into which taxonomy fell during the close of the nineteenth century and early part of the twentieth century is attributable to such excessive splitting.

The most recent phase in the development of systematics has been characterized by study of evolution within species. The rediscovery of Mendel's rules, the rise of genetics and later of population genetics, the use of experimental biology, the introduction of biometrics and, subsequently, of numerical taxonomy, advances in ecology, the study of comparative behavior, and the development of new tools have greatly broadened the outlook of systematics and advanced the field to a bonafide, vigorous science. The backgrounds of these sciences are essential for the solution of many systematic problems. The realization finally emerged that the organisms with which the systematist must work "are neither a cage full of laboratory animals nor a row of dead specimens in a museum, but instead the natural populations of living things in the field" (Myers, 1941). The modern systematist has become even more interested in formulating generalizations, for which the naming and describing of new taxa is merely the initial step.

TASKS OF THE SYSTEMATIST

The ultimate task of the systematist is not only to recognize and describe the infinite array of life around us but also to contribute to its understanding. The necessity for accurate identifications is critical to the work of wildlife managers, geologists, ecologists, and experimental biologists, to name only a few. To become competent in a particular group of animals requires painstaking work and long experience, and authoritative determinations may not be obtainable for some organisms because no specialist of the group is available. The modern systematist is a competent field naturalist who studies the ecology and behavior of species in their natural environment. The breadth of systematics provides opportunities for students with the most diversified interests and talents.

After recognizing and accurately describing species, the systematist must next put these species in order by grouping them into arrays of related taxa and devise a classification that will relate the species and their higher categories. This step involves more speculation than naming species and is more theoretical, since the taxonomist must decide if similarities indicate convergence or true phylogenetic relationship. He has now passed from the analytical to the synthetic stage.

The third task of the systematist is to study how species are formed, and this leads to a consideration of the factors of evolution: how do species originate, how are they related, and what does this relationship mean? In this phase of his work the systematist comes into closer contact with the other biological disciplines, such as genetics and physiology, ecology and cytology, and biogeography and paleontology.

These several stages through which the classification of a particular group passes have been referred to as alpha, beta, and gamma taxonomy. Alpha taxonomy refers to the level at which the species is recognized, characterized and named; beta taxonomy, to the arrangement of species into a natural system of classification; and gamma taxonomy, to the analysis of intraspecific variation and studies of evolution. The three overlap and intergrade, but the trend is real. Although the biologically minded taxonomist may aspire to concentrate on gamma taxonomy, there is still a real need for more refined studies at the alpha and beta levels as well, especially in the lesser-known groups, including fishes.

TAXONOMIC CONCEPTS

In classical taxonomy, organisms were arranged from the species through the higher categories by a given set of characters presumably occurring in every member of a given group. In modern taxonomy, the concepts of variation within a group at a given time and between different periods of geologic time are becoming increasingly influential in characterizing categories.

Classification starts with the individual. Individuals that bear close relationship to each other and breed true are grouped into species. Species are arranged into genera, which may contain one or more species. Related genera are then assembled into families, each family containing one or more genera. Families in turn constitute orders, and orders are arranged in classes.

Present evidence indicates that closely related species are as likely to differ from each other in details of their biochemical functioning as in their external or internal morphology. Newer approaches to deciphering relationships are focusing on chromosomal studies, amino acid sequences, and comparative behavior, among others. Although the study of preserved specimens provides valuable systematic data, it is imperative for the broad view to know the organism in its natural habitat.

Thus, although species are more often than not defined by sets of morphological characters, and by their geographic variation, considerable biological information may in some instances be available (as on behavior, genetics, physiology, and ecology), even including a time element. The species is a concept based on that of populations, and it must be defined according to a particular set of circumstances. Consequently, there may be different kinds of species in different groups of animals or even in the same group. *Monotypic* species are those showing uniform characteristics over the entire range, whereas *polytypic* species are made up of differentiated subspecies. *Sibling* (cryptic) species are those that are morphologically indistinguishable or very similar but are shown to be fully differentiated by genetic, physiological, ecological, or behavioral differences that involve reproductive isolation. Subspecies are a geographically segregated population, or group of local populations, that differ taxonomically from other such subdivisions of a species.

The genus tends more than any other category to retain its classical status and to be defined in practice as a group of species that share certain characters.

The family category includes, in theory at least, several distinct phyletic lines, and hence its assessment takes into consideration variation in time as well as in space.

THE DATA OF CLASSIFICATION (FIG. 13.1)

What kinds of characters are used by ichthyologists in identifying fishes or in proposing a classification? They are many and varied for characters, like gold, are where you find them. Whether they be meristic traits—such as the number of scales, fin rays, or vertebrae—body proportions, life colors, osteological features, behavior patterns, or chromosome numbers they have one common feature: any one of them is likely to have different significance in different groups of fishes. For example, the number of caudal rays is virtually constant in many families and therefore cannot be used in those groups below the family level, but in the Cyprinodontidae their number may distinguish unnamed races, subspecies, or species. Number of vertebrae, very consistent in the majority of spiny-rayed fishes—hence an ordinal character—may be used to distinguish categories from the subspecies to the genus in more primitive groups. Absence of pelvic fins may be a family trait or an individual variation (as in certain desert cyprinodontids). In the minnows (Cyprinidae) relative head size may be used to distinguish species from one another; yet

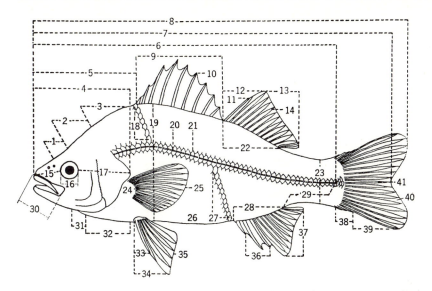

Fig. 13.1   A spiny-rayed fish illustrating parts and methods of counting and measuring: 1—interorbital; 2—occipital; 3—nape; 4—head length; 5—predorsal length; 6—standard length; 7—fork length; 8—total length; 9—length of base of the spinous or first dorsal fin; 10—one of the spines of the first dorsal fin; 11—spine of the second dorsal fin; 12—height of second dorsal fin; 13—length of the distal, outer or free edge of second dorsal fin; 14—one of the soft-rays of the second dorsal fin; 15—snout length; 16—eye length; 17—post-orbital head length; 18—scales above the lateral line or lateral series which are counted; 19—body depth; 20—one of the lateral line pores in a complete lateral line; 21—one of the lateral scales which with the remainder form the lateral series; 22—length of base of the second or soft dorsal fin; 23—least depth of the caudal peduncle; 24—the pectoral fin; 25—one of the soft-rays of the pectoral fin; 26—abdominal region (belly); 27—scales below the lateral line or lateral series which are counted; 28—length of the base of the anal fin; 29—length of the caudal peduncle; 30—length of the upper jaw; 31—isthmus; 32—breast; 33—height of pelvic spine; 34—height of pelvic fin; 35—one of the soft-rays of the pelvic fin; 36—spines of the anal fin; 37—soft-rays of the anal fin; 38—rudimentary rays of the tail (caudal) fin; 39—one of the principal, unbranched soft-rays of the caudal fin; 40—branched soft-ray of caudal fin; 41—the caudal fin with numeral at fork of fin. (Source: Trautman, 1957).

in the sunfishes (Centrarchidae), head size is of family significance. Thus there can be no fixed concept as to what taxonomic level a given character may pertain.

In general, characters that are seen externally and easily are the ones used to separate species and sometimes even genera. These characters are, however, often the most superficial as well as the most variable ones, and hence they must be employed with caution. Features used to distinguish the higher categories of genera, families, etc., are generally more profound and less variable. Certain osteological characters, for example, show little variation and appear to be less subject to environmental influences (such as temperature effects) than are such meristic traits as scale number. Color is often of value as a specific character, but it may also characterize higher groups as does the red color of squirrelfishes (Holocentridae).

The more diverse the data brought to bear on classification the more reliable the resulting arrangement is likely to be. The modern systematist may be able to draw information from such varied fields as biochemistry, genetics, behavior, ecology, physiology, geographical distribution, paleontology, and cytology to supplement and strengthen the more conventional laboratory data of morphology and anatomy. Most classification is accomplished, however, without definite knowledge of the ancestry of the groups involved, their breeding potential, behavior, and much other useful information. It is unfortunate that such data are typically lacking, but we continue to classify without them and to do a reasonably good job, at least at the specific and generic levels.

STUDY COLLECTIONS

Museum collections are the repository for much of the raw material on which systematists' conclusions are based. Such collections are similar to great libraries in that they must be efficiently catalogued to be usable and they undergo continual growth. Unlike a library, the advances in knowledge of groups of living things require continual recataloguing in modern, active museums in order to bring the classification of a particular genus, family, or other category up to date. Whereas in the early days collectors preserved only a few so-called typical representatives of each species, modern practice emphasizes the gathering of large series of specimens from throughout the known range of the species, for the modern taxonomist is a student of populations. Samples must be adequate to give clear pictures of variation, growth, and so on, and to allow for statistical treatment of data.

Museum collections may place emphasis on the fishes of a particular region —for example, the fresh waters of North and Middle America—or they may be cosmopolitan, endeavoring to have some representatives of as many species as possible from throughout the world. Fish collections are represented chiefly by specimens fixed in formalin and preserved in alcohol, but a recent trend to investigate higher classification is resulting in the expansion of skeletal material and of cleared and stained specimens. Many museums now have X-ray machines to assist in working out systematic characters, as well as for studying the effects of radiation on living populations.

Certain essential information should accompany each collection or specimen. The most important is the precise locality, for a specimen not accurately labeled is virtually worthless to the systematist. If the locality is an obscure one, its position relative to a well-known place should be given on the original label. The state or country, the county or district, the date of collection, and the name of the collector(s) should also be recorded. The elevation is often very useful (even if only estimated), and additional ecological information is valuable. The label should be written in the field at the time that the collection is made, should be placed in the fixative (usually 10 percent Formalin, equal to a 4 percent aqueous solution of formaldehyde) with the specimens, and it should not be later replaced. A good quality label is essential, for no matter how carefully the data are recorded if the label subsequently disintegrates, the scientific value of the collection is lost. Labels are best written with waterproof carbon ink, but a soft pencil is satisfactory. Specimens more than a few inches long should be slit along the right side of the abdomen before being dropped in Formalin; fish heavier than a few pounds should also be incised in the muscles on each side of the backbone. If specimens are to be left for several weeks in Formalin, household Borax (1 teaspoon per each ½ gallon of fixative = 5 grams/2 liters) should be added to neutralize the acidic effect, thereby retarding shrinkage, hardening of the soft parts, and preventing decalcification of the bony tissues.

Not all fishes may be preserved whole; large specimens may be skinned and dried or placed in liquid. Material intended to be skeletonized in the laboratory (by the use of dermestid beetles) may be prepared in the field by placement under refrigeration or, if ice is not available, by dry preparation. This is accomplished by scaling, eviscerating (making sure to record the sex and length), and then placing the specimen in a flyproof, screened box to dry. On return to the laboratory a brief soaking in ammonia will soften the dried fish sufficiently for the beetles to do their work.

ZOOLOGICAL NOMENCLATURE

It is essential to distinguish between classification and nomenclature, for their respective roles are often misunderstood. Classification is a biological matter, whereas nomenclature is the tool by which labels are provided for the taxonomic categories so that biologists may communicate with each other. Nomenclature is a means to an end rather than an end in itself. Since zoologists work with animals and use their names, it is essential that they familiarize themselves with the general principles of zoological nomenclature—the language of the zoologist—whether they be systematists or not.

In order to be generally usable, nomenclature must not only be of widespread application but the same words must have the same meaning to all.

The principal aims of nomenclature are stability and uniformity. However, progress in systematics leads to changing concepts of taxonomic categories and of relationships; consequently name changes are inevitable.

Modern scientific nomenclature involves a binomial system, first standardized by Linnaeus. Prior to his time organisms were named, usually in Latin, by descriptive phrases, sentences, or paragraphs—which proved to be a cumbersome and unworkable system. Linnaeus condensed such names to two Latin words—one, the *genus*, to indicate the general kind, the other, the *species*, to designate the particular kind. The binomial system therefore is a two-name system. Since Linnaeus first applied the binomial system consistently to plants and animals in the tenth edition of his *Systema Naturae* (1758), this work is designated in the International Code as the starting point for zoological (and botanical) nomenclature. Names published since then must conform to the principles of the binomial system in order to be accepted.

The two-name system has been expanded to encompass the subspecies— hence we now have a trinomial system. Many of the Linnaean genera now constitute our families, or even orders, for Linnaeus knew less than ten thousand species of animals whereas we now know well over a million. Nevertheless, the Linnaean system still works—which is why it has become universally accepted.

Although the system is basically simple many complications may arise in its practice. Confusion was early wrought by the lack of any accepted code of procedure, for it was not until 1842 that anything resembling an "official" code was proposed; in that year a committee of the British Association for the Advancement of Science published the "Stricklandian Code." With the amendments of 1865, this formed the basis for much of the later codification of the rules of zoological nomenclature. Other codes (French, American, and German) were proposed in subsequent years and the situation was chaotic until a general agreement was reached in 1901 when a code was adopted, and the committee that worked it out was made a permanent organization, the International Commission on Zoological Nomenclature.

At present the International Commission does not have a fixed number of members, but eighteen is the minimum; they represent different geographic regions and disciplines and are elected for 10-year terms at the International Zoological Congress (which meets every 5 years), in classes that terminate at 3-year intervals. The general duties of the Commission are to "study the general subject of the theory and practice of zoological nomenclature." Its powers have been carried out in five main ways: (*a*) by submitting recommendations to the International Congress for amendments to the International Code, or for the addition of "Recommendations" to particular articles of the Code; (*b*) by rendering (since 1907) "Opinions" on questions of zoological nomenclature submitted to it (but which, since 1939, constitute decisions regarding the status of an individual name or book); (c) by compiling (since 1910) the "Official List of Generic Names in Zoology" (nomina conservanda) and other such lists and indices; (d) by issuing (since 1939) "Declarations," which constitute interpretations of the Rules; and (e) by the use (since 1913) of the "Plenary Powers" (names retained by suspension of the rules).

Since 1939, at the instigation of Francis Hemming (England), Secretary of the commission from 1936 to 1957, the Commission has established the *Bulletin of Zoological Nomenclature* (1943) and a series of *Opinions and Declarations Rendered by the International Commission on Zoological Nomenclature* (1939). Proposals submitted to the Commission for decision, comments from and correspondence with zoologists regarding these proposals, and papers on the nomenclatural implications of developments in taxonomic theory and practice are published in the *Bulletin*.

A revision of the rules appeared as the International Code of Zoological Nomenclature (1961. Int. Trust. Zool. Nomen., London), adopted by the XV International Congress of Zoology meeting in London in 1958. The revision is published in parallel English and French texts.

The conventional hierarchy of nomenclature may be summarized as follows for the yellow perch, *Perca flavescens:*

Phylum Chordata
   Class Osteichthyes
      Order Perciformes
         Family Percidae
            Genus *Perca*
               Species *Perca flavescens* (Mitchill)

By using the prefixes "sub" and "super," the foregoing groups may be expanded into fifteen categories that cover most of the needs for expressing group concepts in ichthyology: phylum, subphylum, superclass, class, subclass, superorder, order, suborder, superfamily, family, subfamily, genus, subgenus, species, and subspecies. In some large groups in which the classification is comparatively well known and diverse lines are represented, use of tribes (formed by adding the ending "ini" to the stem of the type genus, e.g., Cyprinini) may be justified. Another useful category is the species group, sometimes called superspecies, which fits into the hierarchy between the species and the subgenus. It may be employed in place of the subgenus where the entire classification of a group has not yet been fully worked out but natural groupings of closely related species are called for.

The generic name is always a noun in the singular, nominative case, and is written with an initial capital letter. The specific part of the scientific name is not capitalized—for example, *Perca flavescens*. It may be: (1) an adjective, which must be of the same gender as the generic name, *Gasterosteus acu-*

leatus; (2) a noun in apposition (a substantive) that is always in the nominative case but need not agree with the generic name in gender, *Cichlasoma maculicauda*; (3) a noun in the genitive singular such as occurs in a patronymic, *Etheostoma jordani* (for David Starr Jordan), or *Epinephelus clarkae* (for Eugenie Clark); or (4) a common name in the genitive plural, usually indicating something about the habitat, *Rivulus paludium* (meaning "of the swamps").

Some of the names of the taxa above the generic level are also formed in accordance with generally accepted principles. For example, the tribe is a subdivision of the family, subordinate to the subfamily, formed by adding "ini" to the stem of an included generic name (such as Fundulini from *Fundulus*); the subfamily, a major subdivision of the family, may be used in large families, and is formed by adding "inae" to the generic stem (Epinephalinae); the family, comprising a group of related genera or, exceptionally, a single genus, is formed by adding "idae" to the stem of an included genus (Percidae); and the order, the major taxon immediately above the family level, is formed in this book by adding "iformes" to the stem of a valid, included genus (Clupeiformes). Unfortunately ichthyologists do not agree on the interpretation or application of ordinal names and the Code does not specify how to form names above the family level.

Since an individual, or occasionally two or more persons, is responsible for the proposal and description of each taxon, the name of the person or persons is placed after the scientific name and is known as the authority of the name. The name is usually written out but may, for well-known authorities (such as Linnaeus), be abbreviated. Citing the authority is not intended as a means of awarding credit to the worker, but rather it fixes responsibility for the name and aids in locating the original description. When a species is transferred to a genus different from that in which it was described, the original author's name (or names) is placed in parentheses.

The original orthography of a name is to be preserved unless there has been an error of transliteration, a "lapsus calami," or a typographical error. The valid name of a genus, species, or subspecies must be the oldest name that fulfills the requirements of the Law of Priority. This important rule was agreed upon in order to avoid confusion in the application of scientific names and to eliminate duplication; in general, names published earlier take precedence over names of the same rank published later. If other names are subsequently published for the same taxon they become synonyms, or invalid names (the same organism can have but one legitimate name, although it may have many synonyms in the literature).

Zoological nomenclature is independent of botanical nomenclature in that an animal name is not to be rejected simply because it is identical with the name of a plant. Generic and subgeneric names are nomenclaturally equal and subject to the same rules; the type species of the genus and the typical subgenus must be the same. Likewise, specific and subspecific names have nomenclatural equivalence and are subject to the same rules.

In the event that two or more names for the same taxonomic unit are published in the same paper (hence simultaneously), it becomes the privilege of the *first reviser* to select one of these names as the valid one and to place the others in synonymy.

Since the identification of a species may not be obvious from the original description, a system known as the type method was devised to tie scientific names to objective taxonomic entities. The type specimen functions as a "name bearer." The author of a new species must designate a certain specimen as the type of that species. Only the primary types are name bearers. These are: (*a*) the *holotype* (or simply *type*), a single specimen designated by the original author in the original description, or the only specimen known at the time the species was described; (*b*) the *lectotype*, a type specimen selected subsequent to the original description from one of the syntypes forming the basis of the original description; (*c*) the *syntype* (formerly often called cotype), one of several specimens forming the basis of the original description when no single specimen is designated as the type; and (*d*) the *neotype*, a specimen selected as the type subsequent to the original description when the type specimen (or specimens) is definitely known to have been destroyed or lost.

Other types in current use are: (*a*) the *paratype* (formerly often called cotype), any specimen other than the holotype used by the original author in describing his new taxon; (*b*) the *allotype*, a paratype of the opposite sex to the holotype; (*c*) the *topotype*, any specimen from the type locality collected there subsequent to the original description; and (*d*) the *paratopotype* or *isotype*, any specimen other than the holotype taken at the same place as the holotype and included in the original description. Types are not required to be "typical specimens" and not infrequently they are quite aberrant. The nomenclatural type of a genus is not a specimen but a species, the one on which the generic name was based. It was formerly called the genotype, but type species is the name currently recommended in order to avoid conflict with the use of the term genotype in genetics.

SPECIAL REFERENCES ON SYSTEMATICS AND NOMENCLATURE

Bailey, R. M., et al. 1970. A list of common and scientific names of fishes from the United States and Canada. 3rd Ed. Special Publication No. 6. American Fisheries Society, Washington, D.C.

Crowson, R. A. 1970. Classification and biology. Aldine Publ. Co., Chicago.

Dobzhansky, T. 1970. Genetics of the evolutionary process. Columbia University Press, New York.

Hennig, W. 1966. Phylogenetic systematics. University of Illinois Press, Urbana.

Huxley, J. S. (Ed.). 1940. The new systematics. Clarendon Press, Oxford.

Mayr, E. 1942. Systematics and the origin of species. Columbia University Press, New York.

———. 1969. Principles of systematic zoology. McGraw-Hill Book Company, New York.

Mayr, E., E. G. Linsley, and R. L. Usinger. 1953. Methods and principles of systematic zoology. McGraw-Hill Book Company, New York.

Myers, G. S. 1941. (Review of Huxley's, The new systematics.) Copeia, 1941 (1): 61.

Schenk, E. T., and J. H. McMasters. 1956. Procedure in taxonomy, 3rd ed. Stanford University Press, California.

Simpson, G. G. 1961. Principles of animal taxonomy. Columbia University Press, New York.

14

Ecology and Zoogeography

The abundance and distribution of fishes in the waters of the earth are the products of interaction among fishes and their chemical, physical, and biological surroundings. The study of the relationships between an organism and its environment is the subject matter of ecology. The study of fish distribution that results from the ecological relationships of fishes is zoogeography, or, more specifically ichthyogeography.

DEFINITIONS

The basic unit of study in ecology is the *individual*. The applications of ecology, however, commonly relate to *species* which are groups of individuals that look more like their own kind than anything else and that produce their own kind (breed true) upon mating. The individuals of one species or of all the species that occupy a given habitat compose a *population*. The members of populations occupy one or more ecological niches. A *niche* can be conceived as the profession of an organism and its *habitat*, as its address in the environment. The *environment* is the total biotic (living, dead) and nonliving (abiotic) surroundings of an organism. In the water environment populations composed of fish and other organisms occur in *communities* in different *habitats*. Finally, the sum of the community plus the habitat that supports it may be regarded as an *ecosystem*, which is an open system, not closed. Obviously the concept of ecosystem may encompass the earth, or *ecosphere*, or it may simply be a small pond or tide pool, or more extensively,

411

again, all the waters of the earth, the *hydrosphere*. An ecologist is concerned, then, with the quantities of matter and energy that pass through a given ecosystem.

WATER

All of the habitats in which fish live and all of the communities in these habitats are continually changing. All are different from one instant to the next, and all tend to evolve gradually in recognized directions, through ecological succession. The dynamics of the aquatic ecosystem depend importantly on the properties of water, among which are the following. Its high heat capacity prevents extremely rapid fluctuations in temperature. Both fresh and sea waters have their maximum densities at temperatures above the freezing point. Were this not so, many water areas would freeze solid, thus, at least seasonally, eliminating their populations of fish and many other organisms. The dissolving power of water is greater than that of any other liquid, and this affords ready transport to a host of materials upon which fish life ultimately depends.

The transparency of water is of prime importance to the amount of solar radiation (light energy) that can be harnessed by phytoplankton and other water plants. The sun is the main source of energy for these primary producers of the entire aquatic ecosystem. The transparency along with the availability of dissolved nutrients for the phytoplankton are features that basically but indirectly determine the quantity of fish any given habitat may support.

ORGANIC PRODUCTIVITY IN AQUATIC ECOSYSTEMS

Energy Transfer

The continuing interaction of components of the aquatic ecosystem is founded on energy transfer. Light is the source of the energy, only a fraction of which in aquatic ecosystems ultimately appears in fish flesh. Energy is first harnessed by water plants, some large, others small. The key to fish production is the harnessing that is done by microscopic algae, desmids, and diatoms—the phytoplankton. Phytoplankton has been termed the real grass of the waters, affording food stuffs which are directly or indirectly converted into fishes. The harnessing of light energy for the manufacture of organic material is the process of photosynthesis. In this process, green plants take carbon dioxide from the water and build carbohydrates and other molecular components. It is possible that more carbon is fixed in primary production (first trophic level) each year in aquatic ecosystems than in terrestrial ones.

Animal consumption and transformation of plant materials follows. Herbivores, including some fishes, are primary consumers. Subsequent grades of consumers are predatory. As there is only about a 10 percent efficiency in the conversion of energy from one trophic level to the next, with only a slight increase in efficiency at higher trophic levels, it follows that the shorter the food chain the greater the likelihood for survival and abundance of a fish species. Ecologists have likened this concept of production to a pyramid either of biomass or numbers. The producers are at the base and exceed in mass and number; primary, secondary, tertiary, etc., consumers are at various intermediate levels in the pyramids; and the consumers with the longest food chains such as highly predatory and/or deepsea fishes are at the top of the pyramids and are least in biomass and numbers. In such nonparasitic food chains the predators not only decrease in number along the various links, they also grow larger than their prey in each succeeding link.

Biogeochemical Cycles

Key components available for synthesis into protoplasm in the hydrosphere are circulated in biogeochemical cycles. Recognized cycles include those for carbon, nitrogen, and phosphorus. Initial sources of the key elements differ, as well as their concentrations throughout the year. Thus in production cycles the atmosphere is the source of carbon (from carbon dioxide) and nitrogen in gases. Phosphorus originates from soil through erosion or leaching. The productive phase of biogeochemical cycles is the fixation of elements into protoplasm by living things. Phytoplankton, for example, is able to fix carbon, nitrogen, and phosphorus. Fish that feed directly or indirectly on phytoplankton convert the plant-fixed compounds of these three elements and others into flesh. The waste products and carcasses of both the phytoplankters and consumers return the key elements directly or through bacterial decay, to their biogeochemical cycles for re-use. The process of return may be very rapid or very slow. Phytoplankters, fish and other aquatic animals may promptly return carbon dioxide by respiration to the aquatic ecosystem. Phosphorus, locked in the skeleton of a fish, may take years for its release after the fish has died and fallen to the bottom. The entire scene, however, is that of a continuum, with the forces of production nourished from the land and the atmosphere consistently managing to exceed those of release. Death is required for balance, but it is the amount of the nutrient raw materials naturally available in an ecosystem that limits the size of the biomass the system can support. For this reason, it is common practice in aquaculture (the intensive production of fish and other useful organisms, most often in ponds), as in agriculture, for fertilizers to be used in order to provide more nutrients than the habitat affords naturally.

The environment of fishes is composed of many factors in addition to nutrients that may limit populations or influence geographic distribution. Any one of these factors may be limiting when it is present in quantities which are either too little or too great.

ECOLOGICAL CLASSIFICATIONS OF FISHES

Aquatic organisms, including fishes, may be classified ecologically in several different ways. According to environmental tolerances, they may be grouped as either narrowly or broadly tolerant. The corresponding expression is prefixed respectively either by "steno" (narrow) or "eury" (broad). For temperature the classification is thus stenothermal or eurythermal, for salinity it is stenohaline or euryhaline, etc. Fishes may also be categorized on the basis of location in aquatic ecosystems—benthic (bottom dwellers or ground fishes), pelagic (free swimming), or planktonic (depending on currents for their movements as do the larval young of many species). In lakes and ponds, littoral zone fishes are those of the inshore waters where light penetrates to the bottom and rooted green plants are often present; limnetic zone fishes are those of offshore waters free of rooted plants, and extend downward to the light compensation level where illumination is inadequate for sustained life of phytoplankton; and profundal zone fishes are those in the darkened waters beneath the light compensation level. Incidentally, the foregoing categorizations illustrate stratification of communities in ecosystems. However, the species composition of the zones may be expected to vary geographically. In the flowing water facies the readiest division of the habitat and its occupants is into two rather subjective zones: pools and riffles. The current of the riffles is generally fast enough to move sand and silt which is then deposited in pools and backwaters.

ECOLOGICAL FACTORS

Environmental forces that impinge on the lives of fishes are many, complex, and interrelated in their effects. We will single out a few for illustration.

Temperature

Temperature is a factor of wide and varied significance, as already mentioned in Chapter 5, on food and growth, and in Chapter 10, on development, among other chapters. Generally, within the range of temperatures that can be tolerated by the fish, the effects are for vital processes to be accelerated by warm temperatures and decelerated by cold ones. Temperature extremes or sudden changes are often lethal. Elevated, sublethal temperatures may induce estivation, and depressed ones, hibernation. Precipitation and evaporation in the hydrologic cycle are thermal phenomena. Freshwater fishes, marine species that use the freshwater habitat for growth, and brackishwater fishes are dependent on this cycle for the supply of water in which they live.

Within bodies of water, temperature may determine success of a species as well as its distribution. Within the geographical range of trouts of the family Salmonidae, waters are commonly classified thermally as cold (inhabited by trout) or warm (intolerable to trout). In lakes of the Temperate zones, when stratified thermally in midsummer, coldwater fishes such as the trouts (*Salmo*), chars (*Salvelinus*), and whitefishes (*Coregonus*) remain in deep water in or beneath the thermocline (the hypolimnion) and the warmwater fishes are restricted to the shallow upper warmwater layer (the epilimnion). The thermocline is a thin layer of rapid temperature decline demarcated in the Temperate zones by a change in temperature of 1°C for each meter of increase in depth. Low temperatures bring covers of ice to lakes in the Temperate and Arctic zones in the winter, and the growth of fishes is slowed or stopped in this season. Just before the ice covers a lake in the Temperate Zone in the fall, and, again, just after the ice leaves in the spring, a condition of homothermy is reached from top to bottom. This condition enables deep thermal and wind-induced circulation of such a lake (and of parts of the oceans) distributing nutrients and flocculent bottom materials from the deeps to the very surface. Dissolved oxygen from the surface is mixed through the water column during this event. In super-cooled streams, ice may form on the bottom to destroy or scour out fish food organisms.

Temperature of water can be an important directive influence in the migrations and movements of fishes. As a result of correlating temperature with the ranges of fishes, zoogeographers have developed isothermal theories of fish distribution. Limnologists have long attempted to classify lakes on the basis of their thermal characteristics. The most widely accepted classification has three major categories: Polar lakes with the surface temperatures never over 4°C (39.2°F); Temperate lakes with the surface ranging both above and below 4°C; and Tropical lakes with the surface temperatures always over 4°C.

It may be concluded that not only are fish temperature sensitive (Chapter 11), but that those capable of movement generally seek what may be termed preferred temperatures. Within the range of tolerable temperatures fish continually move about as if to seek the temperature that is optimum for the vital activity of the moment.

Light

Light is another ecological factor of importance in the lives of fishes. Direct effects are through vision but there are many indirect ones as well. Coloration of the integument at any given time is a direct function of quality and quantity of light (Chapter 4). Light also triggers and directs migrations and movements (Chapter 6), has a timing role in reproduction (Chapter 10), and influences rate and pattern of growth. Light also determines the kinds and amounts of food available for fish and is of course the direct energy source for the first, photosynthetic link in the food chain of all fishes. The region where light intensity is sufficient for photosynthesis is referred to as the trophogenic or euphotic zone and extends downward to some 200 m in clear sea water.

Current

Currents are physical factors in the lives of both flowing- and standing-water fishes. At the surface, water movements tend to equilibrate air and water temperatures. Through circulation, they also tend to homogenate temperature and chemical factors in the water. Forced mixing by wind moving across the surface moves the water molecules and causes turbulence or waves. When turbulence at the surface comes into contact the water layer beneath, a current in the opposite direction is set up the lower layer. Such currents produce turbulent internal eddies which lead to vertical interchange of water particles (intermixing). This interchange is known as eddy diffusion.

The cutting action of currents along stream banks and shores continually alters habitat locally, and dislodged materials are transported for deposition elsewhere to effect changes there. Turbidity resulting from erosion may effect adaptations (as in the eyes) of fishes but it may also be lethal or it may greatly reduce chances for survival by sedimentation of feeding or spawning grounds. It also accelerates the rate with which water absorbs heat from sunlight.

Currents within stream systems typically change from fast in the headwaters to sluggish at base level with accompanying change in fish species. Generally, the stronger the current the more depauperate the fish fauna. The development of streamlining and of holdfast organs among fishes in an obvious response to current. However, even as gradient in streams is an ecological factor for fishes, so are the shape, depth, configuration, and composition of the bottom of standing waters.

Dissolved Oxygen

Dissolved oxygen is required for respiration, the release of energy from food. In unpolluted streams and in the surface of lakes, ponds, and the oceans the water is normally saturated with oxygen for its given temperature. It may, however, fall beneath the minimum few-parts-per-million requirement of fishes at the sources of springs, and seasonally in stagnant bottom waters of standing bodies. Oxygen may also be deficient over highly organic bottom deposits or in waters polluted by organic wastes such as domestic sewage. When large algal blooms die due to a few cloudy days, extensive fish kills (summer kill) may result because of decomposition using up oxygen. Thick snow cover on ice shuts out sunlight and as a result phytoplankton does not produce oxygen, respiration and decomposition use it up and winter kill results when the demand exceeds the supply.

Food

Food is one of several important biological factors in the environment of fishes. Its abundance and varieties are determinants of both the species composition and magnitudes of fish populations. Key foods are the free-floating plankton, the benthos composed of organisms which move over or burrow in the bottom, and the periphyton and perizoon, the "Aufwuchs", made up of the organisms that grow on the free surfaces of submerged objects.

Key chemical factors or nutrients in the first link of the food chains of fishes include carbon dioxide for photosynthesis and such common elements as hydrogen, nitrogen, phosphorus, potassium, and calcium, plus trace elements (oligoelements), including manganese, boron, sulphur, magnesium, zinc, and molybdenum.

The origins of the elements and gases in water are many; the principal processes by which they become available in water are three—slow diffusion, rapid mixing, and internal transformation. Oxygen, carbon dioxide, and nitrogen, but especially oxygen, can diffuse relatively slowly into water from the atmosphere. These same gases plus ammonia may be mixed rapidly with the hydrosphere at times of rain, snow, and wind, or when thermal and gradient currents occur. The internal transformation of nutrients by living organisms yields carbon dioxide to water from respiration, oxygen from photosynthesis, and hydrogen sulfide and organic salts from bodily wastes or from decay of dead organisms.

Social Factors

Among the most subtle in their action of all factors of the environment are the interactions among individuals of the same or different species. Illustrative of these interactions are competition and predation. Of these two, predation is most easily visualized since it involves a prey organism eaten by the predator. In aquatic ecosystems affecting fish, either, neither, or both of the organisms in a predatory relationship may be a fish. Because of replacement through growth of remaining individuals (a process termed recruitment), predation on fishes generally does not eliminate the prey. Rather it may function to increase average size of the residual individuals through reduction of intraspecific competition. Predation also favors the survival of the most fit, both predator and prey. Runts, weaklings, albinos, and diseased individuals are ordinarily the first to disappear.

In a model concept of competition among fishes, as among other animals, the two competitors would be in a perpetual draw. In nature, however, one of the competitors may gain the upper hand and finally eliminate the other; one may cause another to emigrate or disappear, or the forces of evolution may lead to adaptive change. Competition may be among individuals of the same species (intraspecific) or between two or more species (interspecific). Either kind can be a density-dependent factor in determining population size and structure. The most common competitive reactions among fishes are for spawning sites, food, space, and shelter. The crowding of spawners, such as in the Pacific salmons (*Oncorhynchus*), often leads to the destruction of earlier nests by late spawners and may have adverse psychological effects on the spawners as well. Furthermore, competition for nest sites is an underlying factor in the common occurrence of hybrids among sunfishes (*Lepomis*). The resulting hybrids are predominantly male and are apparently sterile or of lower fertility than the parent species, depending on the cross. In addition, since they also have an increased rate of growth, it may be reasoned that they have a shorter life expectancy than the parent species.

The introduction of fishes by planting them in waters where they are not native can lead to competitive displacement of the native species. For example, in New Zealand the stocking of exogenous, predatory salmonid game fish species that have a competitive ability superior to that of the endemic species is eliminating the latter.

In addition to competition and predation, fishes exhibit a variety of other symbiotic relationships—that is cases in which at least one of the species is affected by the presence of the other. Cleaning symbiosis is one of the most remarkable and widespread of these interactions. Cleaner-fishes and cleaner-shrimps feed on ectoparasites, necrotic tissue, scales and perhaps mucus found on the body surfaces, mouths and gill chambers of their host fishes. Dozens of marine and freshwater species of fishes have been found to be cleaners in both tropical and temperate areas. The best known are from the wrasses (Labridae), butterfly- and angelfishes (Chaetodontidae) and gobies (Gobiidae). Many of these cleaners are brightly colored and appear to advertise their presence to attract prospective hosts to their cleaning stations. Their hosts include nearly all of the reef fishes larger than a few centimeters in size including large predators such as groupers (Serranidae) and barracuda (Sphyraenidae), and even some of the pelagic fishes such as certain tunas (Scombridae). Even sharks receive cleaning services from their hitchhiking suckerfish (*Remora*). In some areas cleaning symbiosis appears to be a true mutualism in that both parties benefit—the cleaner has its food "delivered" to its cleaning station by hosts soliciting cleaning while the hosts are rid of potentially damaging ectoparasites. But in other areas ectoparasites appear to be relatively rare and the cleaners feed heavily on the scales and mucus of their hosts becoming a commensal or even parasitic animal in their own right. The apparent paradox as to why hosts continue to solicit cleaning from these commensal or parasitic cleaners has been resolved by the discovery that cleaners continually reward their hosts. This reward or positive reinforcement is in the form of a "tickle" or gentle tactile stimulation delivered by the fins and snout of specialized cleaner fish and the legs and antennae of cleaner shrimp. In this way the cleaning relationship continues under nearly all conditions and is available to fulfill its mutualistic role of ectoparasite removal if and when the fish in the area are subjected to increased rates of infection.

Commensalism occurs among several coelenterates and fishes. The fishes seem either to be immune to the stinging cells (nematocysts) of the coelenterates, or may even inhibit their discharge. The Portuguese-man-of-war (*Physalia*) thus harmlessly harbors the so-called man-of-war fish (*Nomeus gronovi*). Some jellyfish medusae, such as *Crambessa*, give shelter to small fishes, including schools of jacks (*Caranx*). Some damselfishes (anemonefish, *Amphiprion*; *Premnas*) are often associated with large sea anemones (*Stoichactis*). These fishes may even carry or attract food to the anemone and also drive away other fishes. Some cardinalfishes (Apogonidae) stay in the pallial cavity of a large conch (*Strombus*), and some pearlfishes (*Carapus*) take refuge in the cloacal chamber of sea cucumbers (Holothuroidea) or, occasionally, in the stomach of large starfishes (Asteroidea). One pearlfish species takes up its abode in the mantle cavity of a pearl oyster, from whence the common name of the fish genus. The pilotfish (*Naucrates*) schools in front of and beneath large sharks and mantas (Mobulidae) and theoretically in return for table scraps to eat (as well as the chance to clean ectoparasites off the host fish) may warn the host of enemies or alert it to prey. The remoras (Echeneidae), which hitch-hike on sea turtles, whales, sharks, and other fishes, also take feeding leftovers of the host. For swordfishes (*Xiphias*), sailfishes (*Istiophorus*), and the molas (*Mola*), there is reciprocity in that the remoras clean the large fish of parasites and diseased tissue, not only externally but in the gill and mouth cavities. A blind goby (*Typhlogobius*), dwelling in the burrow of a crab (*Callianasa*), has become quite dependent on the crab, both for food and for shelter in an instance of extreme adaptation to commensalism.

In commensal relationships, a fish benefits from the relation but its host is essentially unaffected. Protocooperation in fishes apparently pertains in the schooling of predacious sharks for herding prey to the kill as well as in schooling behavior of small fishes for protection against predators. Protocooperation is defined as the situation in which both species or numerous individuals of one species benefit from cooperation which, however, is not indispensable.

Finally we come to parasitism, a condition in which one species or individual (generally the smallest) inhabits the other and depends on it for subsistence. Many organisms are parasitic on fishes, including fungi, bacteria, worms, crustaceans, larval molluscs, and fish themselves. Only overinfestation by parasites seems to be grossly harmful to the host, so perfectly have the parasitic systems evolved. One fish that can perhaps be regarded as a true parasite is the small candiru, a South American catfish of the family Trichomycteridae, which lives naturally on the gills of larger fishes and sucks blood from them. This fish is a hazard to man and other animals because of its ability to penetrate and become lodged in the urethra (as in human males) or the bladder (as in human females). Other body openings are also entered.

Any of the foregoing social factors in the lives of fishes can vary and interact according to life-history stage, season, and location. As a simple example, consider fry of the brown trout (*Salmo trutta*), which are early prey to minnows (Cyprinidae) with which they dwell. As the trout fry become juveniles, they also become competitors of their former predators. Later, to complete the turn-about, the adult trout are predators on the minnows.

Population Density

The dynamics of specific environments are reflected in the continual change within the fish populations that inhabit them. A basic parameter of population is density as a function of time, volume, or space. Thus, density may be the number of fish in a lake or numbers or weight per unit of its area or of its volume. Changes in population density in fishes are often described in terms of relative abundance because of the difficulty of accurately measuring population size in any but small water bodies. Many aspects of fish behavior and physiology have been shown to be density dependent.

Population density at any instant is the function of natality versus mortality. In a successful species, potential natality normally exceeds potential mortality which affords the species the chance to multiply if circumstances are favorable. If the actual natality is superior to real mortality, the population increases; however, if the two are equal, the population is stationary. Where real mortality exceeds real natality, the population is regressing. The dynamic history of a species population may be summarized as follows:

a.   colonization and expansion
b.   attainment of stability, with some fluctuations in density but with real natality equalling real mortality

c. regression, in which mortality exceeds natality, typically with greatest losses among fishes occurring in the very young

d. giving way to other species

e. disappearance; replacement by other species populations

Population Structure

The structure of the population of a fish species is an indicator of its evolutionary stage as expanding, stable, or regressing. Structure can be learned by determining the age composition of the population. In fishes, age assessment of individuals is facilitated in Temperate Zone waters by the year-marks or annuli that have been identified on scales, bones, ear-stones (otoliths), and in cross-sections of fin rays. From application of such means of age determination it has been learned that young preponderate in expanding populations whereas the aged dominate regressing ones. Fluctuations in abundance of individuals in a stable population are brought about by year-class dominance. In a model of year-class dominance, the same number of eggs may be produced year after year, but only in certain years will there really be substantial survival. As a result, fisheries may often depend for their success on a dominant year-class, decline as it passes, and rise again when a highly successful new one comes along.

Fish species have different characteristics of grouping. Some, like the roving predators may be relatively solitary and randomly grouped. In such nongregarious species, growth and survival are greater as the population density is lesser. For gregarious fishes, such as herrings (*Clupea*) and mullets (*Mugil*), distribution in a population is typically random by schools, small or large. Density of the groups may favor survival and be needed for reproduction and to provide the variants that may favor evolution paced with changes in the environment.

Some populations of fishes are affected importantly by immigration or emigration of individuals. Moderate losses or gains in numbers from migration in or out will be compensated by corresponding changes in mortality rate. For example, massive immigration may reduce growth of all of the individuals through increased competition. However, this negative effect may be offset by increased mortality. Massive emigration may increase growth and survival, but it may also cause a species population to lose its niche. Although the numbers may change in either type of movement, the fixed capacity of the habitat tends to maintain the same biomass.

Communities of fishes are characterized by species dominance. In a stream, the headwater and the baselevel communities have different dominants. Typically, the transition zone, or ecotone, between the two would have representatives of both the zones above and below and might be expected to contain more species than are contained in the zones it separates. Similar streams elsewhere may have the same or different species combinations. However, whether or not the species are the same, the ecological niches of the occupants will be essentially the same; consequently, these occupants will exhibit close similarities in their food-chain relationships.

Succession

In any given part of the earth, the waters and their inhabitants tend to evolve in predictable directions. In so doing there is a succession of environmental conditions and of the dependent fauna and flora. For example, the evolutionary process of glacial lakes is most frequently an orderly one (Fig. 14.1), termed eutrophication, in which the original oligotrophic, nutrient-poor situation becomes more and more nutrient-rich, ultimately filling and eliminating the lake entirely. While the environment is changing, the pioneering coldwater fishes are gradually replaced by warmwater-tolerant fishes, but finally there are no fishes at all. In the evolution of streams, the watercourse cuts ever downward toward baselevel (Fig. 14.2). As the gradient diminishes, pioneering headwater fishes disappear and are replaced successively by others better adapted to the changing environment.

Adaptive evolution of fishes, following evolution of habitats, has left unoccupied few niches in the aquatic ecosystem. In old waters of long evolutionary history this is strikingly true, but even in the relatively new waters, such

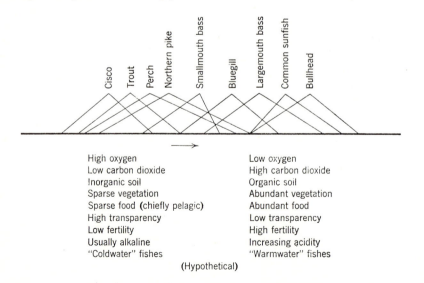

High oxygen
Low carbon dioxide
Inorganic soil
Sparse vegetation
Sparse food (chiefly pelagic)
High transparency
Low fertility
Usually alkaline
"Coldwater" fishes

Low oxygen
High carbon dioxide
Organic soil
Abundant vegetation
Abundant food
Low transparency
High fertility
Increasing acidity
"Warmwater" fishes

(Hypothetical)

Fig. 14.1 Hypothetical evolution of a lake of glacial origin in central North America. (Based on Eschmeyer, 1936).

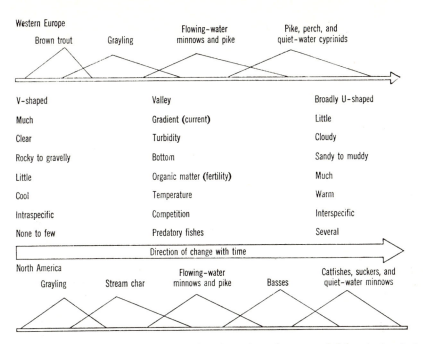

Fig. 14.2 Hypothetical succession of ecological conditions and fishes in trout streams of North America and of Western Europe.

as the rift lakes of Africa, virtually explosive evolution rapidly fills the available niches.

MARINE ECOSYSTEM

To the casual observer, especially to the student whose major experience is in inland waters, the sea is often thought of as a quiet, stable, and uniform environment. However, in actuality the sea is a moving and changing ecosystem with a large variety of biotopes. A biotope may be envisioned as an area in which conditions of the habitat and the organisms present are quite uniform. Oceanic biotopes differ according to regional geology, substrate, depth, currents, and nutrients. They also differ from place to place in salinity, dissolved gases, light, temperature, sound, and, of course, habitation by fishes and other living things. Because of differences among marine biotopes, oceanographers have long felt that a broad, fundamental system of classification could be developed to facilitate the organization and understanding of the growing mass of detailed information. This feeling has led to the proposal of many systems of dividing the marine ecosystem; that of Hedgpeth (1957) serves our needs well. Such a classification is concerned both with the water (the pelagic division of the seas) and with the bottom (the benthic division; Fig. 14.3).

All of the foregoing components of the pelagic and benthic environments may contain one or more biotopes, as previously suggested. In them, uniformity of conditions for life among fishes decreases with nearness to land and to the surface of the water. Consequently, biotopes are most numerous in inshore waters.

Pelagic Environment

The pelagic environment has two major divisions: (1) the neritic zone, water over the continental shelf, extending offshore to a depth of about 200 meters; (2) the oceanic zone, all waters beyond the edge of the continental shelf, thus all those more than about 200 meters of depth. In this environment shelter is generally wanting and pelagic and other fishes, from small to large, frequently gather beneath flotsam, encompassing most anything that drifts such as floating seaweed, parts of trees, pumice, large jellyfish, and dead turtles and whales. Within the oceanic zone, the following vertical divisions are recognized although there are obviously no sharp lines of demarcation among them and their vertical extent differs in the different oceans:

a.   epipelagic, surface waters away from the continental shelf but also to about 200 meters in depth

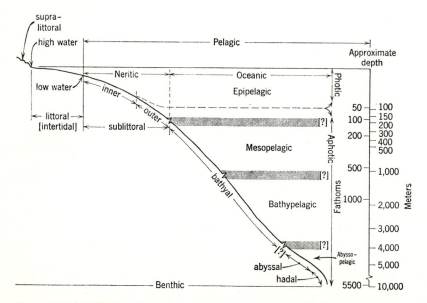

Fig. 14.3   Classification of marine environments. (Source: Hedgepeth, 1957).

b. mesopelagic, approximately 200 to 1000 meters
c. bathypelagic, approximately 1000 to 4000 meters
d. abyssopelagic, roughly below 4000 meters

Neritic Zone. The neritic zone is well lighted but has seasonal variations in light as well as in temperature, salinity, dissolved oxygen, nutrients, wave action, currents, and its biota. Sometimes the abundance of certain components of the neritic biota can change drastically and cause cataclysm for the fish population. For example, blooms of the dinoflagellate alga, *Gymnodinium breve* (called the "red tide") cause mass mortalities on the west coast of Florida. The fish deaths result from the release of a highly toxic metabolite by the algae. This zone also has in it a very considerable drift of detritus. It is the richest in varieties of biotopes and thus has fish of greatest abundance and variety of all waters. Representative habitats include waters of the shores, drifting and stationary kelp beds, and estuaries, as well as those over coral reefs, elevated submarine plateaus or banks, and varyingly narrow or broad expanses of the continental shelf. The most characteristic fishes are the herrings (Clupeidae). Among the many other fishes of the neritic zone are the barracudas (Sphyraenidae), many sharks, and some mackerels (Scombridae), the bluefish (*Pomatomus*), needlefishes (Belonidae), tarpon (*Megalops*), eels (*Anguilla*), tunas (*Thunnus*), marlin (*Makaira*), sailfish (*Istiophorus*), butterfishes (Stromateidae), kelp bass (*Paralabrax*), groupers (*Epinephelus*), snappers (Lutjanidae), grunts (Pomadasyidae), porgies (Sparidae), and sea trout (*Cynoscion*). Based on the analysis of voluminous trawl data, the demersal (bottom) fishes of the continental shelves are organized into communities with depth preferences.

Oceanic Zone. The oceanic zone is less productive than the neritic one, but a wide range of living conditions prevails. In it there is little detrital matter and seasonal variations of temperature, water chemistry, and other environmental factors are not as great as in the neritic zone.

Epipelagic Division. The epipelagic division of the oceanic zone is really a relatively thin, offshore extension of the neritic zone, but it is bottomed by water, not by solid substrate. This division is well lighted at the surface, dimming toward its downward limit of about 200 meters. All wave lengths of visible light enter the water at the surface but blues and violets hold at the lower ranges of greatest depth. Seasonal variations are shown in temperature, light, salinity, oxygen, nutrients, and plant and animal populations. Of these, light and temperature seem most important in determining animal distribution. For example, no mouthfish (*Stomias*) has been taken above 300 meters in daylight trawls but most species migrate to the upper 100 or 200 meters at night. Fish inhabitants include oceanward utilizers from the neritic zone, as well as some mackerels, bonitos, albacores, and tunas (Scombridae), some sharks, dolphins (*Coryphaena*), flyingfishes (Exocoetidae), mantas (Mobulidae), eels (*Anguilla*), marlin (*Makaira*), sailfish (*Istiophorus*), bluefish (*Pomatomus*), molas (*Mola*), lanternfishes (Myctophidae), the mouthfish relative *Cyclothone*, and the mouthfishes (*Stomias*).

Mesopelagic Division.   Occupants of the mesopelagic division of the oceanic zone depend for food on a "rain" of plankton, corpses, and droppings from the overlying epipelagic division and on one another in predatory relationships. There is little seasonal variation of temperature; the water is always cold, mostly around 10°C. The pressure is high and the light extremely dim and in the blue and violet range. This division contains the uppermost aphotic waters of the oceans and is inhabited importantly by dark-adapted, or scotophilic, animals. Many of the fishes are black or red and migrate to feed in the epipelagic division, sometimes at or near the surface at night. The larval stages of such invaders are typically also passed in epipelagic waters. Fish examples include the lanternfishes (Myctophidae), deepsea eels (*Synaphobranchus*), the mouthfish relatives (stomiatoids: *Stomias, Cyclothone, Argyropelecus,* and *Chauliodus*), the stalkeyed fish (*Idiacanthus*), and the deepsea swallower (*Chiasmodus*).

Bathypelagic Division.   In the bathypelagic division of the oceanic zone the key foods also gravitate from the waters above. There are essentially no seasonal variations in physical factors of the environment; the water is very cold (between 2° and 4°C at 2000 meters), pressures are very great, and darkness prevails except for bioluminescence arising from the light organs of many of the inhabitants. Fishes are greatly reduced in numbers and kinds below those of the upper waters but include most ceratioid deepsea anglers, dories (Zeidae), some scorpionfishes (Scorpaenidae), the deepsea swallower (*Chiasmodus*), the mouthfish (stomiatoid) relatives (*Chauliodus* and *Malacosteus*), gulpers (*Eurypharynx*), swallowers (*Saccopharynx*), and deepwater eels such as *Cyema*.

Abyssopelagic Division.   The great deeps of the oceanic zone, the abyss, are extensive and occupy more than half of the earth. In this so-called abyssopelagic division, high pressure (ranging from 200 to more than 1000 atmospheres), coldness (less than 4°C), and darkness—except for rare bioluminescence—prevail along with no seasonal changes of physical conditions. Animals of the abyss are nearly blind or have no eyes; light is of little ecological importance. Primary food needs are met by the fallout from overlying waters or from food chains connected upward. Conditions and fishes at the lower limit are poorly known, and essentially nothing is known of depths greater than 6000 meters. The few fishes present are represented by such kinds as the mouthfish relative, *Chauliodus*, gulpers (*Eurypharynx*), a deepwater eel (*Cyema*), and deepsea anglers, *Melanocetus* and *Borophryne*. From a depth of 3590 meters comes *Gallatheathauma*, an angler of tropical West Africa, with a curious light organ in its mouth—a possible lure to certain prey.

Benthic Environment

In the previous subsections, we have examined fishes of the open waters of the neritic and oceanic zones. On the bottom beneath these zones, there are characteristic bottom-dwellers that compose the benthos. The benthic environment is divided as follows:

a. supralittoral, from hightide mark to landward edge of spray zone
b. littoral, between high and low tidal extremes—that is, intertidal
c. sublittoral, between lowtide mark and edge of continental shelf (at about 200 meters of water depth); the inner sublittoral is subdivided from the outer (offshore) at about 50 meters
d. bathyal, the bottom approximately from water depth of 200 meters to 4000 meters
e. abyssal, about 4000 to 6000 meters
f. hadal, deeper, approximately, than 6000 meters

Supralittoral Zone.   The supralittoral or spray zone of the benthic environment is typically composed of moist surfaces and small pools in the substrate. There are few fishes here because living conditions, especially water supply and temperature, may be extremely variable almost from one instant to the next, let alone among seasons. Indeed it is a most difficult zone for animals and plants to occupy and requires considerable adaptation. The few fishes that have been able to establish niches in this labile environment include some gobies (Gobiidae), eels (Anguilliformes), and clingfishes (Gobiesocidae).

Littoral Zone.   In the intertidal areas of the littoral zone of the benthic environment there are also great hourly, daily, monthly, and seasonal variations in the conditions for life, especially in moisture. In spite of this the productivity is high with the fishes moving in and out with the tides. Most of the fishes are thus also to be found in the sublittoral benthic division. Representative fishes are stingrays (Dasyatidae), several flounders of the families Bothidae and Pleuronectidae, soles (Soleidae), the bonefish (*Albula*), eels (*Anguilla*), morays (Muraenidae), clingfishes (Gobiesocidae), sculpins (Cottidae), searobins (Triglidae), snailfishes and lumpfishes (Cyclopteridae), midshipmen (*Porichthys*), blennies (Blenniidae), gobies (Gobiidae), pipefishes and seahorses (Syngnathidae), and cusk-eels (Ophidiidae).

Inner Sublittoral Zone.   In the inner sublittoral, the productivity remains high, seasonal variations are near maximal, and the lighting is good with yellows and reds near shore and blues offshore. Fish examples include surfperches (Embiotocidae), most skates (Rajidae) and stingrays (Dasyatidae), most flounders and soles (Pleuronectiformes), searobins (Triglidae), dogfish sharks (Squalidae), bonefish (Albulidae), eels (Anguillidae), morays (Muraenidae), seahorses and pipefishes (Syngnathidae), croakers, kingfishes, and drums (Sciaenidae), hakes and pollocks of the family Gadidae, rockfishes and relatives (Scorpaenidae), wrasses (Labridae), butterflyfishes and angelfishes (Chaetodontidae), parrotfishes (Scaridae), filefishes and triggerfishes (Balistidae), trunkfishes (Ostraciidae), puffers (Tetraodontidae), porcupinefishes (Diodontidae), midshipmen (*Porichthys*), and kelpfishes (*Gibbonsia*).

Coral reefs bridge the sublittoral and littoral zones in warm seas. The fishes on them are mostly very colorful. About these reefs are the majority of the some two hundred species of butterflyfishes (Chaetodontidae). Moorish idols (Zanclidae), surgeonfishes (*Acanthurus*), filefishes and triggerfishes (Balistidae), and gorgeous parrotfishes such as *Scarus* are also present. Many species of wrasses (Labridae) are there too, along with damselfishes (Pomacentridae), blennies (Blenniidae), and members of the sea bass family (Serranidae).

Outer Sublittoral Zone. The outer sublittoral remains quite productive with some seasonal variation of physical conditions of life. The light of the bottom here runs to blue and violet. Much of the fauna is bathyal-like; there are also stragglers from the inner sublittoral. Fishes include haddock (*Melanogrammus*), cod (*Gadus*), and hakes (*Merluccius* and *Urophycis*) of the family Gadidae, halibuts (*Hippoglossus*), chimaeras (*Chimaera*), hagfishes (Myxinidae), eels (*Anguilla*), and pollock (*Pollachius*). Occasionally, bottom-dwelling fishes from this and other shoreward zones may be displaced, even cast ashore by giant waves.

Bathyal and Abyssal Zones. In the bathyal zone of the benthic environment, as in all subsequent ones, there is essentially no variation in the physical factors of the environment. The bottom is cold and dark, except for rare bioluminescence among occupants and dim light to about 1000 meters. The primary foods fall from above and are more dense at the bottom than in the water immediately above. Bathyal fishes include halibuts (*Hippoglossus*), chimaeras (*Chimaera*), cods (*Gadus*), and hagfishes (Myxinidae). Abyssal fishes are represented by the eel (*Synaphobranchus*), rat-tails or grenadiers (Melanonidae), brotulas (Brotulidae), the lanternfish relatives, *Ipnops* and *Bathypterois*, and some oceanic mesopelagic and abyssopelagic fishes that probably also come to the bottom here to feed.

Hadal Zone. Few fishes are present in the hadal region. Questionably included are rat-tails (Macrouridae), the deepwater eels, *Synaphobranchus*, the brotulid *Bassogigas,* and snailfishes including *Pseudoliparis.* Only sketchy information exists on hadal communities of the seven great trenches of the seas.

ESTUARINE ECOSYSTEM

An estuary is typically a water body in which fresh water (in which salt cannot be tasted) from a stream mixes with salt water from an ocean (which is chokingly salty at its average of 35 parts of salts per thousand parts of water, i.e., 35‰). Since it is really a transition zone between two environments, it is properly termed an ecotone. Topographically, most estuaries are drowned river mouths, and they range in size from the small one that receives a single stream to the very large one, such as Chesapeake Bay of the American Atlantic Coast, that receives major streams and many small tributaries as well. Known to all is the striking character of estuaries on rocky shores where they are termed fjords, as in Alaska and Norway. Estuaries are characterized by extremes of fluctuation in salinity, tidal and stream-current turbulence, and turbidity and siltation; seasonal fluctuations in these are among the greatest of all environments. Estuaries bordered by cities or industries may also have extremes of pollution. The primary basis for evolution of estuaries is filling. Delta formation is a stage in this process in many of these bodies of water on low coastlines. The effect of filling is seen when the tide is out in the tidal or mud flats that have such pungent, never-to-be-forgotten odors. In the tropics deltaic land building is often accelerated by dense growths of the mangrove trees and shrubs (especially *Rhizophora*). Estuarine shores are often sand, silt, and mud. They may be bordered by switchgrass, one of the panic grasses (*Panicum*), which is densest around the extreme hightide mark. Between the normal hightide mark and the extreme hightide mark, the bottom may also be well vegetated with such plants as rushes (*Juncus*) and marsh grasses (*Spartina*). Between the normals for high and low tide, vegetation may fall off rapidly from stands of marsh grass to disclose the highly organic peaty and mucky bottom. A key dynamic feature of these bodies of water is the rate of flushing or water change. In spite of flushing and other limiting factors estuaries are high in biological productivity.

River-mouth estuaries characteristically exhibit layering, with a freshwater lens overlying the intruding salt water beneath. Both layers may have distinctive fish faunas. Estuaries are travel routes for diadromous fishes and are also places of abode for populations of many coastal species, both marine and freshwater. For others, they may be nursery or feeding grounds. Fish inhabitants or visitors show surprising tolerances for the salinity changes and other environmental rigors of estuarine life. Common estuarine inhabitants include many herrings and relatives (clupeoids) and many killifishes and their relatives (Cyprinodontiformes).

FRESHWATER ECOSYSTEM

Although the marine environment is many times more extensive than the freshwater one, by far the most is known about freshwater fishes and about details of their natural history, ecology, and distribution. Life and conditions in fresh water have been most readily accessible for study. Fresh water occupied by fishes occurs both on the surface of the earth and in the subterranean waters of caves and underground stream channels. Subterranean fishes have evolved from those of the surface waters and are dependent on the life of surface waters, either directly or indirectly, for their food. Consequently, our considerations here will center on surface waters.

The fresh surface waters of the earth are broadly separable into two groups of environments: standing, or lentic, and flowing, or lotic. The lentic habitats include those of natural lakes and ponds along with the many impoundments constructed by man since time immemorial. The lotic environments are those of streams. To these may be added the special conditions found in (a) springs, which may be sources of either lakes or streams, and (b) estuaries and the heads of large impoundments where the lotic conditions of streams grade into the lentic.

Since artificial impoundments, including such areas as the subsealevel polders of the Netherlands, are strongly lacustrine in their nature, no special consideration is given them here. The construction of many of them has, however, significantly altered regional ecology and distribution of fishes. When a dam is built across a river, it completely changes the chemistry and temperature of the water. As a result the local fauna changes and different species become dominant. Dams have become barriers to migratory fishes as is strikingly seen in the Columbia River system of the American West. The frequent location of dams in actively eroding stream courses has furthermore subjected impoundments to relatively rapid rates of filling and ecological change, more so than in most natural lakes; there are downstream effects as well.

Lakes

Although to the casual observer, lakes may seem to differ primarily in size and depth, scientific study has disclosed countless other points of difference. Many of these differences are critical determinants of the presence, absence, or abundance of fishes. For example, the fish-carrying capacity of a lake varies with the character of its basin, fertility of water, the age of the water body, land use in the watershed, and changes in the kinds of fish or in the relative abundance of certain kinds and sizes of fish. Carrying capacity has been related to a morpho-edaphic index and specific conductance of the water.

Origins of Lakes.   Among the distinguishing factors of lakes are their natural origins. Many have come from action by glaciers, especially the great continental glaciers that gouged the earth's crust and left water-damming moraines. In glaciated regions many small circular kettle lakes were formed by ice blocks that were buried, then thawed, leaving a depression that filled with water. Others have resulted from landslides that barred valleys and from crustal movements of the earth that made water-catching depressions. Solution of underlying rocks has formed some lake basins, particularly in limestone country. A few lakes exist in craters of extinct volcanoes. Seashore ones have arisen through formation of lagoons behind offshore bars of sand or coral; their water is sweetened from rain when the bars exclude more and more of the sea. Not to be overlooked is lake origin through river action; oxbow lakes are cut-off stream meanderings in land of low relief. Whatever the origin of a lake an inexorable evolution of aging follows. Regardless of how a lake originated, its fishes are the product of initial drainage connections and survival or adaptation to meet changing conditions. Through fish stocking and other management procedures and through conflicting uses of both the land and water, man has greatly altered the natural direction of evolution of lake (and stream) fish populations.

Classification.   Lakes have been variously classified according to evolutionary state and to direction of evolution. Although they are not all-inclusive, four basic types of lakes are recognized generally in the literature of limnology—oligotrophic, eutrophic, mesotrophic, and dystrophic.

Oligotrophic lakes are most often deep, cold, and low in nutrients and in population by plants—especially seed plants. There are many different species but few individuals of organisms, thus a high diversity results. These species are distributed to great depth throughout the water column. The most characteristic fishes of such lakes in the North Temperate Zone are coldwater and deepwater salmons, trouts, and whitefishes (Salmonidae). Lakes of this type undergo ecological succession into the eutrophic type; this process of aging is termed eutrophication. Eutrophic lakes succeed into ponds, swamps, and marshes. Deepwater salmonids are absent or disappearing, as the lakes become shallower, warmer, and progressively more highly organic and occupied by larger aquatic plants. In the North Temperate Zone, common fishes of eutrophic lakes include many different kinds of minnows (Cyprinidae) and the walleyes (*Stizostedion*), pikes (*Esox*), perch

(*Perca*), and some catfishes (*Ictalurus; Silurus*). A mesotrophic lake is one in the middle stage between eutrophic and dystrophic when the amount of organic material cycling in the lake is increasing. Dystrophic lakes too are shallow but are headed toward becoming peat bogs. Their highly organic nature often renders them at least seasonally devoid of oxygen and thus of fish. Representative of the fishes that may occur are the mudminnows (Umbridae).

To whichever of the four foregoing or other categories a group of lakes may belong temporarily, each lake may still differ greatly in its area, depth, volume, bottom contour, shoreline regularity, and other features. All of these physical features have some relation both to the kinds and the abundance of fishes that may be present.

Zonation.   The environments of fish in lakes can be very diverse. The diversity is, however, conventionally grouped into three major zones: littoral, limnetic, and benthic. Approximating these three terms in the marine ecosystem are the neritic, oceanic (or pelagic), and benthic zones, as already described (Fig. 14.3).

Littoral Zone.   In a lake, the littoral zone extends from the shoreline lakeward to the limit of rooted aquatic plants that ordinarily dominate the zone. This zone is singularly important as the spawning and feeding region of many typical lake fishes. Most often in Temperate waters along the shore are emergent plants including such kinds as the bulrush (*Scirpus*) and cattail (*Typha*). Beyond the emergents, there is often a region of floating-leaf plants which include waterlilies (*Nymphaea, Nuphar*), watershield (*Brasenia*), and some of the pondweeds (*Potamogeton*). Beyond the floating-leaf plants, the submergents may blanket the bottom to the outer limit of the littoral zone. Examples of common American submergents are many pondweeds (*Potamogeton*), watermilfoil (*Myriophyllum*), hornwort (*Ceratophyllum*), and waterweed (*Anacharis*). On the windward sides of lakes rooted vegetation may be sparse or wanting, and the shores may be variably sandy, eroding, or bare rock depending on the geology of the region. The littoral zone is well-lighted but subject to maximum short-term fluctuations of temperature, turbulence from wind action, and turbidity—even when organic, the bottom materials tend to be coarsest near shore. Fishes requiring the warmest water in a lake are usually found in the littoral zone. Many are relatively shortened and compressed in body form, as, for example, the American sunfishes (*Lepomis*) and crappies (*Pomoxis*); this body form is an adaptation to high turning power and short-distance swimming (Chapter 6).

Limnetic Zone.   In a lake the limnetic zone is the lakeward openwater extension of the littoral zone. Upper limnetic waters are well-lighted and subject to relatively rapid change of physical conditions; the downward limit is the light-compensation level. Lower limnetic or profundal waters are comparatively stable in physical conditions or change slowly, with noteworthy exceptions at seasonal turnover.

The limnetic zone is the nursery ground for many open-water spawners including freshwater stocks of certain herrings (Clupeidae) and silversides (Atherinidae). The most abundant fishes of the limnetic zone are plankton feeders such as the emerald shiner (*Notropis atherinoides*) and freshwater stocks of the alewife (*Alosa pseudoharengus*) in the North American Great Lakes. Most often the fishes are silvery in coloration, somewhat darker above than below, and well-streamlined or elongate but slab-sided in body form. Deep waters that lie beneath the thermocline are poor in light and inhabited entirely by heterotrophic organisms. The organisms there depend upon the "rain" of organic material from the waters above for energy. Some zooplankters may inhabit this area but migrate toward the surface to feed. The fish species diversity is usually small.

Benthic Zone.   The benthic zone slopes from the shore of a lake, beneath the littoral waters, to the greatest depths of the profundal region. In so doing it may change in character of substrate from bedrock through particulate inorganic and organic material to a very flocculent organic ooze. The chemical and physical conditions affecting life in and on the bottom change in pace with those of the overlying waters. The bottom organisms (benthos) live in the region of highest nutrients of a lake but are exposed to the greatest hazard of elimination through the development of an oxygen deficiency resulting from decomposition of bottom materials. In northern latitudes, typical lake benthic fishes include sculpins (Cottidae), the burbot (*Lota*), some catfishes (Ictaluridae), and highly adapted bottom feeders such as suckers (Catostomidae) and sturgeons (*Acipenser*). Among many of such bottom dwellers the body form, especially the head, is depressed. The benthic habitat is also at least visited by many other lake fishes for feeding and for nest construction or spawning.

Streams

Natural streams are watercourses that develop from surface run-off or from groundwater sources. Their routes and gradients undergo continual change with time through the forces of solution, erosion, and deposition. Like lakes they have an evolutionary pattern (Fig. 14.2), with associated changes in fish fauna, that they follow from a newly formed, cutting, tumbling stream through to the old-age stage, characterized by sluggishness and meanders near baselevel. Humans have altered the natural pattern of drainage of land masses in many ways, notably by the construction of canals and water diversions. Through pollution and changing practices of land use they have also greatly changed the living conditions in streams, often rendering long stretches fishless.

Characteristics.   Streams or the various reaches of one stream differ from one another in many features. Among the parameters of physical difference are depth, length, width, area, volume of flow or discharge, regularity of shore, gradient, bottom type, and temperature. Streams with these differences may be permanent or intermittent. Some are interrupted, flowing alternately above the bottom materials or through them. Most fluctuate seasonally in discharge and especially downstream may flood widely over the lowlands. Indeed, among streams of all aquatic habitats, there are the most sudden and abrupt changes in living conditions.

Streams are generally shallower than lakes. They have currents moving predominantly in one direction and have less stability of bottom materials and more erosion. They may fluctuate rapidly in temperature but have more uniform temperatures from surface to bottom. They also tend to be saturated with oxygen from the surface to the bottom. Inorganic turbidity is greatest in streams, and light penetration, often minimal. Plankton is greatly reduced, with periphyton being the essential key to productivity. Variations in available nutrients according to substrate are greater in streams than in lakes, ostensibly because of failure of organic stores to accumulate in the bottom. Because of current, there is almost an unending flow-through of nutrients in streams.

Zonation.   Recognition of life zones in streams has been of two principal kinds. In one, the entire course of a stream is separated largely on the basis of gradient into upper, middle, and lower reaches. In the other, working within a reach, riffles, pools, flows, backwaters, and floodwaters are chief among the habitats recognized.

Reaches.   The upper reaches of a stream have the steepest gradient and most V-shaped valleys. An example would be a mountain stream, perhaps one originating in a glacier. The current is swift, slowing only occasionally to pools. The waters are the coolest of all reaches. These are the reaches inhabited by active fishes such as the stream trouts (*Salmo clarki* and *Salvelinus fontinalis*) which find refuge behind boulders or ledges of rock in the stream course. The upper waters grade slowly into the middle reaches of less gradient, slower current, nearly equal numbers of riffles and pools, with some siltation in quiet pools and occasional backwaters—all in U-shaped valleys. Seasonal flooding is largely confined to the stream channel. The fishes increase both in kinds and numbers over the headwaters; the fauna become more diverse, which leads to stability. Often various minnows (Cyprinidae) become conspicuous parts of the fauna. In its lower reaches the stream approaches the baselevel conditions of its mouth, and the valley becomes broadly U-shaped, usually with extensive flood plains. Riffles disappear, and pools elongate into extended, quietly moving flows. The shores may become marshy, and backwaters, sloughs, and bayous appear. Siltation tends to be heavy, and the stream seasonally leaves its banks at flood, often stranding fishes when flood subsides. The waters are warmest of all reaches. Sluggish-water fishes appear, with their greater environmental tolerances and generally less restricted feeding habits. Within their range, suckers (Catostomidae), the larger minnows (Cyprinidae) including the carp (*Cyprinus*), and catfishes (Siluroidei) may abound.

Depending on the continent, one of three families of freshwater fishes has at least some representatives in practically all riverine life zones. These are the minnows (Cyprinidae), characins (Characidae), and cichlids (Cichlidae). Australia is an exception in that it has none of these three.

Riffles and Pools.   A riffle in streams is most often relatively shallow and flows over gravel and rubble. Its surface is broken, in contrast to the smooth aspect of a pool or backwater. The bottoms of riffles are the spawning grounds of many streamfishes which bury their eggs in gravel; included are lampreys (Petromyzonidae), stream trouts and salmons of the family Salmonidae, and several minnows (Cyprinidae), suckers (Catostomidae), and darters (Etheostomatinae). The bottoms of riffles are also important grounds for the production of insect fishfood, especially for many caddisflies (Trichoptera), mayflies (Ephemeroptera), stoneflies (Plecoptera), and crayfishes (Decapoda), due to the physiological richness of the area.

Pools tend to be wider and deeper than the average of the stream course. In them, current is reduced, some little siltation may occur, and aquatic seed plants may form beds. Pools are important to streamfishes as post-hatching nursery areas and as feeding and nesting grounds. Pools with overhanging banks or other shelter seem to support more and larger fishes than ones where shelter is wanting. Fish food in pools includes drift from the riffles along with insects of the land and those of pools. Pool insects include burrowing mayfly nymphs (*Hexagenia*) and dragonfly and damselfly nymphs (Odonata). About the margins in aquatic plant beds, scuds (Amphipoda) may abound. In the quietest pools and backwaters, where silt rich in organic materials accumulates, important fishfood organisms include two-winged fly larvae, such as midges (Tendipedidae), and sludge worms (Tubificidae). Sludge worms are present in most organic bottom areas but are most abundant in locations of high organic pollution. At high levels of abundance they are used as index organisms to such pollution.

Hot Springs

Hot springs issue into both lakes and streams. Such thermal waters are widely distributed over the earth and are hot, clear, highly mineralized, and generally devoid of or very low in oxygen. Fishes generally are not known to be established in waters much warmer than 40°C, although they may survive short spells in water up to 45°C. Although reports of fishes exist for waters from 48° to 70°C, close examination has failed to support any of the claims. Numerous examples of hotwater fishes exist in the American Southwest. Three species, for example, are in Death Valley and live at temperatures up to 45°C. Most of these hotwater fishes are killifishes of the genus *Cyprinodon* and the few related genera. Elsewhere certain minnows (Cyprinidae) and cichlids (Cichlidae) are included.

ZOOGEOGRAPHY

Geographical distribution and ecology are closely related both to each other and to evolution. Some of the early critical evidence that convinced Darwin and Wallace of the fact of evolution was supplied by zoogeography. This interdisciplinary science, through contributions from the fossil record, also attempts to reconstruct and understand the evolutionary history of organisms in space and time and to provide an independent test for conclusions reached by earth scientists about the current exciting theory of plate tectonics (see below). If we attempt to scan the geological past, then our concerns lie with historical biogeography; if our view is contemporary, then geographical ecology becomes the major interest. Both of these approaches have advanced markedly during the past decade, the former because of the revolution in earth sciences, the latter because of noteworthy new ideas in biological theory (MacArthur, 1972; MacArthur and Wilson, 1967; Vuilleumier, 1974).

Conclusions derived from biogeographic studies are only as valid as the taxonomy of the organisms on which they are based and much penetrating work needs to be done on the zoogeography of fishes. We are still not sufficiently knowledgeable about phylogenetic relationships in some animal groups to be able to distinguish fact from opinion.

The science of zoogeography is concerned with the geographical distributions of animals over the world and with the history behind these distributions. Animals are living organisms that are constantly evolving and multiplying in some places, spreading into other places, and dying out in others—all of which contributes to the formation of new geographic patterns. Ranges of species and of groups are therefore pulsating and dynamic, not static as we are forced to visualize them on distribution maps. The aim of the zoogeographer is not only to describe existing distributional patterns but to explain how and why such patterns have been formed. Time as well as space must be reckoned with. In order to do this, zoogeography must draw upon ecology, evolution, and geology (including paleontology and paleoclimatology). Animal distribution can reveal new insights into how evolution takes place; it can assist the geologist in interpreting the past; and, as Darlington properly emphasizes, it is a fascinating study in its own right.

For convenience, zoogeography may be divided into three categories: world-wide distributions, which reveal broad patterns; regional distributions, dealing with particular groups in selected parts of the world; and local distributions, which are concerned with the geography and ecology of species. All of these are treated here but the broad picture is emphasized.

Continental Drift

Scarcely more than two decades ago, most geologists regarded the earth as rigid and believed that the continents formed where they are now positioned. An alternative theory of continental drift had been proposed early in this century, but the seeming lack of a mechanism for moving continents had assured its nearly universal rejection. As has often happened in the history of thought, the unorthodox view has turned out to be correct.

Geophysicists and geomorphologists studying the ocean floors have uncovered a mechanism in the current theory of lithosphere plate tectonics (also called the new global tectonics) and sea-floor spreading, which proposes that the earth's surface is divided into a small number of plates bordered by ridges and subduction zones. New ocean floor is formed at the ridges as older parts of the plates are pushed away. In order to balance this addition, the old plate parts are drawn into the earth's interior at the subduction zones. Plate tectonics is essentially ocean-based, that is, the bulk of intense seismic and thermal activity is associated with plate margins under the oceans and at continental margins. The basic concepts of plate tectonics and sea-floor spreading are securely founded on oceanic magnetic data, in seismological researches, and in the deep-sea drilling that has been (and is still being) pursued relentlessly by the research vessel *Glomar Challenger*.

"The nature of sea-floor spreading and continental drift in Pre-Jurassic time, more than 200 million years ago, has yet to be unraveled. The fossil record may have much to contribute to this problem now that we are learning to interpret the interplay between evolution and continental drift" (*Scientific American*, 1974: 159).

The continents rest passively on the plates and move with them, rather than plowing through solid ocean floor as previously thought. Continental drift, therefore, is but one consequence of plate tectonics. Other important consequences are the occurrence of earthquakes along plate boundaries (e.g., the famous San Andreas Fault of California) and the development of mountain chains where two plates bearing continents collide (as when the Indian subcontinent collided with Asia to form the Himalayas).

In attempting to answer questions whether there are general distributional patterns common to a diversity of organisms and, if so, what these patterns are and how they have changed through time, former students of historical zoogeography (e.g., Wallace, Matthew, Simpson, and Darlington) have assumed that continents and ocean basins have remained relatively stable at least throughout Cenozoic and Mesozoic (if not all of) geologic time.

According to current concepts of earth scientists, all of the continents were fused around 225 million years ago (Paleozoic Era) into one supercontinent, called Pangaea, that subsequently broke up (onset ca. 200 m y BP) into a northern landmass, Laurasia (Eurasia and North America) and a southern landmass, Gondwanaland (Antarctica, the southern continents, and the Indian subcontinent). Gondwanaland later split up as Africa and India broke away, then east Antarctica and Australia, followed by New Zealand, and then west Antarctica and South America (Fig. 14.4).

To determine the historical biogeography of a group such as the vertebrates, a set of working principles has evolved into a biogeographic theory. This theory specifies that a primary goal is to construct the most parsimonious hypothesis about the location of ancestral species, as based on cladistic or phylogenetic classification (Chapter 13). Reconstruction of distributional patterns thus is a deductive inference based on a prior phyletic analysis (Nelson, 1969).

Clues to Geographical History

In attempting to account for the origin and dispersal of organisms, certain criteria have been formulated that may provide valuable clues to their history:

a. The earliest fossil appearance is an important clue to place of origin of a group, particularly where the fossil record is comparatively well known (as it is in mammals). A poor fossil record, however, may be worse than none.

b. A group probably arose in the area wherein it shows the greatest taxonomic differentiation. Genera and species should reveal greater differences, more diversity, and a larger number of endemics in a region where the group has been for a long time than where it has just arrived. The minnows (Cyprinidae) of southeastern Asia provide a good example.

c. The place that contains the largest numbers of genera or species may be taken as the area of origin of the group. However, since most animals are

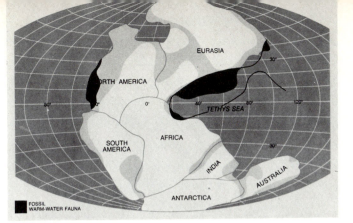

(a) Late Paleozoic

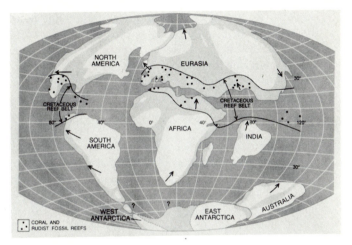

(b) Late Mesozoic

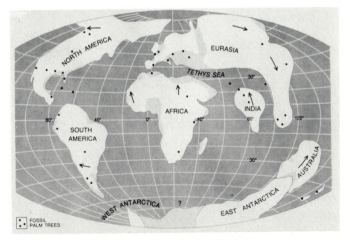

(c) Early Cenozoic

Fig. 14.4 Some effects of the movement of continental plates (plate tectonics or continental drift) and changing invasions of the continents by shallow seas on global geography and climate—and thus on the distribution of fishes. (a) Continental plates gathered more or less into a single land area near the end of the Paleozoic. (b) The seas separate the continental plates by the end of the Mesozoic, and the shallow seas have withdrawn from the continents. (c) Shallow seas reestablish in some areas in early Cenozoic but continental areas are larger than in Mesozoic; seasonal extremes of climate become accentuated but distribution of fossil palm trees show that the range of tropical and subtropical organisms remained quite extensive. (Adapted from Newell, 1972).

probably receding and becoming extinct about as commonly as new groups are increasing and spreading, the numbers clue will often lead to wrong conclusions about the history of a group. For example, the pikes and mudminnows (respectively, Esocidae and Umbridae) are best represented in North America (seven species) but first appeared in Europe (three species) and consequently probably originated in the Old World (Figs. 14.5 and 14.6).

d. Another clue is extent of area, in which the age of a group is supposed to be proportional to the area occupied and the place of origin is presumed to be at the center of the area—the "age and area" hypothesis of the botanist Willis. Cycles of spreading and receding as revealed by the vertebrate fossil record do not support this viewpoint. But the clue does have some value if used with other clues.

e. Continuity of area is an important clue. In the suckers (Catostomidae), the longnose sucker (*Catostomus catostomus*) occurs outside North America only in eastern Siberia, and this is consistent with the idea that it invaded that area recently. In contrast, the very different Chinese sucker genus *Myxocyprinus* is well isolated from the rest of the range of the family and this supports the view that it is a relict of an earlier Chinese sucker fauna which has since receded almost to extinction in southeast Asia (Fig. 14.7).

f. The distribution of related, competing, and associated families provides a very useful clue to family histories. The nearest relatives of the suckers are the minnows (Cyprinidae), from which they were probably derived in eastern Asia; the group subsequently migrated to North America to undergo major differentiation in a region where its closest competitors are comparatively few in number. The probable history of the suckers (Catostomidae) suggests that centers of evolution and dispersal are also likely to be centers of extinction.

GEOGRAPHY OF FRESHWATER FISHES

Freshwater fishes, numbering more than six thousand species, are of major significance in zoogeography even though a consistently clear distinction between groups as based on their salt tolerance cannot always be made. Generally, they can pass from one isolated stream system to another only through the slow physiographic change of the land itself. They typically possess a physiological inability to survive in full sea water. As a result, their dispersal over the world is slow and hence they are most likely to preserve old distribution patterns (Fig. 14.8).

Although the strictly or primary freshwater fishes are singled out as of

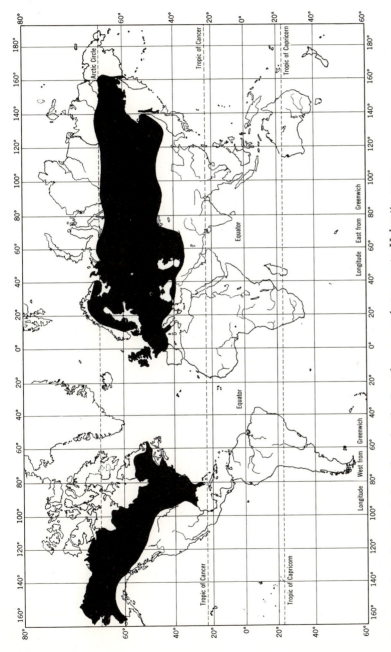

Fig. 14.5 Distribution of the pikes, family Esocidae, a north-temperate Holarctic group.

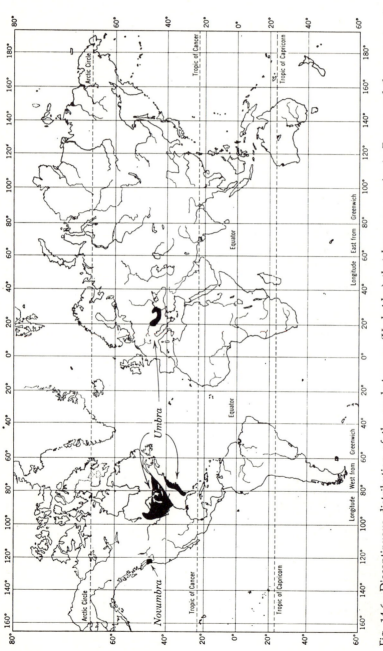

Fig. 14.6 Discontinuous distribution of the mudminnows (Umbridae) that have one species in Europe and three species (in two genera) in North America.

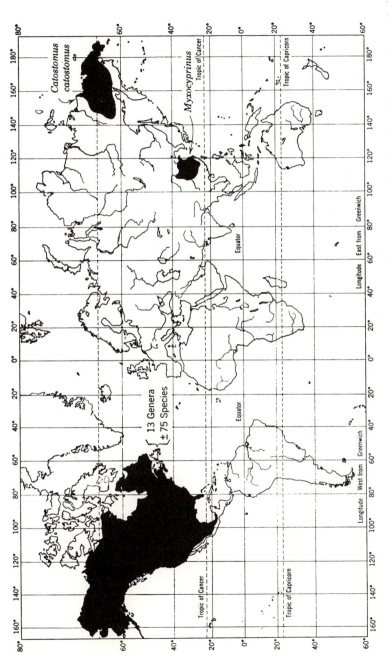

Fig. 14.7 Distribution of the suckers, family Catostomidae, that have only an ancient representative (*Myxocyprinus asiaticus*) and a recent migrant (*Catostomus catostomus*) in the Old World, where the group likely originated.

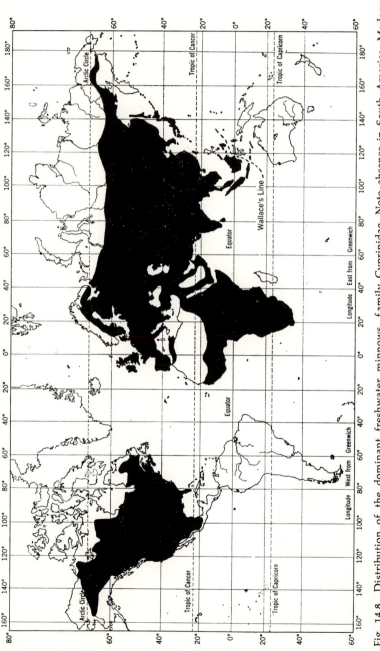

Fig. 14.8 Distribution of the dominant freshwater minnows, family Cyprinidae. Note absence in South America, Madagascar, and the Australian region. (Northwestern American range provided by C. C. Lindsey).

exceptional value in zoogeographic studies, other groups are also important in explaining regional and local distribution patterns. Many of the secondary freshwater fishes, especially those of the topminnow order Cyprinodontiformes (for example, the Mexican family, Goodeidae), are restricted to fresh water and have already proved of value in tracing former hydrographic connections between now isolated drainages.

Groups of Freshwater Fishes

The bony fishes (Osteichthyes) constitute the major groups of freshwater fishes. These may be divided into the non-teleosts (Holostei and Chondrostei) and the teleosts (Teleostei), and the latter may be separated into the Ostariophysi or Cypriniformes (minnows, suckers, catfishes, characins, loaches, etc.) and the non-Ostariophysi. The primitive non-teleosts include groups from the world continents, for example, the African bichirs or polypterids, the Chinese and North American paddlefishes (*Psephurus* and *Polyodon*), and the South American, African, and Australian lungfishes (Dipnoi) that comprise two distinct evolutionary lines. The teleosts are numerous, of diverse origin, and widespread—including, in North America, the mooneyes (*Hiodon*), pikes and mudminnows (*Esox*, and *Umbra* and *Novumbra* respectively), many cypriniforms, trout-perches (Percopsidae) and pirate perches (Aphredoderidae), sunfishes and black basses (Centrarchidae), and the perches and darters (Percidae).

The Cypriniformes, numbering around five thousand species (or about one-fourth of the total number of fishes known), include the hordes of fishes that predominate on all the continents except Australia. Only the perch-like fishes (Perciformes) with about eight thousand species, outnumber them. Of the approximately thirty-five families comprising the order Cypriniformes only two, the ariid and plotosid catfishes, are marine, and they have no doubt been derived from freshwater ancestors; they alone represent the order in Australia and Madagascar, where some of them have reinvaded fresh water.

Continental Patterns

Differentiation of fish faunas on different continents is one of the major distributional patterns of freshwater fishes.

The primary freshwater fish fauna of North America includes twenty-one families comprising about six hundred species. Its center of abundance and diversification lies in the Mississippi River basin (with about two hundred sixty species in thirteen families), southeastern United States, and the Great Lakes drainage—the latter enriched by invasion from the Mississippi fauna during glacial and postglacial times. The fauna becomes depauperate to the west of the Rocky Mountains and to the south of the Isthmus of Tehuantepec and also thins out markedly in Canada, Mexico, and Central America, but with enrichment in eastern Panama due to spillover there from the vast South American fauna. There are no true freshwater fishes among the islands of the West Indies, and but five in Alaska.

North of the southern margin of the Interior Plateau of Mexico the fauna is dominated by the minnows, suckers, and darters, which together comprise about four hundred fifty species or 90 percent of the strictly freshwater fauna.

About 30 percent of all North American freshwater fishes are of North American origin: included are mooneyes (Hiodontidae), cavefishes (Amblyopsidae), trout-perches (Percopsidae), pirate perch (Aphredoderidae), ictalurid catfishes (Ictaluridae), sunfishes (Centrarchidae), and darters (Etheostomatinae). About 55 percent are of Eurasian ancestry, including the pikes (Esocidae), mudminnows (Umbridae), minnows (Cyprinidae), suckers (Catostomidae), and perches (Percidae) other than darters. Approximately 15 percent are of South American affinities and include the characins (Characidae), gymnotids (Gymnotidae), and several catfish families, but chiefly the Pimelodidae.

The South American primary freshwater fishes are extraordinarily limited in number of groups but at the same time is the richest and most varied of all continental faunas, probably comprising close to two thousand species. These were derived from few ancestors, since they include only two major groups, the characins (with derived gymnotids) and catfishes. There are only three other primary (non-ostariophysan) groups: the ancient lungfish (*Lepidosiren*) with one species; the bonytongues (Osteoglossidae) with two species; and two species representing the spiny-rayed Nandidae. These three families also occur in Africa, and the latter two in the Orient as well. The South American fauna developed in isolation during the greater part of the Tertiary period when this continent was separated from North America. The fauna is overwhelmingly dominated by the Ostariophysi, more than half of which are represented amongst twelve endemic or autochthonous families of catfishes, and it is primarily African in its relationships.

The direct relationship between Africa and South America is just as strongly supported by the affinities of the secondary families, the killifishes (Cyprinodontidae; Fig. 14.9) and the cichlids (Cichlidae; Fig. 14.10). The problem of the derivation of this fauna is yet to be solved. Some have argued strongly that the fishes came by way of a land bridge from Africa; others point to the absence of many important African fish groups in South America and feel that the migration was by way of North America, with subsequent

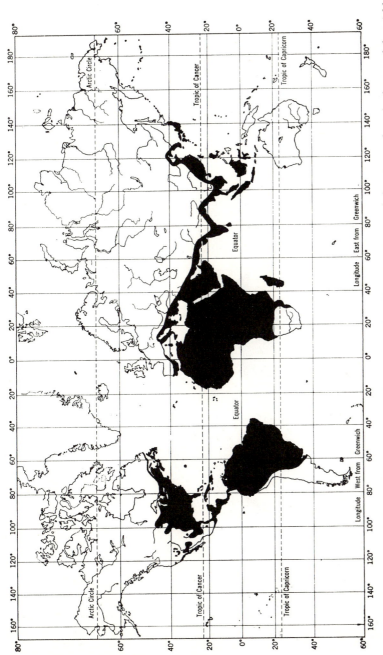

Fig. 14.9 Distribution of the killifishes, family Cyprinodontidae. Note occurrence of these secondary freshwater fishes (able to tolerate salts and cross narrow sea barriers) in the West Indies, Bermuda, Madagascar, and Celebes.

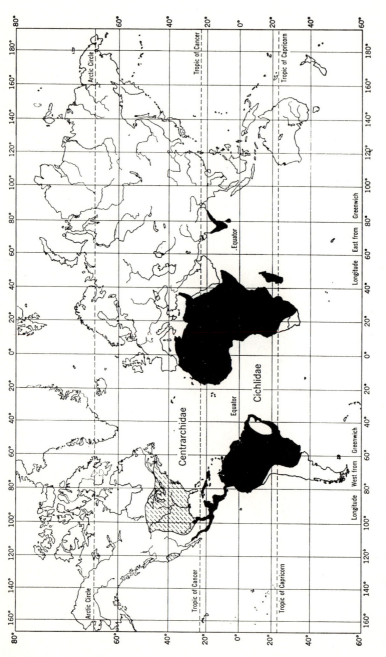

Fig. 14.10 Distribution of the cichlids, family Cichlidae, showing (in black) common occurrence in South America and Africa (including Madagascar) and relatively recent invasion into Middle America and India (including Ceylon). Shown in North America is range (crosshatched) of the ecological counterpart of the cichlids, the endemic sunfish family, Centrarchidae.

extinction in the north. The fossil record is too fragmentary to provide a reliable clue to the riddle.

The freshwater fish fauna of Africa is characterized by great diversity as well as richness and also by the development of extraordinary species swarms and high endemism among the cichlid fishes (Cichlidae) of the great rift lakes. There are numerous archaic and generalized forms—for example, lungfishes (Protopteridae) and bichirs (Polypteridae), and a variety of isospondylous groups including the endemic mormyrids (Mormyridae). There are twenty-three primary families, comprising about fourteen hundred species. Minnows, characins, and catfishes are widespread, and there are also groups of higher, spiny-rayed fishes—for example, the climbingperches (Anabantidae) and mastacembelid eels (Mastacembelidae). The large cyprinid fauna comprises minor diversifications, including a blind form, from but a few types, very probably of rather recent intrusion from Asia. The great Sahara Desert to the north and the Kalahari Desert to the southwest restrict the available areas for fish habitation, but the greater part of Africa represents an ancient land mass that has been continuously above the sea since pre-Cambrian times, thus allowing a long time for the evolution of its distinctive fauna. The long stability of this land mass also permitted the survival of old fish groups. Tropical West Africa is the richest area and has a diversity of ancient stocks unequalled anywhere else in the world. The isolated northwestern part of Africa contains a depauperate fish fauna closely tied in with that of southern Europe.

Little is known of the origin and dispersal of the older groups of African primary fishes, but eight families are presumably autochthonous: the bichirs (Polypteridae), elephantfishes (Mormyridae), Gymnarchidae, African mudskippers (Pantodontidae), Phractolemidae, Kneriidae, Cromeriidae, and electric catfishes (Malapteruridae). The families that include the minnows (Cyprinidae) and loaches (Cobitidae), and probably some of the spiny-rayed forms, apparently entered Africa from Asia during Pliocene and Pleistocene times.

The fish fauna of Eurasia is characterized by scarcity of phylogenetic relicts (only *Psephurus*, the oriental paddlefish, and *Myxocyprinus*, the sucker—Fig. 14.7—of the Yangtze River and vicinity) and great richness and diversity of the cyprinoid fishes in eastern Asia represented by the minnows (Cyprinidae), loaches (Cobitidae), and the hillstream loaches (Homalopteridae). In China, for example, about 80 percent of the nearly six hundred primary fishes belong to the Cypriniformes (cyprinoids and four families of catfishes or siluroids). To the southward, through tropical China, the fauna merges in complex fashion into the fish fauna of the Oriental Region. Europe, recently glaciated, has a depauperate fauna, of which only six families comprising one hundred eighteen species make up the primary freshwater fishes. None of the groups is endemic and three families (mudminnows, Umbridae, pikes, Esocidae, and perches, Percidae) comprise only thirteen species; the minnows

(Cyprinidae) and loaches (Cobitidae) make up the bulk of the fauna, and there are only two catfishes (Siluridae). There is a relative abundance of groups such as lampreys (Petromyzonidae), sturgeons (Acipenseridae), herrings (Clupeidae), trouts (Salmoninae), and gobies (Gobiidae), which tolerate or regularly spend a part of the life cycle in sea water; these also include species tolerant of low temperature. The fauna is well known in Europe, less so in Russia, and poorest known in China, which has by far the greatest number of species.

The Oriental Region includes the main part of tropical Asia, stretching from India and southern China southward through the Indo-Australian Archipelago (Sumatra, Java, and Borneo) to Wallace's Line (Fig. 14.8). It is noteworthy for its rich and diverse cyprinoid fishes and catfishes (nine families, six endemic) and for the absence of archaic groups and almost all of the herring-like families of Africa; it is moderately rich in spiny-rayed fishes. Included are a large variety of lowland fishes, certain of which (such as the Indian Chaudhuriidae) are exclusive to the region, and also certain groups highly specialized for life in mountain torrents (hillstream loaches, Homalopteridae, and suckerbelly loaches, Gastromyzonidae, in particular).

Australia is almost without true freshwater fishes; only the archaic lungfish (*Neoceratodus*) and a bonytongue (osteoglossid) (*Scleropages*) can be placed in this category. The remainder of the depauperate freshwater fauna comprises groups of marine derivation, about one hundred eighty kinds excluding introduced species and stragglers. The principal marine groups represented in the rivers of Australia are lampreys (Petromyzonidae), sawfish (*Pristis*), herrings (Clupeidae), "smelt" (*Retropinna*), galaxiids (Galaxiidae), catfishes (Ariidae), eels (*Anguilla*), silversides (Atherinidae), mullets (Mugilidae), serranids (Serranidae), theraponids (Theraponidae), cardinalfishes (Apogonidae), and gobies (Gobiidae and Eleotridae).

GEOGRAPHY OF MARINE FISHES

Unlike the fishes that inhabit continental fresh waters, marine fishes are not primarily restricted in their distribution by physical barriers, aside from the continental borders themselves. The important factor limiting their dispersal is temperature. Other major distributional influences are the existence of broad expanses of deep oceanic water (as between America and Polynesia), the nature of coastline configuration and submarine contours, variations in water salinity, and the effects of ocean currents (not only their direction of flow but their temperature). Random dispersal of pelagic life history stages by way of the equatorial counter current, for example, may account for the common occurrence of closely related or identical species on the Atlantic coasts of Africa and the Americas.

For zoogeographic purposes, marine fishes may be classed into three main categories: (a) the shore or shelf fauna, comprising species that occur inshore, along the continental shelves, in depths of 100 fathoms (600 feet, or about 200 meters) or less; (b) the pelagic or open sea fishes, generally living near the surface of tropical and warm-temperate seas, often far from shore, in 100 fathoms or less; and (c) deepsea or abyssal forms, inhabiting depths greater than 100 fathoms—these also may be divided into bathypelagic and benthic forms, the latter bottom fishes occurring to depths of almost 6000 fathoms (Challenger Deep, near Guam).

The shore forms, which include well over half the known species of marine fishes, occupy the continental shelves (including the sea floor) from the intertidal zone (the littoral fauna, as restricted by many workers) to their outer edges. Within the limits set by temperature, their ranges are largely limited by the extent of available coast line and archipelagos. In contrast, the open sea is generally a highway of dispersal for the pelagic fishes, many of which can cross areas having markedly different water temperatures, for example, the whale shark (*Rhincodon*), the dolphin (*Coryphaena*), the swordfish (*Xiphias*), and the tunas (*Thunnus*). Circumglobal fishes range from primitive lancelets (Branchiostomidae) and sharks (Squaliformes) to specialized puffers (Tetraodontidae) and anglerfishes (Ceratiidae), with nearly 85 percent of such widespread species being typical of the open ocean. However, the barrier between Polynesia and America has prevented more than a third of the widely ranging pelagic and bathypelagic fishes from attaining worldwide distribution through their failure to cross into the eastern Pacific.

We know least about the deepsea fauna, and presently held ideas about its distribution will be modified as new information is made available through advances in oceanographic research. Many of the species seem to obey distributional laws of their own. Not infrequently they are widely distributed, evidently because the barriers to dispersal found in shallower regions are absent. Or, put in another way, the greater physical conformity within the ocean water masses—as compared with shore waters—has enabled species to spread widely. Although we have much to learn about the benthic deepsea forms, their distribution may generally be said to be rather restricted; only three such fishes are known to have a circumtropical distribution. The deepsea fauna is a depauperate one, in evident correlation with the severity of environmental conditions and the paucity of food material at great depths. In general it can be said that with increasing distance from the continents the deepsea floor tends to be occupied by fewer individuals belonging to fewer species.

Although deepwater fish faunas (below 1000 fathoms) are generally thought to be rather undiversified, exploration of the Kurile-Kamchatka Trench reveals nearly fifty species belonging to twenty-two families; fifteen of these families were previously unknown from Russia. These deepwater fishes are believed to represent older species that differ in origin and distribution from the younger fish populations in the upper water layers.

The tropical shore faunas are by far the richest and most diverse, not only in regard to fishes but to other marine organisms as well. Because of their special need for warm water (mean annual surface temperatures of 20°C or more), these fishes hardly penetrate the subtropical regions to the north and south. Consequently, tropical faunas exhibit a high degree of endemism. The subtropical faunas may be generally characterized as diluted tropical faunas, limited by a yearly minimum temperature of about 16° to 18°C; they have a good deal in common with the tropical fauna, with much less resemblance to the neighboring temperate fauna. Many tropical fishes occur throughout the tropics on both sides of the Atlantic and Pacific, and are spoken of as circumtropical.

Tropical shore faunas may be divided into four major regions: Indo-Pacific, West African, West Indian, and Pacific American or Panamanian. All are relatively distinct in species and have at least some distinctive genera, but there are broad similarities that link all four more closely together than each is to the temperate shore faunas immediately to the north and south.

The Indo-Pacific fauna is by far the richest and contains essentially all of the families and a considerable number of the genera that are found in the other three regions. It also harbors families and genera not found elsewhere. The complex configuration of continental shores and numerous islands has afforded great opportunity for evolution by providing a diversity of habitats. This fauna covers a vast expanse, extending from East Africa and the Red Sea to northern Australia, southern Japan, the Hawaiian Islands, and the other islands of Polynesia—in all, some half way around the world. The other three regions are, in effect, depauperate geographic segregates of this great world fish fauna. The distribution of coral reefs or of parrotfishes (Scaridae) may be taken as delimiting the boundaries of the Indo-Pacific fauna. The Hawaiian Islands harbor the most distinctive element of the fauna, for, although there are only three known endemic genera, more than 30 percent of the reef species are endemic. A striking characteristic of this inshore fauna is the virtual absence of groupers (subfamily Epinephelinae, family Serranidae) and snappers (Lutjanidae); elsewhere in the tropical Pacific these two important groups are dominant reef fishes.

The West African tropical fauna is comparatively small and shows relationships with the tropical Mediterranean fishes and also with those that occur in the West Indies.

The West Indian or Caribbean fauna, distributed from southern Florida and the mainland shore of central America south of Yucatan southward to Brazil (Bahia), is a very rich one, containing around two-thirds of the truly tropical shore families that occur in the Indo-Pacific. Most of the species of the two regions are distinct. Some northern elements in the northern part of the Gulf of Mexico represent glacial relicts that were trapped after the recession of continental glaciers. That the Gulf Stream also provides a barrier is reflected in the restriction of certain fishes to the Gulf and Caribbean respectively.

The Pacific American or Panamanian fauna extends from the Gulf of California to Peru; in the upper Gulf, conditions have led to limited endemism and some entrapped elements of the northern cool-temperate fauna. The fishes of this fauna are largely of New World affinities, well cut off from the great Indo-Pacific fauna by a broad expanse of deep oceanic water, the Eastern Pacific barrier. They are essentially West Indian, with numerous species common to the two sides of the Isthmus of Panama which was under the sea in late Tertiary time. The distinctness between the eastern Pacific and the Indo-Pacific faunas has been reduced by very recent (geologically speaking) invasions of Indo-Pacific forms along counter currents.

There are four major barriers to the dispersal of tropical shore fishes:

a. The Eastern Pacific which is not crossed by 86 percent of the species. Some fifty-six species are trans-Pacific, occurring on both sides of this great barrier; but thirteen of these are worldwide species.

b. New World land, the importance of which as a faunal barrier appears to be increasing as more groups in the eastern Pacific and western Atlantic are compared; that is, there seem to be fewer species in common to both sides than has been thought, so that the barrier (at the species level) is around 98 percent.

c. The mid-Atlantic, which is crossed by about one hundred forty tropical species, and shows an estimated barrier effectiveness of about 72 percent.

d. Old World land, linking Eurasia and Africa, eliminated the ancient Tethys Sea. The Suez Canal breakthrough is reshuffling Red Sea fishes into the Mediterannean (seven species recorded by 1976) and some of the latter (five to date) into the Red Sea.

It is therefore seen that the Eastern Pacific barrier is only one and perhaps not the most important deterrent to the mixing of tropical shore fishes.

The temperate and boreal faunas lying to the north and south of the Tropical zones are characterized by much less diversity and fewer species than is the tropical fauna, and often by spectacular numbers of individuals, for example of cods (Gadidae), herrings (Clupeidae), and flatfishes (Pleuronectiformes). In these regions occur the most important commercial fisheries of the world. However, numerous complications in ocean currents, temperatures, and seasonal changes make the separation of temperate and boreal regions extremely arbitrary.

The north Pacific is hereby defined as that area of ocean lying between Eastern Asia and Western America north of the tropics. California and Japan form transition zones, but the situation is complicated by the cold California current and the warm Kuroshio current. In general, therefore, the coldwater fauna extends farther southward along the American than along the Asiatic coast. This southward extension of boreal elements is further abetted, in northern Baja California, by upwelling of deep, cold water along the coast, so that species not occurring in California farther south than Point Conception reappear in Baja California, after a nearly 483-km break in their range. The north Pacific fauna has a great diversity of rockfishes (Scorpaenidae), sculpins (Cottidae), snailfishes (Cyclopteridae), and poachers (Agonidae) and several endemic families, including the viviparous surfperches (Embiotocidae) and the greenlings (Hexagrammidae). Endemic genera are numerous. It possesses the richest fauna of any outside of the Indo-Pacific. This may be the result of the extent and variety of the habitat, former connections between the American and Asiatic elements, past connections between the western element and the Indo-Pacific, and survival of the coolwater fauna during Interglacial periods. The north Pacific fauna shows only a weak resemblance to the fauna of the Arctic, with but a handful of species in common, whereas the north Atlantic and Arctic faunas have many similarities. This is explained by the narrow connection between the Pacific and Arctic, the past history of the Bering land bridges that repeatedly isolated the two faunas, and the relative dissimilarity of oceanographic conditions in the two regions.

The temperate south Pacific includes southern Australia, New Zealand, southern South America, and South Africa. In all of these areas there are many north temperate genera and sometimes northern species also appear; the latter are cited as evidence for bipolar distribution (see later treatment of bipolarity). The currents south of the tropical region are not deflected by land masses to nearly the same extent as they are to the north; hence the distribution zones are easier to define and the isotherms are more nearly parallel. The Australian–New Zealand fauna contains a number of endemic genera and a few peculiar families, but the relationship to Chilean and Antarctic forms is clear cut, especially in New Zealand.

The cold Antarctic region is bounded on the north by the isotherm of 6°C. Its waters are capable of producing large populations, in terms of numbers of individuals, during the summer months. The fauna is limited but peculiar, consisting largely of the Antarctic blennies (Nototheniidae) and their relatives. The relative sparseness of the fauna reflects the severity of the environment (as with the benthic deepsea fishes), with low temperatures, reduced productivity of the waters, ice-action in the littoral region, and a virtual absence of macroflora providing limiting factors.

In the north Atlantic, the seasonal variations in the temperature of the sea are marked; the great ocean currents are greatly spread out and cause the isotherms to be widely separated. The meeting of the cold Labrador Current with the warm Gulf Stream produces an abrupt change from frigid to sub-tropical conditions along the coast of North America, with a corresponding change in the fish fauna. The 6°C isotherm, for example, runs from about 45° N. lat. on our coast to about 68° N. lat. on the coast of Norway, making a difference in latitude of more than 1,200 nautical miles. The southern boundary of the temperate north Atlantic is approximated by the 12°C isotherm, which runs roughly from the mouth of the English Channel west and southward to Cape Hatteras. Here occurs a transition from the typical northern types, like the herrings (Clupeidae), cods (Gadidae), and flatfishes (Pleuronectiformes), to the warmwater forms such as the sardine (*Sardinella*) and anchovy (*Anchoa*). Many Mediterranean species reach their northern limit at about this latitude, which also very nearly marks the southern limit, as marine fishes, of such cold-loving groups as the salmon family (Salmonidae). The fauna has much in common on both sides of the North Atlantic.

In the south Pacific, the Chilean fauna has a few endemics, a noteworthy Australian element (among the lampreys, Petromyzonidae, and galaxiids, Galaxiidae, for example), and an admixture of tropical forms to the north and Patagonian ones to the south, in addition to those groups common to all south temperate faunas.

The Arctic fauna, lying north of the 6°C isotherm, is a very depauperate one for the same reasons that the Antarctic fauna is so poor (see above). The important groups are the sculpins (Cottidae), cods (Gadidae), flatfishes (Pleuronectiformes), and eelpouts (Zoarcidae), the latter group also being represented in the Antarctic but with distinctive genera in each region.

BIPOLARITY

Species that occur in the northern and southern hemispheres but have a pronounced distributional gap in between are said to show the phenomenon of bipolarity. For example, the deepsea anglerfish, *Ceratias holboelli*, has been taken off Greenland and also in the Antarctic, but not in intervening tropical waters. Such submergence also occurs along the West African coast. During the lowered sea temperatures of Pleistocene time, surface forms were presumably able to cross over. These two regions, moreover, are still very incompletely investigated, and species thought to be bipolar may, in fact, occur in one or the other (or both) of the intervening regions. Bipolarity is somewhat higher among genera and taxonomic categories above the genus than it is among species—for example, the lamprey family (Petromyzonidae) exhibits bipolar distribution. Upwelling of cold water in the eastern Pacific has favored the passage of northern and southern forms across the tropics.

SPECIAL REFERENCES ON ECOLOGY AND ZOOGEOGRAPHY

A. Freshwater Ecology

Carpenter, K. E. 1928. Life in inland waters. Sidgwick and Jackson, London.

Chapman, R. N. 1931. Animal ecology. McGraw-Hill Book Company, New York.

Coker, R. E. 1954. Streams, lakes, ponds. University of North Carolina Press, Chapel Hill, N.C.

Hutchinson, G. E. 1957. A treatise on limnology. Vol. 1: Geography, physics, and chemistry. John Wiley and Sons, New York.

Needham, J. G., and J. T. Lloyd. 1930. The life of inland waters. The Comstock Publishing Company, Ithaca, N.Y.

Reid, G. K., and R. D. Wood. 1976. Ecology of inland waters and estuaries. D. van Nostrand Company, New York.

Ruttner, F. 1953. Fundamentals of limnology. University of Toronto Press, Toronto. (Translated by Frey and Fry.)

Ward, H. B., and G. C. Whipple. 1918. Fresh-water biology. John Wiley and Sons, New York. Second edition, 1959, edited by W. T. Edmondson.

Welch, P. S. 1951. Limnology. McGraw-Hill Book Company, New York.

B. Marine Ecology

Coker, R. E. 1954. This great and wide sea. University of North Carolina Press, Chapel Hill, N.C.

Colman, J. S. 1950. The sea and its mysteries. G. Bell and Sons, London.

Halstead, Bruce, W. 1959. Dangerous marine animals. Cornell Maritime Press, Cambridge, Md.

Harvey, H. W. 1928. Biological chemistry and physics of the sea. Cambridge University Press, Cambridge.

Hedgpeth, J. W. (editor and contributor). 1957. Treatise on marine ecology and paleoecology. Mem. Geol. Soc. Amer., 67.

Idyll, C. P. 1964. Abyss: the deep sea and the creatures that live in it. Thomas Y. Crowell Company, New York.

Johnstone, J. 1928. An introduction to oceanography. Hodder and Stoughten, London.

Kuenen, P. H. 1955. Realms of water. John Wiley and Sons, New York.

MacGinity, G. E., and N. MacGinity. 1949. Natural history of marine animals. McGraw-Hill Book Company, New York.

Moore, Hilary B. 1958. Marine ecology. John Wiley and Sons, New York.

Nicol, J. A. C. 1967. The biology of marine animals. Interscience Publishers, New York.

Pettersson, H. 1954. The ocean floor. Yale University Press, New Haven, Conn.

Romanovsky, V. 1953. La mer. Librairie Larousse, Paris.

Russell, F. S. 1928. The seas. F. Warne and Company, London.

Sears, Mary. (Ed.). 1961. Oceanography. Publ. 67, Amer. Assoc. Adv. Sci., Washington, D.C.

Smith, F. G. W. 1954. The ocean. Charles Scribner's Sons. New York.

Sverdrup, H. U., M. W. Johnson, and R. H. Fleming. 1942. The oceans. Prentice-Hall, Englewood Cliffs, N.J.

Sverdrup, H. U. 1942. Oceanography for meteorologists. Prentice-Hall, Englewood Cliffs, N.J.

Tait, J. B. 1952. Hydrography in relation to fisheries. E. Arnold, London.

C. Fish Geography

Barbour, C. D., and J. H. Brown. 1974. Fish species diversity in lakes. *Amer. Nat.*, 108: 473–489.

Briggs, J. S. 1974. Marine zoogeography. McGraw-Hill Book Company, New York.

Darlington, P. J., Jr. 1957. Zoogeography: The geographical distribution of animals. John Wiley and Sons, New York.

de Beaufort, L. F. 1951. Zoogeography of the land and inland waters. Sidgwick and Jackson, London.

Ekman, S. 1953. Zoogeography of the sea. Sidgwick and Jackson, London.

Gosline, W. A. 1975. A reexamination of the similarities between the freshwater fishes of Africa and South America. Mém. Mus. Nat. D'Hist. Nat., n.s., Ser. A., Zool., 88: 146–155.

Gosline, W. A., and V. E. Brock. 1960. Handbook of Hawaiian fishes. University of Hawaii Press, Honolulu.

Grey, M. 1956. The distribution of fishes found below a depth of 2000 meters. *Chicago Nat. Hist. Mus. Fieldiana: Zool.*, 36 (2): 75–337.

Hubbs, C. L., and R. R. Miller. 1948. Correlation between fish distribution and hydrographic history in the desert basins of western United States: *In:* The Great Basin, with emphasis on glacial and postglacial times. *Bull. Univ. Utah*, 38: 17–166.

MacArthur, R. H. 1972. Geographical ecology. Harper and Row, New York.

MacArthur, R. H., and E. O. Wilson. 1967. The theory of island biogeography. Princeton University Press, Princeton, N.J.

Marshall, N. B. 1954. Aspects of deep sea biology. Philosophical Library, New York.

Miller, R. R. 1959. Origin and affinities of the freshwater fish fauna of western North America. *In:* Zoogeography. Amer. Assoc. Adv. Sci., Publ. 51 (1958): 187–222.

Myers, G. S. 1938. Fresh-water fishes and West Indian zoogeography. *Ann. Rept. Smithsonian Inst.*, 1937: 339–364.

———. 1940. The fish fauna of the Pacific Ocean, with especial reference to zoogeographical regions and distribution as they affect the international aspects of the fisheries. *Sixth Pacific Sci. Cong.*, 3: 201–210.

———. 1949. Salt-tolerance of fresh-water fish groups in relation to zoogeographic problems. Bijd. Tot Dierk., 28: 315–322.

———. 1951. Fresh-water fishes and East Indian zoogeography. *Stanford Ichthyol. Bull.*, 4: 11–21.

Nelson, G. J. 1969. The problem of historical biogeography. Syst. Zool., 18 (3): 243–246.

Scientific American, Inc. 1972. Continents adrift. Readings from *Scientific American*, with introductions by J. Tuzo Wilson. W. H. Freeman and Co., San Francisco.

Scientific American, Inc. 1974. Planet Earth. Readings from *Scientific American*, with introductions by Frank Press and Raymond Siever. W. H. Freeman and Co., San Francisco.

Seyfert, C. K., and L. A. Sirkin. 1973. Earth history and plate tectonics. An introduction to historical geology. Harper and Row, New York.

Sterba, G. 1959. Süsswasserfische aus aller Weit. Verlag Zimmer und Herzog, Berchtsgaden.

Vuilleumier, F. 1974. Zoogeography. Vol. 4. In press.

Systematic Index

This index includes the common and scientific names of fish-species and other fish groups used in the text. The subject matter of the book is treated in a separate index. Italic numbers indicate that the entry appears in a figure on that page. Insofar as it is practical, common names follow those of *Special Publication No. 6* of the American Fisheries Society.

fathead, 167, 282, 287, 330, 366, 375
Old World, 134
stoneroller, 141
see also Carp
Mirapinnidae, 40
Mirapinnids, 40
Misgurnus, 188, 391
fossilis, 240, 347
Mistichthys luzonensis, 304
Mobulidae, 32, 419, 426
Mochokidae, 38
Mojarras, 44
thick-lipped, *138*
Molas and *Mola, 54,* 181, 271, 323, 419, 426,
301, 302, 323
Molgula, 11
Molidae, 49, 181, 301, 302
Molly, Amazon, 268, 381, 389
Monacanthidae, 49
Monacanthus, 183
polycanthus, 124
Monocentridae, 42, 125
Monocirrhus, 121, 181
Monodactylidae, 44
Monopterus, 197
Monorhina, 19
Mooneye, 36, 176, 445, 446
Moorish idols, 428
Morays, 34, 134, 141, 153, 190, 210, 227,
366, 427, 428
Mordacia mordax, 356
Mormyridae, 36, 223, 237, 320, 368, 374,
375, 449
Mormyrids, 174
Mormyroidei, 36
Mormyrus caballus, 37
Morone, americana, 305
chrysops, 275
saxatilis, *43,* 143, 192, 193
Mosquitofish, 276, 277, *278,* 288. *See also*
Gambusias and *Gambusia*
Mouthfishes, 246, 425, 426
Moxostoma, 86, 135
anisurum, 135
erythrurum, *135,* 140
macrolepidotum, 135
Mrigal, 199
Mudminnows, 36, 176, 188, 238, 432, 442,
445, 446, 449
Mudskippers, 99, 190, 202, 209, 233, 261,
262, 265, 343, 354

African, 36
Mugil, 143, 149, 190, 375, 421
cephalus, *144,* 322
Mugilidae, 44, 265, 342, 450
Mugiliformes, 42
Mugiloidei, 42
Mugiloididae, 46
Mullets, 44, 143, 149, 190, 265, 342, 375,
421, 450
striped, *144,* 322
Mullidae, 44, 114, 147, 153, 227
Mummichog, 156, 231, 292, 294, 297, 305,
342, 344, 346
Muraena, 210
Muraenesocidae, 34
Muraenidae, 34, 134, 141, 153, 190, 227, 427,
428
Muskellunge, 133
Mustelus, 122, 207, 232, 288, *320*
antarcticus, *206, 208*
canis, 213
Myctophidae, 40, 242, 243, 248, 426
Myctophiformes, 40, 125, 126
Myctophum, 121
Mylinae, 193
Myliobatidae, 32, 126, 175
Myliobatis, 140
californicus, 126
Myoxocephalus, 194
Myxine, 57, 80, 92, 105, 207, 328, 364
glutinosa, *31*
Myxini, 12, *23*
Myxinidae, 22, 30, 56, 64, 67, 80, 99, 102,
136, 138, 200, 211, 261, 293, 300, 428
Myxiniformes, 15, 30, 220, 320, 364
Myxocyprinus, 440, 449
asiaticus, 443

Nandidae, 46, 446
Naso, 127
Naucrates, 419
Needlefishes, 40, *54,* 124, 141, 303, 425
Nemichthyidae, 34, 53
Neoceratodus, 82, 86, 89, 97, 233, 236, 450
forsteri, *33*
Neopterygii, 12, 86
Neostethidae, 44
Nettastomidae, 34
Nocomis biguttatus, 167
Noemacheilus, 352
Noestethus amaricola, 39

Subject Index